AF601315

Big Data Management

The big data paradigm presents a number of challenges for university curricula on big data or data science related topics. On the one hand, new research, tools and technologies are currently being developed to harness the increasingly large quantities of data being generated within our society. On the other, big data curricula at universities are still based on the computer science knowledge systems established in the 1960s and 70s. The gap between the theories and applications is becoming larger, as a result of which current education programs cannot meet the industry's demands for big data talents.

This series aims to refresh and complement the theory and knowledge framework for data management and analytics, reflect the latest research and applications in big data, and highlight key computational tools and techniques currently in development. Its goal is to publish a broad range of textbooks, research monographs, and edited volumes that will:

- Present a systematic and comprehensive knowledge structure for big data and data science research and education
- Supply lectures on big data and data science education with timely and practical reference materials to be used in courses
- Provide introductory and advanced instructional and reference material for students and professionals in computational science and big data
- Familiarize researchers with the latest discoveries and resources they need to advance the field
- Offer assistance to interdisciplinary researchers and practitioners seeking to learn more about big data.

The scope of the series includes, but is not limited to, titles in the areas of database management, data mining, data analytics, search engines, data integration, NLP, knowledge graphs, information retrieval, social networks, etc. Other relevant topics will also be considered.

Senlin Luo · Limin Pan

Big Data Analysis

Theory and Technology

Senlin Luo
Cyberspace Science and Technology
Beijing Institute of Technology
Beijing, China

Limin Pan
School of Information and Electronics
Beijing Institute of Technology
Beijing, China

ISSN 2522-0179 ISSN 2522-0187 (electronic)
Big Data Management
ISBN 978-981-95-5724-0 ISBN 978-981-95-5725-7 (eBook)
https://doi.org/10.1007/978-981-95-5725-7

Jointly published with Beijing Institute of Technology Press, Beijing, China
The print edition is not for sale in the mainland of China. Customers from the mainland of China please order the print book from: Beijing Institute of Technology Press, Beijing, China

This Springer imprint is published by the registered company Springer Nature Singapore Pte Ltd.
The registered company address is: 152 Beach Road, #21-01/04 Gateway East, Singapore 189721, Singapore

Preface

Big data analysis can reveal internal laws from data resources, facilitate the extraction of useful information, and help people make scientific decisions based on large-scale data. Its strategic significance lies not only in the accumulation of data resources but also in the analysis and processing of these resources. Big data analysis has become the new normal of data science. It is an interdisciplinary field of research. At present, universities, research institutes, and enterprises should conduct in-depth research and application in this field. The teaching of big data analysis has become a part of various disciplines, including cybersecurity, information and communication engineering, computer science and technology, and software engineering. These subjects are currently offered at almost all major universities. There is an urgent need to develop a comprehensive theoretical and technological framework for big data analysis, along with a core knowledge map for its incorporation into academic curricula.

This book presents a systematic overview of big data theory and technology, along with a detailed discussion of the system framework and core knowledge map associated with them. It also incorporates the author's latest research findings. The main contents include fundamental knowledge of big data analysis, the core architecture of big data analysis, big data analysis computing mode, big data and information security confrontation, big data and natural language processing, big data and biological information processing, etc. It aims to comprehensively develop students' big data theoretical and technical capabilities, emphasize understanding of basic theories and applications of data science, and guide the training of competitive data scientists. Furthermore, the book facilitates students' capacity to employ computer and mathematical knowledge to analyze and process large-scale datasets, rapidly obtain information and discover relationships within complex data, and model specific problems in the real world.

In terms of system structure, this book places significant emphasis on the systematic and hierarchical nature of knowledge, while simultaneously highlighting the key points that can be observed at both the tree and forest levels. The content is comprehensive yet not overly burdensome. The book is divided vertically into theoretical technology, engineering practice, and horizontally emphasizes multidisciplinary applications such as cyberspace security, natural language processing, and life information management. While grasping its essentials, the knowledge points are as comprehensive as possible, covering multiple individual technologies while maintaining a clear structure and systematic knowledge. This will help readers quickly master its core technologies.

In terms of content scope, this book focuses on the depth, advancement, timeliness, and characteristics of interdisciplinary integration of theory and practice. The book adheres to the dynamic development of big data theory and technology while maintaining the fundamental and timely nature of its theoretical and technical core knowledge. It introduces new theories, new technologies, new methods, and new cases, incorporates the author's research results, and ensures the advanced, applicable, and forward-looking nature of its theoretical and technological content.

In terms of flexible use, this book is based on research-oriented teaching methodology, focusing on readers' interests and the flexibility of content learning, as well as the indirect effects of learning, which can meet the long-term training needs of diverse talents in various universities. This book can be used for regular learning and comprehension of the fundamental principles and methodologies associated with the field of big data, as well as the practical application of big data technology in multidisciplinary fields. Furthermore, it can serve to support the sustainable development of learning. In addition, this book integrates the cultivation of humanistic literacy, discusses non-technical and engineering issues in the research and practice of data science, and addresses the availability of algorithms, ethics, privacy, law, intellectual property protection, economics, and social influence related to big data and data analysis, social impact, etc.

Luo Senlin and Pan Limin co-authored the book. Pan Limin was responsible for Chaps. 4–6, while Luo Senlin was responsible for the remaining chapters. Luo Senlin was also responsible for the overall design, content planning, and drafting of the book.

In the process of writing this book, I have received invaluable assistance from Liu Xiaoshuang, Yan Han, Li Yu, Zhang Hanqing, and other students from the Institute of Information Security and Confrontation Technology of Beijing Institute of Technology. I would like to express my profound gratitude here. In addition, I would like to express my sincere thanks to Ms. Ni Wenhui of Higher Education Press for her comprehensive and serious editorial revisions and her enthusiastic help with this publication. I would also like to extend my sincere thanks to Higher Education Press for its invaluable support and help in numerous aspects.

Given the constraints of time and resources, we kindly invite readers to offer constructive criticism and corrections for any identified areas that may have been overlooked or any errors that may have inadvertently slipped through. This feedback is invaluable as we strive for excellence and continuous improvement. Finally, we extend our sincere gratitude to all who have supported us in this endeavor.

Beijing, China
January 2025

Senlin Luo

Contents

Chapter 1
Fundamentals of Big Data Analysis

1.1 Introduction

The term "Big Data" is used to describe a large and complex set of information (i.e., data) that is beyond the scope of traditional data processing applications and algorithms [1]. In contrast to traditional data, the implicit patterns and rules embedded within big data offer more effective guidance. Consequently, it is essential to dig deep into data through corresponding data analysis techniques. Peter Drucker, the Founder of Modern Management, stated in his book "Management Challenges in the 21st Century" that the world is undergoing an information revolution, not in technology, machinery, software, or speed, but a revolution in the field of "concept." Over the past few decades, the focus of information technology has primarily been in enhancing the term "technology" itself, striving to broaden its scope, increase capacity, and efficiently disseminate information. In contrast, the current revolution of information is based on the "information" itself. Consequently, in the "seventh information revolution" wave, big data and big data analysis technologies play a vital role.

This chapter aims to present the big data analysis concept, an overview of existing research history and current situation, and present the big data analysis applications. Additionally, it addresses existing issues and presents prospective avenues for future research.

S. Luo and L. Pan, *Big Data Analysis*, Big Data Management,
https://doi.org/10.1007/978-981-95-5725-7_1

1.2 Knowledge Base

1.2.1 Key Concepts

The term "Big data" is used to describe the vast quantities of the generated data that cannot be collected, managed, and processed using conventional software tools and traditional data analytics due to their sophisticated characteristics. The analysis of such massive data volumes requires the development of novel analytical and processing techniques to enhance decision-making, insight, and process optimization capabilities [2]. As a consequence of the exponential growth in the quantity, diversity, and velocity of data, IBM employs the "5 V" framework to summarize the characteristics of big data, as presented in Fig. 1.1, and is described as follows:

1. Volume: This refers to the quantity of collected, stored, and calculated data. The size and scale of the dataset determine whether the data is big data or not. If measured in Petabyte (PB), the data is considered big data, which requires significant computing power and an effective computing architecture.
2. Variety: This refers to the type and source of data, including structured data, semi-structured data, and unstructured data. Structured data is organized correctly and efficiently stored, unstructured data is random and difficult to process, while semi-structured data uses organizational properties such as tags to make it easy to analyze. The rapid expansion of the Internet and the Internet of Things (IoT), including the proliferation of web pages, social media, healthcare, smart cities, and perception data covering audio, images, video, and analog signals, has led to a significant increase in the diversity of collected data, which has in turn necessitated the development of more sophisticated data processing capabilities.

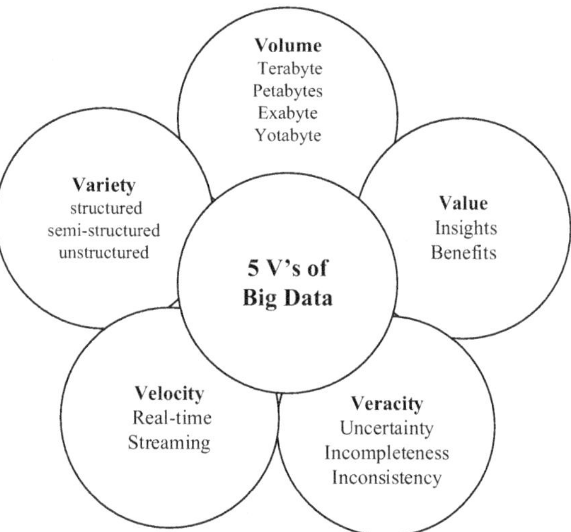

Fig. 1.1 Common big data characteristics

3. Value: This refers to the value of the insights and benefits derived from data. For example, if we draw a parallel between the petroleum industry and big data analysis, the emphasis is on extracting high-quality crude oil rather than merely refining the vast quantities of data produced within the Internet industry. The primary objective is to dig out the most valuable information. Given the low density of data value in big data, it can be likened to finding precious gems in a sea of sand. With the wide application of the Internet and IoT, information perception is ubiquitous, and the amount of information is massive, yet the value density remains low. The extraction of useful insights from data by combining business logic and using powerful machine algorithms remains the most challenging task in solving data problems.
4. Velocity: This refers to the velocity of processing data within a short duration, such as real time or streaming. For instance, in a healthcare context, sensors will continuously generate a vast quantity of data. In this case, any delays in the processing phase by monitors may result in patient injury or death. Similarly, the search engine requires that the news generated a few minutes ago can be queried by a user, and the personalized recommendation algorithm requires that the recommendation be completed in real time whenever possible. One of the defining characteristics of big data is its velocity, which is not present in traditional data mining.
5. Veracity: This indicates the accuracy and reliability of data. It is also referred to as the quality of data. The big data content is closely related to events in the real world. Consequently, it is essential to guarantee the accuracy and reliability of the data. Nevertheless, the extraction and interpretation of diverse events from massive network data present challenges to the establishment of accuracy and trust in big data analytics.

The strategic significance of big data technology is not to merely master vast data information, but rather to professionally process these meaningful data. In other words, in the industry area, the key to profitability is improving the "processing capacity" of the data, while realizing the "value-added" of the data through "processing." Typical processing methods can be divided into two categories: data analysis and data mining. Data analysis is the process of using appropriate statistical analysis methods and tools to the collected data in order to achieve the desired analytical outcome. This outcome may involve the extraction of valuable information, which then play the role of data. It mainly achieves three significant functions: status analysis, cause analysis, and predictive analysis (quantitative). The objective of data analysis is to identify and verify assumptions, which are then used to draw conclusions. The most common analytical techniques employed include comparative analysis, group analysis, cross-analysis, and regression analysis. Furthermore, data analysis typically obtains an indicator statistic result, such as a sum or average. In order to ascertain the value and role of data, it is necessary to interpret the aforementioned indicators in conjunction with the business.

Data mining is the process of mining unknown and potentially valuable information and knowledge from a large amount of data through the application of statistics, artificial intelligence, machine learning, and other methods. Data mining is concerned with the resolution of four distinct types of problems: classification, clustering, association, and prediction (both quantitative and qualitative). The significance of data mining lies in the identification of previously unknown patterns and laws. This is exemplified by the oft-cited examples of data mining, such as the cases of beer and diapers, condoms and chocolate, and so forth. These are patterns that are not known in advance but are highly valuable. The primary tools employed in data mining are statistics, artificial intelligence, machine learning, and other mining methods such as decision trees, neural networks, association rules, and cluster analysis. The output models, rules, corresponding model scores, and labels can be obtained using various inputs, including churn probability value, sum score, similarity, predicted value, and labels such as high-, medium-, and low-value users, as well as non-churn, excellent, and poor credit.

In terms of technology, the relationship between big data and cloud computing is as inseparable as the front and back of a coin. A single computer cannot process the big data as the latter is built on a distributed architecture. Its characteristic lies in the distributed data mining of massive data. However, it must rely on cloud computing's distributed processing, distributed database, cloud storage, and virtualization technology to achieve correct performance and results.

For data collection, the Internet webpages' search engine must download all the webpages of the entire Internet. This is an inherently impossible task to complete with a single device, as it requires multiple machines to form a web crawler system to work simultaneously. Each machine must be capable of downloading massive data in a limited time (as shown in Fig. 1.2, the Nutch search engine obtains distributed data through an injector, generator, crawler, and parser).

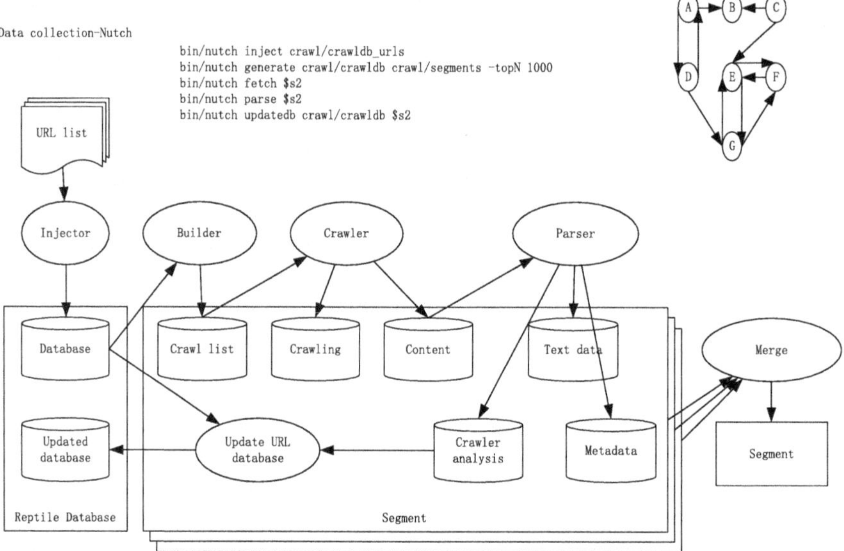

Fig. 1.2 Distributed data acquisition

Due to the sheer amount of data, the queue in the memory of a single device must overflow during data transmission. This requires the use of a hard disk-based distributed queue (Fig. 1.3). It is possible for distributed queues to be transmitted by multiple machines simultaneously, provided that the number of queues is sufficient. Consequently, there is no need to concern oneself with the possibility of data overflow.

In order to store vast quantities of data, a distributed file system is required. This is achieved by using the hard disks of multiple machines to form a unified file system. For example, the Hadoop Distributed File System (HDFS) stores distributed data. As shown in Fig. 1.4, HDFS is designed to be a distributed file system that can be run on common hardware. It relaxes certain constraints imposed by the Portable Operating System Interface of UNIX (POSIX) to achieve the purpose of streaming file system data reading.

Another example is the analysis of data. The computing power of a single device is quite limited, and it is often challenging to deal with massive data. The distributed computing method can effectively address this problem. The approach employs the "divide and conquer" strategy to divide a substantial dataset into small components. Each machine processes a specific subset, and multiple machines process

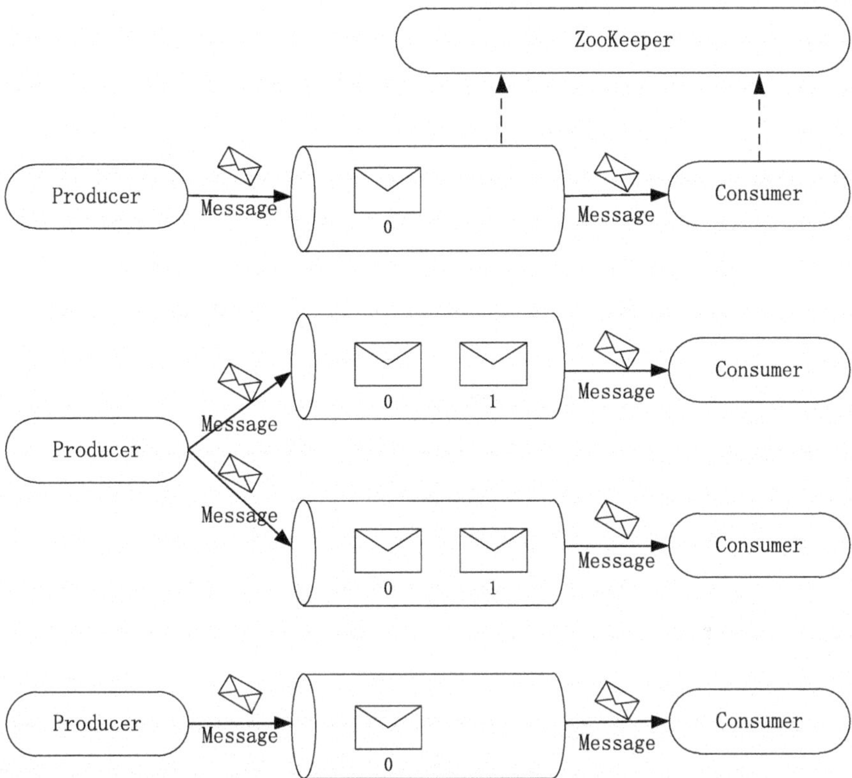

Fig. 1.3 Distributed data transmission

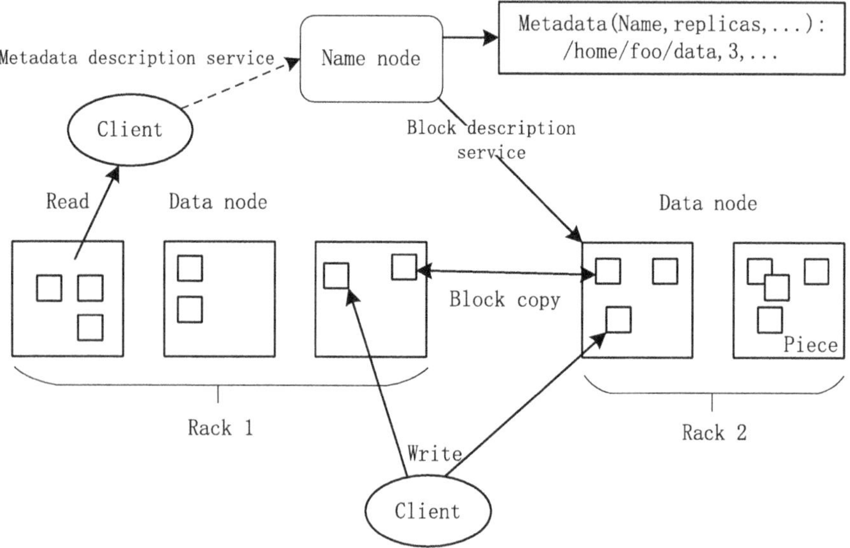

Fig. 1.4 Distributed data storage

the data in parallel, which enables the completion of the task in a relatively short time (MapReduce, as shown in Fig. 1.5).

From the above analysis, it can be seen that the fundamental principle of big data processing is the "many hands make light work" adage. By distributing tasks to a large number of distributed nodes, the performance and speed of computing can be improved, and finally massive computing tasks can be completed [3].

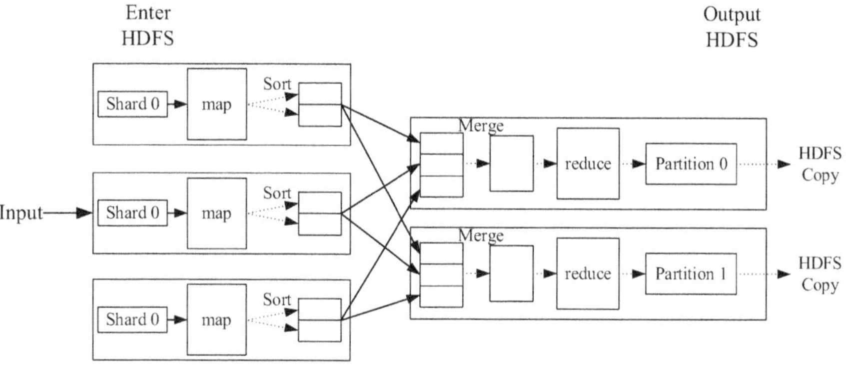

Fig. 1.5 Distributed data analysis

1.2.2 Interdisciplinary Connections of Data Science

The term "big data" is used to describe data collections with "5 V" characteristics that cannot be captured, managed, and processed using conventional software tools. The discipline that studies information extraction techniques based on massive data is called data science. Formally, data science refers to the emerging disciplines based on theories and technologies of computer science, statistics, information systems, and other disciplines. The objective of data science is to collect and collate research data and analyze and process it from massive data to obtain adequate knowledge and apply it. In contrast, data engineering is the application of engineering perspectives to the data management and analysis, the development, and application of data systems [4].

In contrast, the computer science discipline is the study of algorithms, while data science encompasses a much broader range of topics. Data science represents an interdisciplinary foundation for the research and application of big data. Its theoretical foundation is derived from many different academic disciplines, including computer science, statistics, artificial intelligence, information systems, and information sciences. The objective of data science is to explore various scientific and technical issues in a systematic manner, with a particular focus on the challenges encountered in the implementation of big data. This includes issues related to the full lifecycle management of data, data management, and analysis technologies and algorithms, data system infrastructure construction, and the broad application of data and the promotion of data-driven initiatives. Consequently, the integration of disciplines is a defining feature of data science. Figure 1.6 shows the first Venn diagram on the subject of "data science," created by Drew Conway in 2010. The central part of the diagram is data science, which is defined as a combination of hacking techniques, mathematics, statistics, and other substantial expertise.

1. **Data Science and Computer Science**

Computer science is a discipline that regularly studies the theoretical foundations of information and computing, as well as the way that they are implemented and applied in computer systems. It is often described as a systematic study of algorithms that create, describe, and transform information. There is an overlap between computer science and data science. Both areas use computational processes, and both require a practical understanding of programming languages and algorithms. However, the way this understanding is applied is the main difference between these two areas. Computer science focuses on the "how," whereas data science focuses on the "why." Computer science is a fundamental discipline, whereas data science is an applied discipline.

Furthermore, computer science is concerned with the principles of algorithms and the specific details of the calculation process, with less attention paid to the specific logical results of function implementation. Computer scientists may engage in the development of applications, the creation of new programming languages, or the design of systems for the generation and sorting of data streams. Nevertheless, for

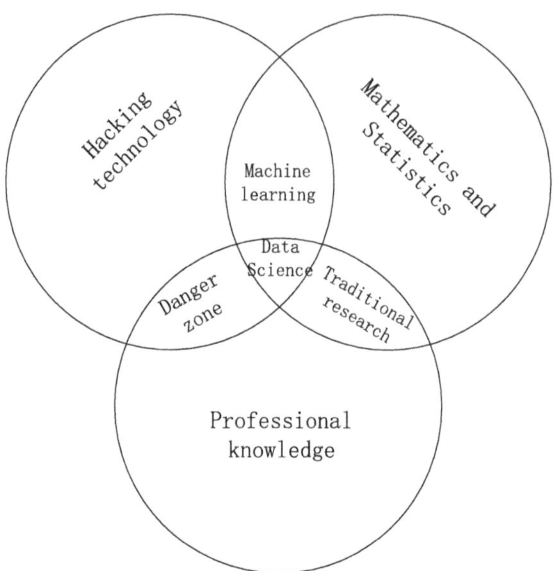

Fig. 1.6 The data science concept venn diagram (by Drew Conway)

computer scientists, these processes are typically based on voltage-to-bit symbolic logic, and the results are predictable.

In data science, the principles of algorithms are applied to more uncertain fields, often resulting in probabilistic answers to interdisciplinary questions such as those pertaining to business. Modern data scientists are typically proficient in computer science, but they may also have backgrounds in mathematics, statistics, or business. Data scientists are responsible for designing algorithms, refining datasets, and parsing large amounts of data through mathematical models in order to mine actionable knowledge. To achieve this process, data scientists must adopt an interdisciplinary approach, accepting and dealing with uncertainty.

2. **Data Science and Software Engineering**

Software engineering is the application of engineering methods in the field of software development in a systematic manner. In 1993, the Institute of Electrical and Electronics Engineers (IEEE) provided a more comprehensive definition: "The systematic, standardized, and measurable methods used in the process of software development, operation, and maintenance will apply engineering to software." Data science is typically required to process data collections that cannot be captured, managed, and processed using conventional software tools. The objective of software engineering is to design and implement systems and architectures that can handle massive amounts of data [5].

3. **Data Science and Artificial Intelligence**

Artificial intelligence is a relatively new technical science that studies and develops theories, methods, technologies, and application systems with the objective of simulating, extending, and expanding human intelligence. The primary application of artificial intelligence is to achieve its objectives through training vast quantities of data. Consequently, artificial intelligence frequently requires a considerable dataset. Nevertheless, although the principal methodologies of data science and artificial intelligence are founded upon the analysis of big data, the objectives of both disciplines diverge to a certain extent. In particular, data science is concerned with the generation of "insights," whereas artificial intelligence is focused on the development of "behaviors." Another concept frequently mentioned is machine learning, which is usually designed to produce "predictions" [6].

Consider a scenario in which we are developing a driverless car. One of the specific challenges we are addressing is the ability of the car to automatically park at a location with a parking sign. To achieve this, we must draw upon the knowledge and skills derived from three distinct fields: machine learning, artificial intelligence, and data science. In the realm of machine learning, our vehicles are equipped with cameras designed to recognize stop signs. A dataset comprising millions of street view sign images was constructed, after which an algorithm was trained to predict the location of parking signs. In the field of artificial intelligence, once the vehicle has successfully identified a parking sign, it must then determine the optimal moment to apply the brake. Braking too early or too late could pose a significant risk. Our car must be able to deal with different road conditions (for example, identifying a smooth surface that does not decelerate rapidly). This is a quintessential control theory problem [7]. Shifting our focus to data science, during on-street testing, we observed that the car's performance was suboptimal, with numerous factors contributing to misidentification of parking signs. A thorough analysis of the test data revealed a correlation between the time of day and the rate of missed judgments: the early morning hours before sunrise and the evening hours after sunset were particularly problematic. This insight led us to recognize a gap in our training data, which predominantly featured daytime scenes. Consequently, we enhanced our dataset to include images captured at night, enabling us to revisit and refine our machine learning models with this expanded perspective [8].

1.3 Historical Overview

1.3.1 Evolution of Big Data

1. **The Emergence of "Big Data" (1980–2008)**

In 1997, NASA researchers Michael Cox and David Ellsworth used the term "big data" for the first time to describe the challenge of the 1990s related to simulating the airflow around an aircraft, which could not be processed or viewed [9]. These datasets are usually large, exceeding the capacity of the main memory, local disks, and even remote disks. Consequently, this issue has been designated the "big data problem." In the aftermath of the 9/11 attacks in 2002, the US government has been engaged in extensive data mining operations to prevent terrorism. Former National Security Adviser John Poindexter directed the Department of Defense to integrate existing government datasets and form an extensive database for the purpose of identifying individuals who may be engaged in suspicious activities. This database would be used to screen records pertaining to communications, crime, education, finance, medical care, and travel. One year later, Congress terminated the project due to concerns regarding civil liberties.

In 2004, the 9/11 Commission recommended that anti-terrorism agencies establish a unified "network-based information sharing system" to facilitate the rapid processing of overwhelming data. In 2006, Google first proposed the concept of cloud computing. It was not until the advent of cloud computing that the term "big data" became prominent.

From 2007 to 2008, the advent of social networks saw technology bloggers and professionals inject new life into the concept of "big data." According to Chris Anderson of Wired magazine, "Some other tools currently available worldwide will be replaced by large amounts of data and applied algorithms," which is "the end of theory." At the same time, a number of U.S. government agencies and leading computer scientists claimed that "we should be deeply involved in the development and deployment of big data computing because it will directly benefit the implementation of many tasks." Hence, in September 2008, the journal Nature launched a cover column called "Big Data."

2. **Popular Stage of "Big Data" (2009–2011)**

From 2009 to 2010, the term "big data" became a popular term in the Internet technology industry. In 2009, the Indian government launched the Aadhaar project, which has become the world's largest biometric database project. It is estimated that over 95% of the country's population, or approximately 1.1 billion individuals, have had their names, addresses, mobile phone numbers, and highly sensitive personal information, including fingerprints, photos, and iris scans, included in the central government database as a unique identification code for citizenship (UIDAI). Similarly, the United Nations Global Pulse Project has studied the utility of data sources from mobile phones and social networking sites in order to analyze and predict issues such as spiral prices and disease outbreaks. Consequently, the U.S. government launched a new website, Data.gov, which further opened the door to data. This website contains more than 44,500 datasets, which are used to ensure that some websites and smartphone applications track information. This action has inspired a similar range of actions from Kenya to the United Kingdom. Likewise, Governments have launched similar initiatives in rapid succession. Some of the most prominent research libraries, scientific, and technological information research institutions in Europe have established partnerships with the objective of enhancing the accessibility of scientific data on the Internet.

In 2010, Kenneth Cukier published the special report, "Data, Data Everywhere." In 2011, the scanning of 200 million years of page information and four terabytes of disk storage required only a few seconds. In a televised quiz show "Dangerous Edge," IBM's Watson computer system defeated two human competitors, prompting the New York Times to describe this as "a big data computing victory." In the same year, in June, McKinsey published a report on "Big Data" entitled " Big data: The next frontier for innovation, competition, and productivity," which formally defined big data and gradually received attention from all walks of life. In December, the Ministry of Industry and Information Technology released the IoTs for the 12th Five-Year Plan. The report proposed information processing technology as one of the four critical technological innovation projects, including mass data storage, data mining, and intelligent analysis, which are the most critical components of big data.

3. **"Big Data" Has Become the Characteristics of the Times (2012–2016)**

In 2012, numerous countries, institutions, and researchers proposed and issued various reports pertaining to the big data concept. Viktor Mayer-Schönberger (one of the earliest data scientists to recognize trends in the era of big data) and Kenneth Cukier in their book "Big Data: A Revolution That Will Transform How," divided the impact of big data into three different levels: thinking change, business change, and management change. The concept of "big data" plays a pivotal role in all aspects of life in the Internet age. Consequently, the term big data is becoming increasingly prevalent in discourse. The term "big data" is employed to describe and define the vast quantities of data generated in the era of the information explosion, as well as to designate the related technological developments and innovations. The quantity of data is rapidly expanding and becoming more massive, and it is therefore a determining factor in future developments. With time, it is likely that individuals will become increasingly aware of the significance of data.

In the same year, the World Economic Forum published the report "Big Data, Big Impact," in which they declared that data had become a new economic asset class. Respectively, the Obama Administration published the "Big Data Research and Development Initiative" on the White House website. They considered that big data had become an essential feature of the contemporary era. Consequently, the United States government invested US$200 million in big data. The field of data technology has grown from a commercial phenomenon to a pivotal element in national science and technology strategy.

Similarly, as the United States declared the "Big Data Research and Development Initiative," the United Kingdom announced the "UK data capability strategy." Likewise, Japan issued the "Declaration on Creating the Most Advanced IT Country." Hereafter, South Korea proposed the "Big Data Center Strategy," and other countries in the world have also formulated corresponding strategies and plans. The United Nations published a white paper on Big Data Government Affairs, entitled "Big Data for Development: Challenges & Opportunities." This document provides an overview of how governments are leveraging big data to enhance service delivery and protect their citizens.

The first year of big data in China was 2013. Although the concept of big data has existed for a long time, it only recently attracted significant attention due to the development of the Internet and the information industry. This year, big data began to gradually develop in China, gaining a foothold in people's minds and influencing various social exploration and land practice fields. On January 1, 2013, Alibaba Corporation transformed and reshaped its three businesses—"platform," "finance," and "data"—becoming the first enterprise to propose data-based operations.

In 2014, big data was included for the first time in the "Government Work Report." This report pointed out the need to establish an entrepreneurial innovation platform for emerging industries, surpassing the advances in big data and other aspects, and leading future industrial development. The State Council passed the "Interim Regulations on Enterprise Information Disclosure (Draft)," which requires the establishment of an interconnected and shared information platform among enterprise departments and the use of big data and other means to improve the level of supervision. Subsequently, the topic of big data has become a key focus in China.

In 2015, the "Guiyang Big Data Exchange 702 Convention" and the "2015 China Big Data Trading White Paper," both of which were unveiled at the China Big Data Trading Summit Forum, were launched by the Guiyang Big Data Exchange. The white paper systematically elaborated on the status of big data transactions, while the convention introduced and analyzed from the global to the country and from the country to the industry. The report played a pivotal role in advancement and direction of the big data industry and transactions.

In 2016, the Ministry of Industry and Information Technology issued the "Big Data Industry Development Plan (2016–2020)," which outlined a comprehensive strategy to advance the big data industry during the 13th Five-Year Plan period. The plan aimed to accelerate the construction of data capacity and provide robust industrial support for the realization of manufacturing capacity and network capacity. The "China Big Data Development Survey Report" stated that the size of China's big data market was 16.80 billion yuan in 2016, representing a growth rate of 45%. It is expected that the growth rate will remain above 30% between 2017 and 2020.

1.3.2 Current Research Landscape

In the field of big data, numerous novel technologies are developed annually, which have become effective means for big data acquisition, storage, processing analysis, and visualization. Big data technology can dig out the hidden information and knowledge in large-scale data, provide a basis for human socio-economic activities, improve the operational efficiency of various fields, and even intensify the entire social economy.

1. **Research Status of Big Data Analysis Architecture**

The fundamental analytical process of big data and the traditional data analysis process are largely similar. However, the former involves a significant amount of unstructured data, which can be processed in parallel at each stage. Distributed processing methods, such as Hadoop, MapReduce and Spark, have become the preferred choice for big data processing for various aspects of big data processing.

Hadoop stands out as a robust data management system, enabling users to efficiently construct and utilize applications that swiftly process enormous volumes of data [10]. As the core of data analysis, Hadoop brings together structured and unstructured data, distributing it across the layers of the traditional enterprise data stack. Hadoop, an open-source community-driven project, primarily offers tools and software designed to tackle big data challenges. It also provides a massively parallel processing framework endowed with supercomputing capabilities, ideally positioned to facilitate the execution of enterprise-level applications. Despite these numerous functions of Hadoop, it is more accurately described as a Hadoop ecosystem that consists of multiple components, including but not limited to data storage, data integration, data processing, and a suite of specialized tools for data analysis. The Hadoop ecosystem comprises core components such as HDFS, MapReduce, Hbase, Zookeeper, Oozie, Pig, and Hive. Furthermore, the ecosystem is continually expanding, incorporating frameworks like Sqoop and Flume to ensure seamless integration with other enterprise systems. It also embraces new additions such as Mahout, Ambari, Whirr, BigTop, and other elements, thereby enriching the ecosystem with ongoing updates.

Based on the dynamic and real-time demands of businesses, the big data analysis architecture has evolved to include processing architectures such as Storm, Cloudera Impala that supports online processing, Spark that supports iterative calculation, and the stream processing framework S4. Storm, a distributed and fault-tolerant real-time computing system, was originally developed by BackType and later acquired by Twitter. As a stream processing platform, Storm excels in performing real-time calculations and updating databases instantaneously. It can also be employed for "continuous calculation." The continuous query of the data stream generates a results stream for the user during the calculation. It is also used for "distributed remote procedure call (RPC)" to run expensive operations in parallel. Cloudera Impala is an open-source massively parallel processing query engine developed by Cloudera. The same metadata, Structured Query Language (SQL) syntax, Open Database Connectivity (ODBC) driver, and user interface as Hive can directly provide fast and interactive SQL queries on HDFS or HBase. Impala was developed under the inspiration of Dremel, rather than using the slow Hive plus MapReduce batch processing. However, through a distributed query engine and similar to the commercial parallel relational database (Query Planner, Query Coordinator, and Query Exec Engine), it can directly use SELECT, JOIN, and statistical functions query database HDFS or HBase. This significantly reduces the delay.

2. **Research Status of Big Data Acquisition and Preprocessing Technology**

Data collection represents the initial stage of the development cycle of big data. According to the classification of the application system for MapReduce's data, there are four primary sources of big data collection: management information system, Web information system, physical information system, and scientific experiment system. The data in datasets can be structured with different modes, such as files, Extensible Markup Language (XML) trees, and relational tables, which contribute to the heterogeneity of the datasets. Consequently, further integration processing is required to collect, sort, clean, and convert the data into a new dataset to provide uniformity for subsequent query and analysis processing views. Various research works have been proposed to integrate heterogeneous databases in the management information system, the entity recognition technology in the Web information system, the DeepWeb integration technology, and the data fusion technology for sensor networks. A variety of data cleansing and quality control tools have been introduced, including Data Flux from the Statistical Analysis System (SAS), Data Stage from IBM, and Informatica Power Center from Informatica.

3. **Research Status of Big Data Storage and Management Technology**

Most data storage and management systems are structured, which has led to the development of the Relational Database Management System (RDBMS). This system is capable of meeting the diverse needs of various applications in the world. However, the rise of big data has led to the emergence of semi-structured and unstructured data, which is often combined with structured data. Big data applications are often designed for diverse tasks such as content retrieval across different types of data, cross comparison, in-depth data mining, and comprehensive analysis. Given these evolving application requirements, traditional databases are becoming increasingly inadequate in terms of both technical and functional capabilities. Consequently, in recent years, there has been a coexistence of three distinct categories of databases: oldSQL, NoSQL, and NewSQL.

According to different data types, big data storage and management generally use different technical routes, which can be roughly divided into three categories. The first category mainly deals with large-scale structured data; hence, new database clusters are usually used. These clusters use the technology of column storage, row-column hybrid storage, and coarse-grained indexing, combined with Massive Parallel Processing (MPP) architecture efficient distributed computing model, in order to achieve the storage and management of PB-level data. This type of cluster has high-performance characteristics, high scalability, and has been widely used in enterprise analysis applications. The second category mainly faces semi-structured and unstructured data. The system platform based on the Hadoop open-source system is more reliable for such application scenarios. They expand and encapsulate the Hadoop ecosystem to achieve the storage and management of semi-structured and unstructured data. The third category is the big data mixed with structured and unstructured. The MPP parallel database employs a mixed cluster and Hadoop cluster to manage and store hundreds of petabytes and exabyte (EB) data. On one hand, MPP is used to manage and calculate high-quality structured data and provide robust SQL and online transaction processing (OLTP) services. On the other hand, Hadoop is used to process semi-structured and unstructured data to support applications such as content retrieval, in-depth analysis, new applications, data mining, and comprehensive analysis [11]. The hybrid model is set to be the future trend in big data storage and management.

4. **Research Status of Big Data Computing Models and Systems**

The term "big data computing mode" refers to a variety of high-level abstractions or models that have been refined and established based on various data and computing needs associated with big data. For example, there is MapReduce parallel computing abstraction, the "distributed memory abstraction RDD" in the famous Spark system of the University of California, Berkeley, and the "graph parallel abstraction" in CMU's famous graph computing system GraphLab.

The advent of computing models has significantly promoted the development of big data technology and applications, making it the most successful and widely accepted mainstream big data computing model for big data processing. However, the reality of big data processing is complex and diverse, making it challenging to find a single computing model that can address all different big data computing needs. Research and practical applications have demonstrated that MapReduce is mainly suitable for offline batch processing of big data. It is highly incompatible when confronted with low-latency and big data problems, particularly those involving complex data relationships and calculations.

Consequently, in recent years, academia and industry have been continuously researching and introducing a variety of big data computing models. In response to the diverse needs of big data processing and the above different feature dimensions, a variety of typical and crucial big data computing models have emerged. In response to these calculation modes, numerous corresponding big data calculation systems and tools have emerged. A more abstract and detailed description of the calculation mode can be provided by providing the corresponding typical calculation systems and tools, as shown in Table 1.1. This will help you to understand the computing model and grasp the status of technology development [12].

5. **Research Status of Big Data Analysis and Visualization**

In the application process of big data analysis, visualization plays a crucial role in helping people explore and understand complex data through interactive visual representation. Visualization and visual analysis can streamline and optimize data flow, facilitate data filtering, accelerate discovery from complex data, enhance data comprehension, and serve as a vital tool for in-depth analysis. The visualization of large-scale data is primarily based on the technology of parallel algorithm design, the

Table 1.1 Typical big data computing model and typical system

Typical big data computing model	Typical system
Big data query analysis calculation	Hbase, Hive, Cassandra, Impala, Shark
Batch calculation	Hadoop MapReduce, Spark
Streaming computing	Scribe, Flume, Storm, S4, Spark Steaming
Iterative calculation	HaDoop, iMapReduce, Twister, Spark
Graph calculation	Pregel, Giraph, Trinity, PowerGraph
Memory calculation	Dremel, Hana, Spark

rational use of limited computing resources, and the efficient processing and analysis of specific characteristics of datasets [13]. Large-scale data visualization technology will combine multi-resolution representation and other methods to ensure sufficient interactive performance under normal circumstances [14]. The parallel visualization of scientific large-scale data is based on four fundamental technologies: data streamlining, task parallelization, pipeline parallelization, and data parallelization.

Microsoft has developed a large-scale machine learning visualization platform (Azure Machine Learning) on its cloud computing platform Azure. The platform displays big data analysis tasks as directed acyclic graphs and presents them to users in the form of data flow graphs [15]. In China, Alibaba's big data analysis platform Yushanfang has adopted a similar approach to provide interactive big data analysis platforms for business personnel.

1.4 Major Applications

1.4.1 Applications in the Internet Industry

1. **Application in Online Advertising Push**

   (1) Establishing user database: Firstly, the system employs collection technology in big data to gather user behavior data from various social platforms, including online media and communities. This data is then analyzed to identify user needs, which are subsequently utilized to implement discount policies. For example, users with a higher posting volume may be granted access to certain virtual privileges, thereby encouraging them to register as members and expand the user database. Secondly, the system utilizes data mining technology in big data to conduct member recruitment, sales transactions, social surveys, and other activities, thereby obtaining user information data. For instance, China Unicom applies Hadoop and big data technologies in its user record, analysis, and query support system, as well as in the analysis of user online behavior, with the objective of providing the user with the most appropriate business offering.

   (2) Changing the form of broadcasting online advertisements: At this stage, the prevalence of advertisements on Internet browsing pages has the potential to significantly affect the user's online experience. In this case, it is imperative that advertisers leverage big data to change the form of online advertising. Firstly, the system employs big data collection technology to gather user click behavior, analyze it to determine user focus, and then formulate the size, form, and location of online advertising. Secondly, the pre-processing technology is used to transform the manual window closing method in an automatic timing closing process, which significantly attracts the user's attention.

(3) Timely adjustment of advertisements. The use of big data allows for the multi-dimensional positioning of online advertisements by regions, netizen (citizen of networks) characteristics, keywords, and investment time. Simultaneously, real-time optimization dynamic technology is used to set the coverage and delivery time of advertisements, identify more real concerns of the audiences, and dig deep into the Internet. This enables the potential consumers to be updated and have their advertisements adjusted promptly.

2. **Application in Network Security**

(1) Application in the risk assessment system: Risk assessment is a vital part of the Internet defense system. When applied to big data, the risk assessment technology and level analysis and quantification technology are used to effectively analyze and evaluate the Internet security complexity, thus eliminating the need for excessive dependence. Furthermore, D-S evidence and gray theories can also be used to optimize the network security assessment system in order to meet the needs of network security adequately.
(2) Application in the security audit system: Data analysis represents the core content of the security audit system. The application of big data technology provides access to genetic algorithms and BP neural network algorithms. In this process, big data can be employed to construct a genetic neural network, thereby significantly improving the network security accuracy of defense system identification data.
(3) Application in the active defense system: Relevant personnel can combine detection technology, protection technology, and early warning technology, as well as utilize big data to construct a simulated dynamic anti-virus system. Hence, the system can automatically monitor network program words and concurrently analyzing the logical relationship between the various actions of network programs. In this way, the independent analysis function of big data enables the rapid identification of viruses, the automatic blocking of intrusions, and the automatic repair of network loopholes, thereby providing comprehensive network security.

3. **Application in the Portal**

(1) Data mining: The most critical technology in big data is the data mining technology, which uses this technology to extract the user's concerns about browsing information from the Internet database. Firstly, the data is calculated using genetic algorithms. The calculation target may be one or several, and the objective is to extract more valuable information from it. Secondly, the data is visualized, with a greater number of variables displayed on the user's surface, and a variety of graphics provided to the user.
(2) User behavior analysis: The analysis of user behavior is mainly conducted using cluster analysis technology in big data, which enables the identification of various clusters and classes based on the characteristics of users, and the subsequent classification of various featured data. Subsequently, the wavelet clustering algorithm is used to compute the high-frequency signal and the

user feature space. The data object vector serves as the input value for the generation of the clustering object, thereby meeting the diverse needs of the user.

1.4.2 Applications in Healthcare

Apart from Internet companies that started the use of big data at an early stage, the medical industry is one of the first traditional industries to allow big data analysis. The medical industry has many case studies, pathology reports, treatment programs, and drug reports, etc. The organization and application of these datasets will significantly benefit both doctors and patients. The number and variety of germs, viruses, and tumor cells with which we are confronted are in the process of continuous evolution. It is found that the diagnosis of the disease and the determination of the treatment plan represent the most challenging aspects of this process. In the future, the use of a big data platform will facilitate the collection of different cases and treatment options, as well as the essential characteristics of patients. Furthermore, the establishment of a database for disease characteristics will also be possible.

In the event of the mature development of gene technology in the future, it can be classified according to the characteristics of the patient's gene sequence, and a patient classification database in the medical industry can be established. When a doctor diagnoses a patient, he can refer to the patient's disease characteristics, laboratory, and test reports, and refer to the disease database to quickly confirm the diagnosis and locate the disease. When formulating treatment plans, doctors can call for effective treatment plans with similar genes, ages, races, and physical conditions based on their genetic characteristics. This allows them to develop treatment plans that are suitable for patients and to treat more people in a timely manner. At the same time, these data are conducive to developing more effective drugs and medical devices in the pharmaceutical industry.

The field of data mining has made significant contributions to the field of medical big data research. The main applications are as follows:

1. **Early Warning of Disease**

The medical field often needs more precise real-time early warning tools. The establishment of disease early warning models based on data mining can improve the early diagnosis, warning, and monitoring of diseases. Simultaneously, it can also help medical institutions take preventive and control measures to reduce the deterioration of the disease and the occurrence of complications. In the early warning of disease, it is first necessary to collect data related to disease or risk factors. This data can be used to establish a model that identifies the relationship between disease pathogenesis and the underlying conditions revealed by the data.

Forkan et al. collected a variety of vital signs data, including heart rate, diastolic blood pressure, systolic blood pressure, average blood pressure, respiration rate, and blood oxygen saturation, as well as other daily monitoring data. The researchers used the J48 decision tree, random forest tree, and sequence minimum optimization algorithm to establish disease early warning models for remote home monitoring. These models were designed to identify the occurrence of undiagnosed diseases, to send the monitoring results to medical emergency institutions, to achieve complete integration of vital signs big data patients and medical institutions, and to reduce the incidence of sudden illness and death. In a related study, Easton et al. used the Bayesian classification algorithm to establish a death prediction model for stroke sequelae. They believed that the death probability of stroke sequelae is a function of the time elapsed since the initial stroke. This insight facilitates the follow-up monitoring of these patients. Tayefi et al. established a coronary heart disease prediction model based on the decision tree algorithm. The model demonstrated that hs-CRP, a novel marker for coronary heart disease, exhibited greater specificity than traditional markers, such as Fasting Blood Glucose (FBG).

2. **Chronic Disease Research**

Chronic diseases like diabetes, high blood pressure, and cardiovascular diseases pose significant threats to public health. Identifying chronic disease risk factors and establishing early warning models are crucial steps in reducing the occurrence of chronic disease complications. In a pioneering study, Alagugowri et al. established a sophisticated early warning system for heart disease. This system was developed through the meticulous extraction of key indicators from a comprehensive heart disease database. The K-means clustering algorithm was employed to identify the underlying risk factors of heart disease. Furthermore, the Apriori algorithm was leveraged to mine association rules between high-frequency risk factors and heart disease risk levels.

Ilayaraja et al. used high-frequency item sets to identify risk factors for heart disease and to determine patient risk levels. This method can avoid the generation of meaningless item sets, thus resolving the issues of a large number of item sets and the required storage space in previous studies. Jen et al. studied the identification of chronic disease complications, which is divided into three steps. First, the data from healthy individuals and those with chronic disease are selected. Then, linear discriminant analysis with sequence prior selection is employed to identify the characteristic variables of related diseases. Subsequently, the K-NN algorithm is used to classify the characteristic variables. Finally, the results of the K-NN algorithm's classification are applied to the establishment of chronic disease early warning models.

Aljumah et al. used regression analysis and SVM to predict and assess the optimal match between different diabetes treatments and distinct age groups, thereby providing patients with a foundation for choosing the best treatment. Perveen et al. used the Adaboost integration algorithm, Bagging algorithm, and decision tree algorithm to establish a prediction model for diabetes using patient demographic and clinical index data. The Adaboost integration algorithm is believed to be more accurate.

3. **Assisted Medical Diagnosis**

Medical data is not only astronomical in volume, but also intricate and interrelated. The analysis of a substantial quantity of medical data and mining of valuable diagnostic rules will serve as a reference for disease diagnosis. Yang et al. investigated the correlation between lung cancer pathology reports and clinical information, employing the decision tree and the Apriori algorithms. This study provided a basis for the diagnosis of lung cancer pathology, thereby eliminating the need for surgical methods to obtain pathological tissue in the diagnosis.

Becerra-Garcia and other algorithms use SVM, K-NN, and CART algorithms to perform signal preprocessing, pulse detection, and classification on the electrooculogram, thereby providing a foundation for the study of non-spontaneous saccade eye movement recognition in clinical electrooculogram examinations. Peng et al. conducted an association rule mining analysis on a 5-year dataset of breast ultrasound data from a hospital. They established association rules between pathological breast diagnosis and ultrasound diagnosis. Additionally, they developed a breast ultrasound database for a data retrieval system, which facilitates the rapid retrieval of ultrasound diagnosis, pathological diagnosis, various diagnostic information, and case information by medical professionals.

1.4.3 Applications in Finance

Big data is widely used in the financial industry. Typical cases include: Citibank uses IBM Watson computers to recommend products to wealth management customers; Bank of America uses customer click datasets to provide customers with unique services, such as competitive credit lines; China Merchants Bank uses a customer's credit card, deposits, withdrawals, electronic bank transfers, WeChat comments, and other behavioral data for analysis, and sends targeted advertising information to customers every week, including products and discounts that customers may be interested in.

On the other hand, big data can improve the controllability and management of risks by reviewing and evaluating massive data. It finds and resolves possible risk points in time by having an accurate grasp of the regularity of risk occurrence, encourages financial institutions to require deeper and more thorough data analysis, and supports refined business management. Although banks have a lot of payment pipeline data, the various departments do not intersect, the data cannot be integrated, and the big data financial model has encouraged banks to effectively use the deposited data. Big data will promote financial institutions to innovate brands and services, achieve refined services, personalize customers, use data to develop new forecasting and analysis models, and implement the analysis of customer consumption patterns to increase customer conversion rates. Big data will definitely bring more up-to-date data-based business and internal management optimization opportunities to financial enterprises.

1.4.4 Applications in Transportation

The application of big data in transportation is mainly classified into two aspects. On the one hand, big data sensors can be used to understand vehicle traffic density and reasonable road planning, including single-line route planning. On the other hand, live big data can be used to implement real-time semaphore scheduling and improve the operational capability of existing lines. Scientific signal arrangement is a sophisticated system engineering and a big data computing platform must be used to calculate a more reasonable plan. A scientific signal arrangement will increase the capacity of existing roads by about 30%. In the United States, the government added traffic lights based on traffic accident information on a specific section, reducing the traffic accident rate by more than 50%. Relying on big data for take-off and landing at airports will improve the efficiency of flight management. Airlines can use big data to increase the number of passengers and reduce operating costs. Railways can use big data to effectively schedule passenger and freight trains, improve efficiency, and reduce costs.

1.4.5 Applications in Education

In the classroom, data can help to improve education and teaching, but big data is more useful in making major educational decisions and reforms. The United States uses data to diagnose students at risk of dropping out of school, explore the relationship between education expenditure and students' academic performance, explore the relationship between education spending and student achievement, and explore the relationship between student absenteeism and achievement. For example, data analysis of public elementary and secondary schools in a particular state in the United States shows that teachers' college entrance exam scores and students' scores have a significant positive correlation in terms of language performance. That is, teacher's college entrance exam scores have a clear relationship with the students' academic performance in the language classes they teach. The better the teacher's college entrance exam scores, the better the students' language scores. This relationship allows us to further explore the real reason behind it. In fact, the level of teachers' college entrance examination scores is to some extent a teacher's characteristic at work, and this characteristic plays a vital role in teaching students well. Teachers' college entrance examination scores can be used as an indicator for selecting teachers. If we have sufficient data, we can explore the relationship between teachers' characteristics and students' performance, thus providing a better reference for selecting teachers.

Big data can also help parents and teachers to identify children's learning gaps and effective learning methods. For example, the McGraw-Hill Education Publishing Group in the United States has developed a predictive assessment tool to help students in identifying existing knowledge gaps and the required level of compliance tests, as well as pointing out areas for improvement. The assessment tool allows teachers to track students' learning progress, identify students' learning characteristics and methods. Some students are suitable for a step-by-step learning approach, while others may benefit from a nonlinear schema and integrated information approach. These can be effectively identified through the use of big data collection and analysis, providing a solid basis for education and teaching.

In China, particularly in Beijing, Shanghai, Guangdong, and other cities, big data has been used in the field of education, including MOOC, online courses, and flipped classrooms. Among these applications, a lot of big data tools are used.

1.5 Key Challenges

1.5.1 Data Storage

The capacity for data storage has continued to increase over time, progressing from the TeraByte (TB) level to PetaByte (PB) level and beyond, reaching the ExaByte (EB) level. In 2016, Google processed over 63,000 search requests per second, while Facebook handled nearly one million posts per minute and sent 400 million emails simultaneously. Baidu, for its part, processed tens of billions of requests per day. The number of WeChat users exceeded one billion. There are more than one hundred billion online human relationship chains, and the amount of newly added data is close to 500 TB per day. Considering this vast quantity of data, the traditional structured data storage method is no longer capable of supporting the storage of large amounts of data, thereby posing new challenges to related technical architectures.

Firstly, traditional databases cannot handle PB-level data. The exponential growth in data volume has exceeded the management capabilities of conventional databases. Consequently, it is necessary to add servers and construct a distributed data warehouse to expand capacity. Secondly, the traditional database technology does not consider the diversity of data categories, especially the compatibility with structured, semi-structured, and unstructured data. Finally, the processing of massive data requires the provision of excellent network architecture support and a robust data center as a foundation. Ensuring data stability has also become a significant problem.

Moreover, the cost of data storage represents a significant limitation in the development of big data. In 2012, Tianjin's security system projected the construction of a network with 4.6 EB storage capacity at a cost of over 50 billion yuan, which was equivalent to the total GDP of Tibet that year. Therefore, the exponential growth of data volumes and the associated high costs have become significant constraints on the development of big data. The majority of cities in our country use shorter data storage time limits and lower data storage quality to reduce storage costs, and use dynamic data processing techniques to alleviate data storage issues. Nevertheless, the aforementioned approach does not fundamentally address the data storage issue, so it is still the bottleneck of the development of big data technology. The advent of cloud databases offers a potential solution to the challenges posed by storage limitations.

A cloud database is a method of sharing infrastructure based on the development of cloud computing technology. Databases should be deployed and virtualized in a cloud computing environment. Cloud databases are highly scalable, highly available, multi-tenant, and support efficient resource allocation. From the perspective of the data model, cloud database is not a new database technology, it just provides database functionality in the form of services in the cloud environment. The data model used by cloud databases can be a variety of data models, such as the relational model used by the relational database (Microsoft SQLAzure cloud database uses the relational model).

1.5.2 Information Security

The pervasiveness of big data security threats has extends across the entire big data industry chain, encompassing data production, collection, processing, and sharing. The underlying causes of these risks are complex and intertwined. They include external attacks and internal leaks, technical vulnerabilities and management defects, and the introduction of new technologies, which have triggered new security issues within the context of the evolving big data model. This section introduces the security problems faced by big data itself; analyzes three aspects of big data platform security, data security, and personal information security; and expounds on the security problems faced by big data analysis.

1. **Big Data Platforms**

Firstly, the Hadoop open-source model lacks comprehensive security planning for big data platforms. Furthermore, the security mechanisms employed are inherently limited. Hadoop was initially designed to manage a large number of public web data based on the assumption that the cluster is always in a trusted environment. Consequently, Hadoop was not initially designed with a security mechanism, nor did it have a security model or an overall security plan. As the use of Hadoop has expanded, a number of malicious actions have emerged, including the submission of jobs without authorization, the modification of JobTracker status, and the tampering of data. The open-source community of Hadoop has started to consider security concerns by introducing a suite of security features, including Kerberos authentication, file ACL access control, and network-layer encryption. Although these functions can address some security problems, they also present limitations. In terms of identity management and access control, it relies on the Linux identity and permission management mechanism.

Identity management is designed to support users and user groups, but not roles. The system offers three permissions: readable, writable, and executable. However, these permissions are not sufficient to satisfy the demands of role-based security, such as those related to identity management and fine-grained access control. In terms of security auditing, there are only log records distributed in various components in the Hadoop ecosystem that lack a native security auditing function. Instead, log records are distributed across various components, necessitating the use of external tools for log analysis. In addition, the open-source development model also brings potential security risks to the Hadoop system. In the process of tool development, enterprises tend to prioritize the realization of functions and the improvement of performance, with less focus on the quality of code and data security. Consequently, open-source components often lack strict test management and security certification, and are inadequately protected against component vulnerabilities and malicious backdoors.

Secondly, the big data platform has many users and diverse scenarios, and the performance of traditional security mechanisms is challenging to meet the demand.

In a big data scenario, data is aggregated from multiple channels, data types, user roles, diverse application requirements, and access control that faces many new problems. Firstly, the massive aggregation of multi-source data increases the difficulty of access control policy formulation and authorization management. Over-authorization and insufficient authorization are serious issues. Secondly, the data diversity, the refinement of user roles, and requirements increase the difficulty of describing objects. Traditional access control schemes often use data attributes (such as ID numbers) to describe the objects in the access control strategy. However, unstructured and semi-structured data cannot be described in the same way, which hinders the ability to specify the range of data that users can access accurately. This makes it difficult to adhere to the principle of minimum authorization. The sophisticated data storage and flow scenarios of big data make the realization of data encryption extremely difficult. The key management of massive data is also an urgent problem to be solved.

Furthermore, the large-scale distributed storage and computing mode of the big data platform has multiplied the difficulty in the security configuration.

The functions of authentication, authority management, encryption, and audit of the open-source Hadoop ecosystem are all completed through the configuration of related components, without configuration inspection and effect evaluation mechanisms. Simultaneously, the large-scale distributed storage and computing architecture also increases the difficulty of security configuration work. It has higher technical requirements for security operation and maintenance personnel. If an error occurs, it will affect the regular operation of the entire system. According to an analysis conducted by Shodan Internet device search engine, improper configuration of the big data platform server has resulted in the leakage or leakage risk of 5120 TB data worldwide. The countries with the most leakage cases are the United States and China. At the beginning of this year, the ransomware attack on the Hadoop platform did not involve conventional vulnerabilities during the entire attack. Instead, it used the insecure configuration of the platform to manipulate the data efficiently.

Finally, in response to the new features of the network attack methods of the big data platform, traditional security monitoring techniques have exposed their deficiencies.

The development of big data storage, computing, analysis, and other technologies has led to the emergence of a range of new advanced network attack techniques, making traditional detection and defense technologies exposed to serious deficiencies and unable to resist external intrusion attacks effectively. The traditional detection approach relies on real-time matching detection based on threat characteristics at a single point in time. The Advanced Persistent Threat (APT) against big data adopts a long-term hidden attack implementation method, which does not have obvious characteristics that can be detected in real time, making it difficult to find. Additionally, the low density of big data's value makes it challenging for security analysis tools to focus on the value point. Hackers can hide attacks in big data, and traditional security policy detection has great difficulties. Consequently, APT attacks against big data platforms occur from time to time. Large-scale Distributed Denial-of-Service (DDoS) attacks suffered by big data platforms are not uncommon. According to Verizon's 2018 data breach survey, 48 percent of data breaches were related to hacking. It can be seen that traditional security monitoring technologies are no longer able to provide stable and reliable security protection for the current big data platforms.

2. **Data Security**

The characteristics of big data, including large volume and various types, present a novel challenge to data security in a big data environment.

In the data collection process, the characteristics of big data, including large volume, various types, and complex sources, present challenges to the verification of data authenticity and integrity. Currently, there are no strict identification and monitoring methods for data authenticity and credibility, which cannot identify and eliminate false or even malicious data information. It is possible for hackers to use network attacks in order to inject malicious data into the data collection end, thereby destroying the authenticity of the data and leading data analysis results in a predetermined direction. This ultimately serves to achieve the purpose of attacking and manipulating the analysis results.

In addition, with the advent of the digital economy era, more and more enterprises and organizations need to collaborate to integrate the industrial chain and carry out production activities based on data flow and cooperation. In an application scenario where an enterprise or organization uses data resources to engage in collaboration, the data flow extends beyond the organizational and system boundaries, resulting in cross-system access or multi-party data aggregation for joint operations. Ensuring the confidentiality of personal information, business secrets, or unique data resources during the cooperation process is a prerequisite for enterprises or organizations to participate in data flow and data cooperation. Furthermore, this issue must be addressed in order to ensure the secure and orderly interconnection of data.

The application system of big data is complex and the frequent data sharing and exchange make the data flow path become interlaced and complex. From generation to destruction of data, it is no longer a simple one-way, single-path flow model, nor is it limited to the organization's internal flow, from one data controller to another controller. In this process, it is challenging to achieve full-path data tracking and tracing across data controllers or security domains in a heterogeneous network environment. This is particularly evident in the credibility of data tags in data tracing, the binding between data tags and data content, and security issues. In the "Cambridge Analysis" incident in March 2018, Facebook lacked adequate supervision and effective accountability mechanisms for the use of third-party APP data. This resulted in the misuse of 87 million user profiles, a serious drop in stock prices and a loss of credibility.

3. **Personal Privacy**

Initially, the methods and capabilities of big data collection, processing, and data analysis have posed significant challenges to the traditional personal privacy protection framework and technical capabilities. In the context of big data, companies can perform multi-source and multi-type dataset correlation analysis, and in-depth mining to restore anonymous data, thereby obtaining personally identifiable information and valuable, sensitive information. Consequently, the traditional concept of defining a "fixed range" for personal information is no longer applicable in the era of big data. In traditional privacy protection technologies, data collectors select privacy parameters in isolation for a single dataset to protect private information. In the context of big data, the correlation between datasets has increased significantly due to the wide distribution of individuals and other interrelated individuals and groups. This increases the risk of privacy leakage after dataset fusion.

In addition, traditional privacy protection technology is not well suited to the big data non-relational database. In the context of big data technology, data is characterized by dynamic change, semi-structured and unstructured data. For unstructured data, which accounts for over 80% of the total dataset, a non-relational database (NoSQL) is typically used for storage, capture, and management. However, non-relational databases lack strict access control mechanisms and comprehensive privacy management tools. Existing privacy protection technologies, such as data encryption and data desensitization, are mostly used in relational databases. However, they are not effective in coping with the evolution of non-relational databases and are prone to privacy leakage risks.

1.5.3 Data Sharing

The development of the big data industry must address the free flow and sharing of data information. Currently, many enterprises face the challenge of data fragmentation, which hinders the sharing of data between enterprises. This data-sharing problem has become a development barrier for the big data industry. There are two main reasons for this problem. First, the chaos in data collection and management hinders the timely and standardized delivery of data updates. The same type of data is subject to conflicts and contradictions between different departments and industries, which results in a lack of practical application value.

Many medium-sized and large enterprises generate a great deal of data on a continual basis. However, many companies have chaotic management during the big data pre-processing stage, which results in the irregular data processing. The big data preprocessing stage requires the extraction of data, conversion of the data into a data type that is easy to handle, and cleaning and denoising of the data to obtain valid data. Due to the lack of standardization in data processing, the quality of enterprise data is low, and its availability is reduced. This makes it difficult to extract valuable information, and even leads to data-derived knowledge and decision errors. According to a 2007 survey conducted by Gartner, an internationally renowned technology-consulting agency, more than 25% of global Fortune 1000 companies have data missing and inaccurate problems. In 2013, in the United States, approximately 98,000 patients died annually as a result of medical accidents caused by data errors. Furthermore, annual losses of US$611 billion were incurred by US industrial companies due to production accidents and decision-making errors produced by stale data.

On the other hand, data is a new type of commodity, given the existence of privacy and sensitivity concerns. Consequently, some sensitive information is inevitably incorporated into data, which makes data transactions a significant challenge. The absence of unified planning for government construction, enterprise, and industry information systems, and the lack of unified standards between systems have led to the formation of numerous "information islands." Moreover, the limited degree of data openness is constrained by administrative monopoly and commercial interests, which impede the utilization of data. Another important factor restricting the opening and sharing of our country's data resources is the imperfect policies and regulations, and the lack of corresponding legislation for big data mining, which cannot guarantee the sharing and prevention of abuse simultaneously.

Consequently, the establishment of a beneficial data-sharing ecosystem is challenging, and China's big data development requires crossing the hurdle. Concurrently, the balance between openness and privacy represents the most significant challenge in the process of big data opening. The effective protection of the privacy of citizens and enterprises while promoting the comprehensive opening, application, and sharing of data and the gradual strengthening of privacy legislation will be a major challenge in the era of big data. However, with improved performance and lower threshold of technologies such as homomorphic encryption, differential privacy, quantum ledger, and tools such as blockchain and secure multi-party computing, we can expect to see further integration with data circulation scenarios. This integration will likely lead to a reduction in the barriers to data sharing and data circulation.

1.6 Emerging Trends

1.6.1 Development Trend of Big Data Technology

The ongoing advancements in big data technology are driving new developments and posing challenges across various sectors. This section outlines the key trends in big data technology, focusing on five pivotal areas: edge computing, quantum computing, dark data migration, visualization technology, and multidisciplinary integration.

1. **Edge Computing Accelerates Data Analysis**

The concept of edge computing was first proposed in 2003 by cloud service providers AKAMAI and IBM. As the IoT continues to expand, the limitations of traditional central server processing mode are becoming increasingly apparent. Edge computing, which involves the processing and analysis of data at the network edge, is gradually becoming a more viable solution [16]. Edge computing is a methodology that involves the processing and analysis of data by the edge nodes of a network. This approach decentralizes the processing and analysis of the data, thereby reducing the load on the central server. The term "edge node" is used to describe any node situated between the data source and the cloud center with the capacity for computing and analysis. For instance, a mobile phone represents the edge node between the user and the cloud center, while a gateway inside the home serves as the edge node between the furniture and the cloud center. In an optimal scenario, edge computing entails analyzing and processing data in the vicinity of the data source, without data flow, thereby reducing network traffic and response time. This calculation model does not require transmitting data to the cloud center for calculation, reducing network load, accelerating data analysis, and ensuring user data privacy.

In 2016, Huawei Technologies, Shenyang Institute of Automation, and the Chinese Academy of Sciences, in collaboration with six other companies and scientific research institutions, established an edge computing industry cooperation platform. The objective was to facilitate collaboration within the edge computing industry, promote the development of the IoT, and improve the automation level of the industry. Since then, edge computing has gained significant traction, becoming a pivotal area of focus within the broader field of big data technology.

2. **Quantum Computing Improves Computing Power**

In the context of a substantial increase in data volume, edge computing realizes decentralization through the improvement of its architecture, thereby improving the performance of big data processing. The enhancement of computing power through the application of quantum computing techniques represents a significant improvement in performance and provides a fundamental solution to the problem of data explosion.

Quantum computing is the result of the convergence of quantum mechanics and computer science. In contrast to the binary 0/1 mode of classical computing, quantum computing adopts the qubit mode. It has two unique quantum effects. One is quantum entanglement, which enables quantum bits to exist in a state of quantum superposition, simultaneously occupying both the 0 and 1 states. The other is quantum superposition, which allows quantum bits to share their state with other quantum bits that are spatially independent, thus creating a kind of superposition. Due to the unique nature of quantum computing, it is capable of performing computations in a parallel manner that is not possible with classical computing. This enables the acceleration of certain traditional algorithms. In recent years, there has been a convergence between big data and quantum computing, resulting in notable advancements in the fields of data integration and data search.

3. **Dark Data Migration Expands the Dataset**

The term "dark data" is used to describe information that has not been converted into a digital format. This is considered to be an extensive repository of data that remains largely untapped. "Dark data" can be conceptualized as a subset of big data, including structured data, log files, and even unstructured data from social media stored in the Customer Relationship Management (CRM) data warehouse. In fact, it is estimated that more than 80% of the data currently available in the market is being utilized or developed. Consequently, it is possible to treat all these datasets as dark data. The potential for mining in these dark datasets is considerable. The full utilization of these resources will rapidly promote the development of big data and artificial intelligence.

4. **Visualization Technology Promotes the Popularization of Big Data**

In recent years, the concept of big data has gained significant traction in popular culture, with the development outcomes of big data often presented in a visual format. The term "visualization technology" is used to describe the process of presenting complex data in a concise and easily comprehensible manner using an interactive graph. This approach not only facilitates a deeper understanding and analysis of the data but also uncovers its inherent patterns and hidden knowledge. It has considerably bridged the gap between big data and the general public. Looking ahead, with the continuous progress of visualization technology, the continuous innovation of visualization methods, and the constant changes in visualization tools, the presentation of big data analysis outcomes will become increasingly intuitive for both researchers and ordinary people. Consequently, this will enable a more profound understanding and application of big data, allowing the full potential of its value to be realized and harnessed effectively.

5. **Multidisciplinary Integration Is a Key Driver of Data Science Development**

Big data technology is a multidisciplinary fusion product that draws on multiple disciplines, including statistics, computer science, and management science. Big data applications are often cross disciplinary, such as in the security and medical fields. The interdisciplinary and multidisciplinary integration has led to the continuous rise of data science, which has in turn led to an increased focus on the importance of data and a corresponding increase in research into data science, including data analysis and data mining. The advent of data science has led to a shift in focus toward finding practical solutions from large volumes of raw and structured data, uncovering insights within the data, and optimizing the use of big data.

1.6.2 Development Trend of Big Data Application

1. **Big Data and Network Security**

With the increase in data volume and the application of technologies such as cloud computing and virtualization, the boundaries between the host and network are becoming dynamic and blurred. At the same time, the prevalence of covert and continuous attacks is gradually increasing. The traditional network security and intelligence analysis techniques are limited by a single data source and limited processing capacity. The inability to respond promptly to threats in the network represents both a challenge and an opportunity. The combination of big data and network security has led to the emergence of a novel approach to security analysis, specifically tailored to the unique characteristics of big data.

On the one hand, big data processing technologies—including batch data processing, streaming data processing, and interactive data query—have tackled challenges in real-time restoration, and analysis of high-performance network traffic, rapid analysis and retrieval of vast historical log data, and the real-time processing and retrieval of massive text data. These advancements have also enhanced the capabilities for network security and intelligence analysis.

On the other hand, the application of big data technology to visual security analysis, security event correlation, and user behavior analysis has led to the development of interactive visual analysis tools and multi-source events. This has expanded the scope of big data security analysis to encompass a range of research areas, including association analysis, user entity behavior analysis, and network behavior analysis, all of which are crucial for addressing the evolving challenges of network security.

Since 2001, big data security analysis technology has been applied in the detection of APT attacks, the identification of network anomalies, the monitoring of network security situations, and the analysis of network threats. However, with the development of technology, current network protection technology still faces significant challenges. These include a lack of sufficient detection methods for advanced network threats and attacks, a lack of predictive capabilities for unknown complex network attacks, the absence of an evaluation system to measure the results of network security posture assessment, and an imperfect situation assessment index system for both assets and the network as a whole [17]. In the future, big data will be integrated with network security to improve the current network environment and maintain a more harmonious and secure network environment.

2. **Big Data and IoT**

The concept of the IoT was first proposed in the 1990s and has since undergone rapid development. In fact, the core of the IoT is the Internet, which is a network that continuously evolves and expands based on the Internet. A key distinction is that the client can be extended to any item. The development of the IoT is inextricably linked to big data. The IoT can be conceptualized as the Internet's sensory and motor nervous systems, while big data and cloud computing serve as the nerve centers. The combination of these two factors represents an inevitable trend for future development.

The combination of the IoT and big data has achieved promising results, including the development of smart homes and smart roads, as well as the implementation of intelligent traffic management systems for both home and road traffic. In this process, the generation of big data by IoTs assists hardware manufacturers in exploring the usage behaviors of their users. This enables manufacturers to gain insight into their users, optimize product and market strategies, learn from their users, build user profiles, and provide personalized intelligent experiences for different users. In the forthcoming years, researchers will further promote the integration of big data and the IoT to facilitate the development of smart cities and to the benefit of all.

3. **Big Data and Artificial Intelligence**

The advent of big data has ushered in a new era for intelligent applications driven by artificial intelligence, which are now being integrated across various sectors. AI is making particularly significant strides in areas such as unmanned driving, image recognition, and voice recognition. Essentially, artificial intelligence is the application of big data computations. The capacity to execute these computations and the outcomes are the basis of artificial intelligence. Take unmanned driving as an example: it necessitates the collection of comprehensive data on intersection and road condition, which is then analyzed to determine subsequent actions. The foundational structure of data analysis and decision-making process is based on the logic algorithm derived from big data. It starts with the initial storage of extensive datasets, such as road condition and surface data. These datasets are subsequently analyzed to meet specific human requirements, encoded into logic programs, and finally submitted to the system for further processing. Hence, the future development of artificial intelligence is inextricably linked to big data, highlighting the indispensable role it plays in powering the intelligence of tomorrow.

4. **Big Data and Deep Learning**

Deep learning is a learning process that uses deep neural networks to resolve feature expressions. The deep neural network can fully use the input data to simulate and construct the corresponding model structure, exhibiting a high degree of flexibility and an inherent capacity for self-optimization. This significant advantage entails a significantly increased number of calculations. Big data analysis-related tools for processing can effectively address the challenges posed by the volume and power of computing resources. Therefore, the technology associated with big data has considerably promoted the development of deep learning. Concurrently, the primary challenge in the context of vast data volumes is to develop effective analytical and processing techniques that can extract meaningful insights from the data. Deep learning methods play a vital role in the processing of big data. It can adaptively extract its internal representation from the data, reducing the necessity for human involvement. The deep model used to extract features can be applied to various scenarios, exhibiting robust generalization performance.

At present, with the support of large-scale labeled datasets, deep learning based on supervised features has achieved favorable outcomes. However, there remains a need to address unlabeled datasets. Consequently, future research will prioritize deep learning based on unsupervised features.

5. **Big Data and Medicine**

The rapid development of big data technology has facilitated the widespread use of health and medical informatization, generating massive datasets in medical services, healthcare, health management, and the formation of health, and big data medical care. The development and application of health and big data medical play an essential role in improving medical and health services and promoting the development of the health industry. Many countries have reached a consensus on this, and some developed countries have adopted it as a major national strategy and put it into practice. In future development, big data medicine will bring creative changes to clinical treatment, drug development, health monitoring, public health, policy formulation, and implementation, and will lead to comprehensive improvements in governance capacity and healthcare level.

The application of big data in the medical field will facilitate the process of clinical diagnosis, treatment management, and decision-making. By comparing and researching effects, analyzing large datasets, including patient signs, costs, and efficacy data, doctors can determine the most effective and cost-effective treatment methods. At the same time, the clinical decision support system can effectively expand clinicians' knowledge base, reduce human error, and help doctors improve work efficiency, quality of diagnosis, and treatment. By integrating data from diagnosis, treatment operations, and performance datasets, visual flow charts and performance charts can be created to identify abnormalities in the medical process and provide a basis for optimizing business processes.

Medical big data will provide invaluable support for drug development. By analyzing clinical trial registration data and electronic health files, we can optimize the clinical trial design and recruit suitable clinical trial participants, thereby improving the efficiency and effectiveness of the clinical trial process. Furthermore, analysis of clinical trial data and electronic medical records can assist in the analysis of drug utility and the reduction of the impact of drug interactions. The prompt collection of data on adverse drug reactions can enhance the monitoring and prevention of adverse drug reactions. By analyzing disease prevalence and development trends, we can simulate market demand and costs, predict the clinical outcomes of new drug development, and help determine new drug development investment strategies and resource allocation.

The application of medical big data will facilitate the monitoring of public health. The application of big data-related technologies can expand the scope of health surveillance, from partial case sampling to full sample data, thereby improving the timeliness and accuracy of the disease transmission situation's judgment. The speed of identification, processing, and response to public health events can be improved through the integration and analysis of demographic information, disease data from various sources, and risk factor data from various sources. The comprehensive tracking and processing of data enables the effective scheduling of resources, rapid response to crises, and informed decision-making.

The analysis of medical big data will facilitate the management of public health. Wearable medical devices can collect personal health data, assist in health management, improve people's health, and provide an effective way of communication between doctors and patients. At the same time, doctors can take timely intervention measures or make recommendations for diagnosis and treatment based on the patient's health data. The integrated analysis of individual signs, diagnosis, treatment, behavior, and other data can predict the individual's disease susceptibility, drug sensitivity, etc. Thus, it is possible to achieve early detection, treatment, personalized medication, and personalized care of individual diseases.

The application of medical big data will provide a scientific basis for the formulation and implementation of medical and health policies. Integrating and mining health data and network public opinion information at different levels and in different business areas helps to comprehensively analyze the characteristics of both medical supply and demand, for specific service provisions and utilization and its influencing factors. This is achieved through statistical analysis of population and individual health status indicators in the region and individual health status and its influencing factors, which in turn forecast future demand and supply-side development trends. Additionally, disease risk factors are discovered, and a scientific basis for medical resource allocation, medical security system design, population and individual health promotion, and population macro-decision-making is provided. Concurrently, the integration of data from population health departments and medical service institutions at all levels enables the identification and comparative analysis of key performance indicators, thereby facilitating the rapid understanding of the implementation of policies in various locations, the timely identification of problems, and the prevention of risks.

1.7 Summary

The term "big data" is used to describe datasets that cannot be captured, managed, or processed by conventional software tools within a certain period. To achieve more reliable decision-making, insight, and process optimization capabilities, a new processing mode is required. The characteristics of big data, which include a massive volume, high growth rate, and diversified information assets, can be encapsulated by the "5 V" model. This model stands for Volume, Variety, Velocity, Veracity, and Value.

The field of data science is the study of information extraction technology based on massive data. It is a cross-disciplinary field that supports the research and application of big data. Its theoretical basis is drawn from a variety of disciplines, including computer science, statistics, artificial intelligence, information systems, and information science. Data science has played a pivotal role in the development and evolution of computer science, software engineering, communication engineering, and artificial intelligence.

The origins of big data technology can be traced back to the 1980s. Since then, a number of related technologies have emerged, including big data processing architecture, data collection, data storage, calculation mode, and visual analysis. Even though these technologies have already been put into production practice, there is still considerable room for further development. The "Outline of Action to Promote the Development of Big Data," issued by the State Council of China, provides a framework for the development of big data applications, industries, and technologies in the country.

The emergence of big data technology has precipitated profound changes in many fields. Its applications include targeted marketing in the realm of the Internet and precision medicine in the medical area. In fact, considerable quantity of untapped industry-related data exists across all sectors of society. The emergence of data science has unearthed the intrinsic value of these data, thereby propelling the development of diverse industries.

In terms of data storage, the scale of big data often reaches petabyte levels, presenting new challenges to traditional storage technologies. Existing structured databases frequently struggle to accommodate such vast datasets and exhibit limited scalability, failing to adapt to the dynamic nature of big data. The security of data, including the threat of penetration in production/data acquisition, processing and sharing of large datasets, such as the industrial chain of each link, is facing new challenges in the era of big data. Data security, privacy, and platform security are all vulnerable to new threats and risks. As a result, these platforms, data and personal privacy are all at risk. Regarding data sharing, the lack of standardized data processing and the absence of comprehensive relevant legislation and regulations have resulted in uncertainty about the extent to which data can be shared across departments and made available to the public. This has led to the formation of data islands within and between enterprises, preventing the full exploitation of big data's potential.

Technological advancements, including margin calculation, are driving the expansion of cloud computing, narrowing the gap between data and the user. This accelerates data analysis and breaks through traditional computational barriers with quantum computing. Parallel computing further enhances computational capabilities, while dark data migration facilitates further mining. If we fail to leverage the full potential value of our data and expandable datasets, and fail to integrate visualization technology with big data in a way that benefits production and living, we will fail to foster common and multidisciplinary data integration, thus fail to promote the development of science and big data to new heights. There is considerable potential for the application of big data, network security, IoT, artificial intelligence, deep learning, and medical field.

1.8 Exercises

1. What is big data? Briefly describe the basic characteristics of big data.
2. Briefly describe the relationship between data science and other disciplines.
3. What stages can the development of data processing be divided into? What are the characteristics of the data at each stage?
4. Briefly describe the current status of big data analysis research.
5. Briefly describe the main applications of big data analysis.
6. What are the main problems in big data analysis?
7. What is edge computing?
8. What is dark data?
9. Briefly describe the development trend of big data analysis from both technical and application perspectives.

References

1. Baidu Baike. Big Data entry. 2019. https://baike.baidu.com/item/%E5%A4%A7%E6%95%B0%E6%8D%AE/1356941
2. Wikipedia. Big Data entry. 2019. https://zh.wikipedia.org/wiki/%E5%A4%A7%E6%95%B8%E6%93%9A
3. Databanker. An Article to Clarify: The Relationship Between Big Data, Data Mining, Data Analysis, and Mathematical Statistics. 2017. http://www.databanker.cn/point/175156.html
4. EDUCBA. Computer Science vs Data Science. 2019. https://www.educba.com/computer-science-vs-data-science/
5. OnLine Engineering. CS vs DS. 2019. https://www.onlineengineeringprograms.com/faq/computer-vs-data-science
6. David R. The difference between data science, machine learning and artificial intelligence. 2019. https://zhuanlan.zhihu.com/p/33036628
7. Runqi Z. Talking about internet big data application. Comput Fans. 2019;01(01):63.
8. Sun X, Wang X. Common data mining algorithms and their applications in medical big data research. Chinese Digital Medicine. 2018.
9. Zhongguancun Online. 1983 to 2012: A Chronicle of Big Data Evolution. 2013. https://server.zol.com.cn/376/3768036_all.html#p3768036_param
10. Zuo X. Technical route for the development of big data systems. 2019. http://www.360doc.com/content/17/1101/20/48786200_700107893.shtml
11. cdadata. Big Data Era: Implementing Technologies and Applying Analytics. 2015. http://www.cdadata.com/4951?from=singlemessage
12. Shubangke. Analysis of Big Data Industry Outlook in the Next Three Years. 2019. http://www.databanker.cn/point/248688.html
13. Yunqi Community. BigData: Ten data development trends worth knowing. 2018. https://yq.aliyun.com/articles/641749
14. Cheng X, Jin X, Yang J et al. Big data technology progress and development trend. Sci Technol Rev. 2016; 34(14): 49–59. https://doi.org/10.3981/j.issn.1000-7857.2016.14.006.
15. Redman, Thomas C. The impact of poor data quality on the typical enterprise. Communications of the ACM. 1998; 41(2):79–82.

16. The Edge Computing Industry Alliance was formally established. 2016. https://www.huawei.com/cn/press-events/news/2016/11/Edge-Computing-Consortium-Established
17. Schultz MG, Eskin E, Zadok F et al. Data mining methods for detection of new malicious executables. IEEE Symposium on Security & Privacy. IEEE. 2001.

Chapter 2
Core Architecture of Big Data Analysis

2.1 Introduction

The big data analysis core architecture addresses the challenge of traditional data analysis systems being unable to store massive data, which makes calculating and analyzing large and heterogeneous datasets more difficult. In light of these challenges, a big data analysis architecture is proposed. This architecture provides an abstract description and modeling of overall structure and components of a big data analysis system. It addresses the challenges of massive data storage and analysis through the use of distributed storage, computing, retrieval, and storage technologies. An instance of a big data analysis architecture is referred to as a big data analysis framework, which is a fundamental tool in the field of big data development. Frameworks for data processing include (1) Apache Hadoop, a batch processing framework for big data analysis using the MapReduce programming model. Hadoop was originally designed for computer clusters constructed from commercially available hardware, which remains the predominant approach. (2) Apache Spark, a hybrid processing framework designed for big data analysis. It supports both batch processing and stream processing, offering a flexible and scalable solution for big data analysis. It provides an interface for programming entire clusters with implicit data parallelism and fault tolerance. With regard to data storage, there are a number of frameworks that may be employed, including: (1) Hadoop Distributed File System (HDFS), a distributed, scalable, and portable file system written in Java for the Hadoop framework. HDFS provides shell commands and Java application programming interface (API) methods that are analogous to those of other file systems. (2) Apache HBase, a non-relational distributed database running on top of HDFS. HBase provides a fault-tolerant method of storing large quantities of sparse data, which is defined as small amounts of information caught within a large collection of empty or unimportant data. (3) Apache Hive, a data warehouse software project developed on the foundation of Apache Hadoop for providing data query and analysis. Hive provides an SQL-like interface for querying data stored in various databases and file systems that

S. Luo and L. Pan, *Big Data Analysis*, Big Data Management,
https://doi.org/10.1007/978-981-95-5725-7_2

integrate with Hadoop. This chapter aims to present the core concepts and features of the most widely used big data analysis frameworks, which include Hadoop, HDFS, Spark, HBase, and Hive.

2.2 Foundational Concepts of Data Analysis Architecture

2.2.1 Software Architecture

Software architecture is an abstract description and modeling of the software's overall structure and components. It can be divided into logical, physical, and system architectures. The logical architecture describes the functional composition of a software. The physical architecture describes the manner in which the software's various functional components are deployed on the hardware in the real network. The system architecture describes the non-functional characteristics such as performance, reliability, and scalability. A software architecture [1] provides a high-level abstraction of structure, behavior, and attributes of a software system. It comprises a description of the components that make up the system, the interaction of the components, the modes that direct component integration, and the constraints associated with these modes. The software architecture specifies the system's organizational structure and topology, and shows the alignment between the system requirements and the components that constitute the system. It provides the fundamental principles of design decisions and the basis for the development of computer software practice.

The software architecture can be regarded as a collection of components. In the context of the UML 2.0, a component is defined as a modular unit that can provide a well-defined interface, be replaceable in its environment, and pass between components' interface. In object-oriented programming ideas, this feature is usually summarized as "reusable." The software architecture components can be classified into three main categories: processing components, which are responsible for data processing; data components, which are responsible for storing the processed information; and connection components, which are responsible for combining and connecting different parts of the architecture. Abstraction is a communication mechanism that facilitates the passing of messages between components, enabling them to coordinate and collaborate with each other.

The software architecture exhibits a variety of patterns. Each pattern represents a common organizational structure for a specific application domain. It reflects the structural and semantic characteristics shared by many systems within the domain, guiding each module and subsystem into a complete system effectively. More specifically, the software architecture model can be regarded as a set of constraints on the architecture, which define the restrictions on each component type and operation control/data transmission mode. These constraints determine a group or a series of architectures that can satisfy them. The elements of the software architecture model include

1. A set of component types that perform certain functions while the system is running.
2. A topology that can show the interrelationship of components while the system is running.
3. A set of semantic constraints.
4. A set of connectors that provide an intermediary for communication between components.

Common software architecture patterns include layered architecture, event-driven architecture, micro-core architecture, microservice architecture, etc. [1]. The layered architecture (see Fig. 2.1 Architecture Layers) is the most common and standard software architecture. The software is divided into several horizontal layers, with each layer providing services for the upper layer and serving as customers for the next layer. In a layered architecture, each layer is clearly defined and has a distinct role and division of labor. It is unnecessary to understand the specifics of other layers, rather, layers communicate with each other through interfaces. The most common layered architecture comprises four layers: the presentation layer, which is responsible for the user interface; the business layer, which is responsible for business logic; the persistence layer, which is responsible for providing data; and the database, which is responsible for storing data. The layered architecture offers several advantages, including a simple structure, independent development, and the ability to test different layers. Additionally, it is well suited to the organizational structures of most software companies. However, when the environment undergoes changes, necessitating code adjustment, or the addition of functionality, the layered architecture is typically more time-consuming. The majority of existing mainstream big data analysis frameworks, such as Hadoop, are designed based on a layered architecture model.

Event-driven architecture (see Fig. 2.2 Event-driven architecture) is a software architecture that communicates through events (e.g., notifications issued by the software when the state changes). This architecture mainly includes four components: event queue, distributor, event channel, and event processor. The event queue represents the initial point of reception for events. The distributor is responsible for distributing different events to different business logic units. The event channel represents the contact channel between the distributor and the processor. The event processor is responsible for implementing business logic. Upon completion of the processing, an event will be sent to trigger the next operation. The event-driven architecture is a distributed, asynchronous architecture. The event processors are highly decoupled, so it has the advantages of strong scalability, wide applicability, and straightforward deployment. Nevertheless, it is also subject to certain constraints, including high development and testing costs and the challenge of supporting atomic operations.

The micro-core architecture (see Fig. 2.3 Micro-core architecture), also known as the plug-in architecture, refers to the relatively small core of a software program. The main functions and business logic are implemented through the plug-in, while the

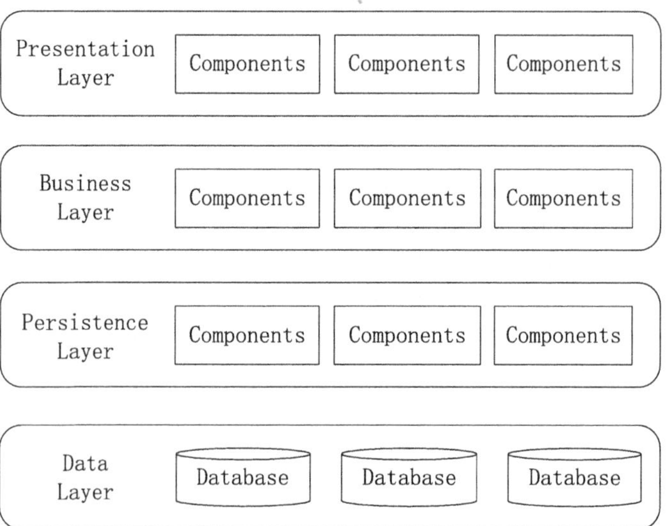

Fig. 2.1 Architecture layers

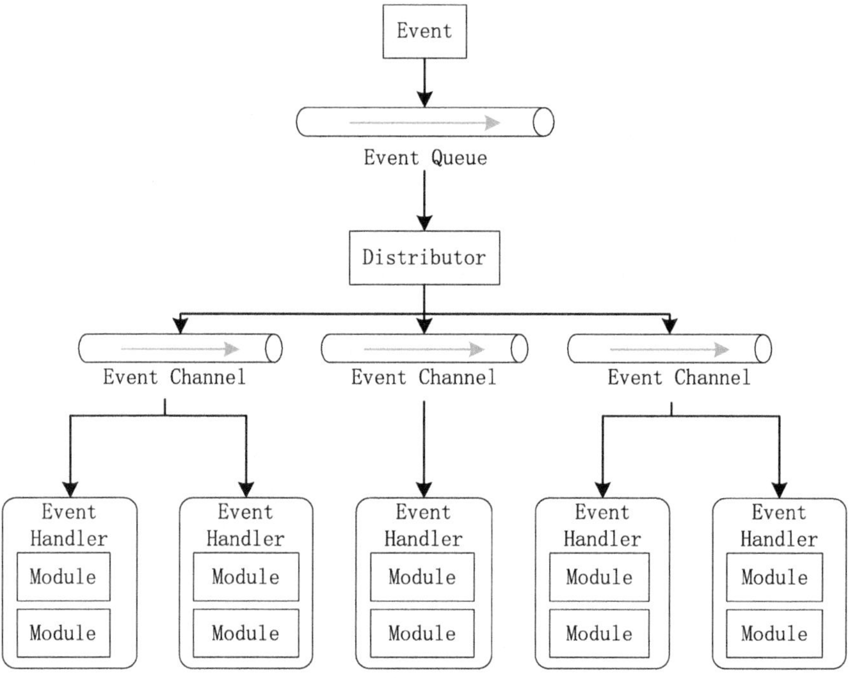

Fig. 2.2 Event-driven architecture

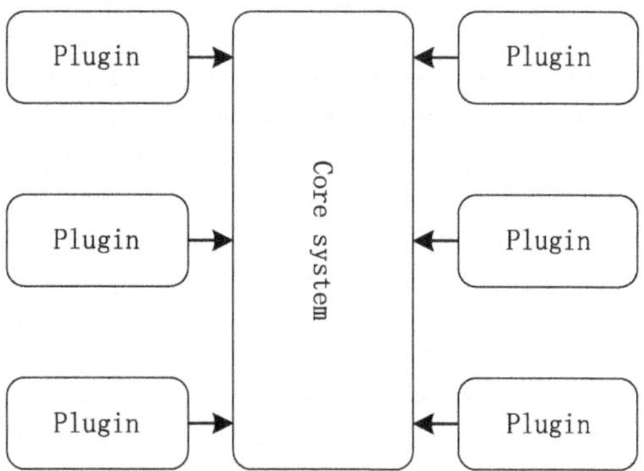

Fig. 2.3 Micro-core architecture

kernel contains only the minimum functions necessary for the system to run. Plugins are designed to be independent of each other, and communication between them should be reduced to a minimum to avoid interdependence. The advantages of the micro-core architecture include good functional extensibility, high customizability, support for progressive development, and easy deployment. However, there are also some disadvantages, including the difficulty of achieving distributed cores and high development costs.

Microservice architecture (as depicted in Fig. 2.4) is an upgrade of Service-Oriented Architecture (SOA). In microservices architecture, each service is an independent deployment unit. These units are distributed and decoupled from each other, and are connected through remote communication protocols (such as REST (Representational State Transfer) and SOAP (Simple Object Access Protocol)). The microservice architecture is comprised of three primary implementation modes: RESTful API mode, RESTful application mode, and centralized message model. In the context of microservice architecture, RESTful API mode is used to describe services that are provided through APIs. Cloud services can be classified within this mode. RESTful application mode is used to describe services that are provided through traditional network or application protocols. In this mode, a multi-functional application is often utilized, which is a common practice in enterprises environments. Centralized messaging is used to describe message brokers that can implement message queues, load balancing, unified logging, and exception handling. The advantages of microservice architecture include strong scalability, easy deployment, and testing. The disadvantages include an emphasis on independence and low coupling, which result in the creation of highly divided services that contribute to a complex and unwieldy system organization. Besides, the distributed nature of this architecture presents challenges in supporting atomic operations.

The software architecture is used to describe the logical organization and interactions of different software components. Upon deployment of each system's software component on a physical machine, the same system may exhibit different deployment methods. The final instance of software architecture following deployment is designated the system architecture. Common system architectures include centralized architecture non-centralized structure and hybrid architecture. A centralized

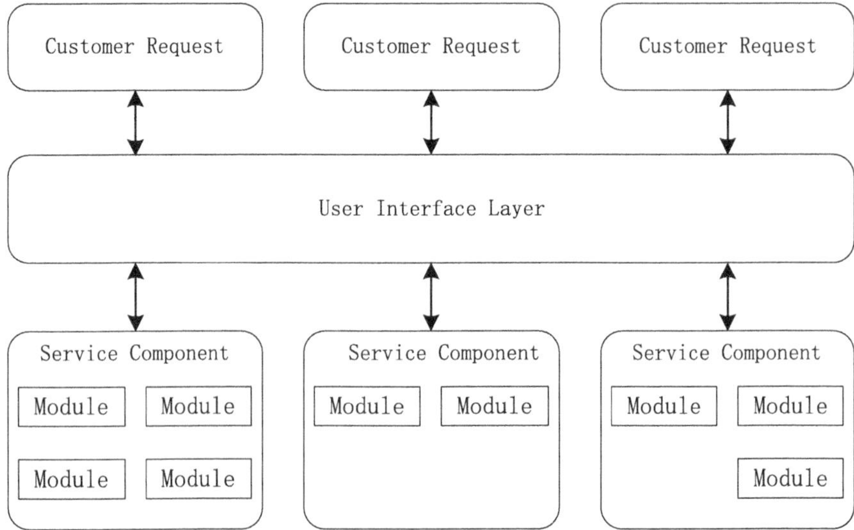

Fig. 2.4 Microservice architecture

architecture is one in which service functions are concentrated on the server side. Clients only maintain the process of sending requests to the server and waiting for the server to reply. This is exemplified by the C/S structure. A non-centralized structure includes both a vertical distribution structure and a horizontal distribution structure. The vertical distribution structure refers to the deployment of different layers of components in a layered architecture on different machines. The horizontal distribution structure, in contrast, refers to the client's physical division into logically equal parts, such as a P2P structure. Finally, a hybrid architecture is a combination of a centralized structure and a decentralized structure, such as a border server system (see Fig. 2.5) or a collaborative distributed system.

2.2.2 Database and Management System

Database technology emerged in the late 1960s as a pivotal branch of computer science, serving as the core and foundation of information systems. Its advent has significantly accelerated the integration of computer applications across various sectors of society. Key metrics such as the scale of database construction, the volume of information contained within database, and the frequency of database utilization are now critical benchmarks for gauging a nation's level of informatization.

Fundamental concepts in databases include data, database, Database Management Systems (DBMS), and Database Systems (DBS). Data, which describes transactions through symbolic records, is the fundamental object stored within a database, and can include text, graphics, images, audio, and videos. The relationship between data and

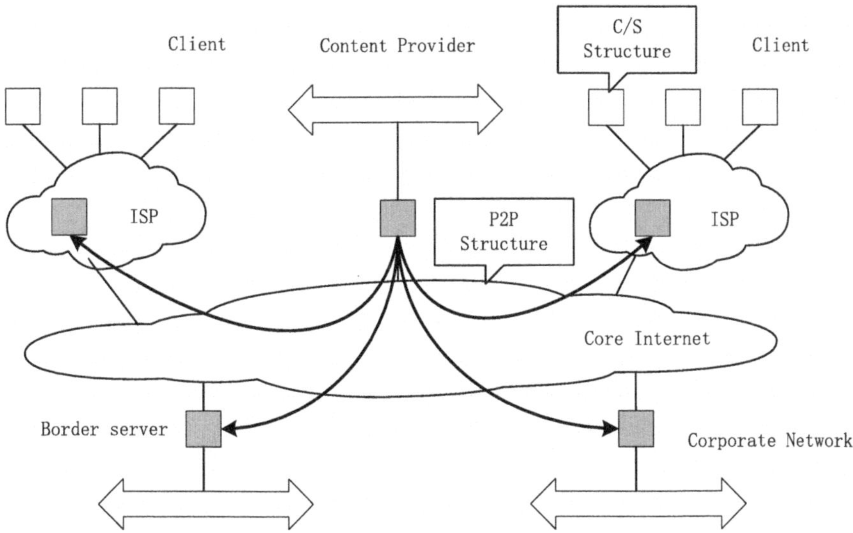

Fig. 2.5 Border server system

semantics is intrinsic, with data often being discussed alongside the concept of information. There is a nuanced distinction and interplay between the two: data represents the raw, qualitative, and quantitative descriptions of a subject, while information is the knowledge contained in the data and messages. Essentially, data are objective representations, whereas information is the interpretive meaning that data conveys, becoming impactful when it influences actions and behaviors. Data, recorded as recognizable symbol recorded, can take various forms and be transformed, yet the inherent information it contains remains constant. Information can exist independently from an information system or separately from different components and stages of such a system. The data format is frequently related to the computer system and changes with the physical medium in which it is stored.

A database is a collection of large amounts of organized, shareable data maintained in a device for a long period of time. The data within a database is interconnected rather than isolated, meaning a database must represent both the data and its interconnections within the data domain. Database exhibits a low degree of data redundancy, typically achieved by the database management system (DBMS). The DBMS is responsible for the unified management and organization of the data within the database, as well as all related operations. Databases support data independence, meaning that data organization, storage methods, and applications operate independent of each other.

Ensuring the security and reliability of data within the database is imperative for database technology. Actively, data should be effectively protected against unauthorized use or modification. Passively, in the event of data destruction, it must be fully and immediately restorable. Databases must support concurrent access, ensuring that during use by multiple users, no conflicts arise and data consistency and accuracy are maintained.

A database management system is a software system that facilitates the acquisition and maintenance of data storage, acting as an intermediary between users and operating systems. Some examples of DBMSs include ORACLE and MySQL. The primary functions of a DBMS include

1. Data definition function provides a data definition language.
2. Data query function provided index search, hash search, sequential search, and other access methods.
3. Data manipulation function provided a data manipulation language for data insertion, deletion, and modification operations.
4. The database transaction management and operation management functions are unified to ensure data security, integrity, and concurrent use of data by multiple users. Furthermore, the unified control of these functions enables system recovery after a failure.
5. Database construction and maintenance functions include database initialization, data entry and conversion functions, database storage and retrieval, database reorganization functions and performance insights, and analysis functions.
6. Other functions include the communication function of the database management system and other software systems within the network, a database management system and other database management data conversion functions of the system or file system, mutual access and interoperation between heterogeneous data, etc.

A database system contains four primary elements: the database itself, related software, related hardware, and Personnel (see Fig. 2.6). Related software encompasses operating systems, database management systems, and application programs. Related hardware includes various physical devices and external storage devices that constitute computer systems. Personnel includes system analysts, database designers, application developers, and users.

2.2.3 Parallel Computing

Parallel computing is defined as the utilization multiple computing resources to simultaneously address computational challenges. It is an effective means to improve the computing speed and processing capacity of computer systems. Furthermore, parallel computing represents a standard solution for computing tasks involving large data volumes, such as big data analysis. Parallel computing can be classified into two categories: time parallel computing and space parallel computing. The field of parallel computing places greater emphasis on the exploration of spatial parallel computing. The fundamental concept is the utilization of multiple processors to collectively address a singular problem. This entails the decomposition of the problem into independent components, each of which is then processed by an independent processor. Finally, the solution to the problem is obtained by aggregating all the processing

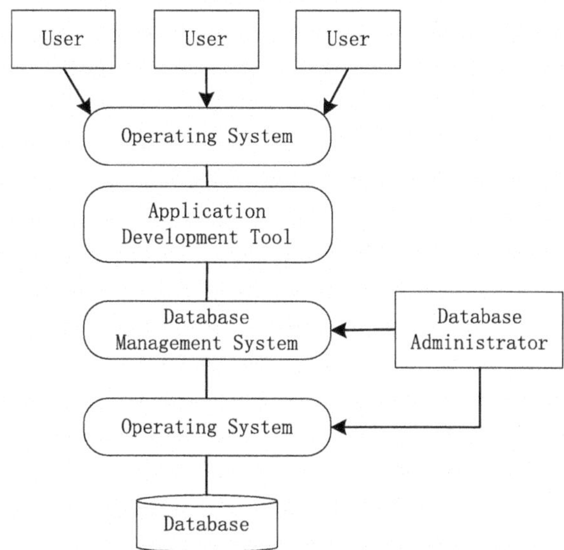

Fig. 2.6 Database system

results. A parallel computing system may be either a specially designed supercomputer containing multiple processors or a cluster of several independent computers interconnected in a certain way. In the latter case, data is processed by the parallel computing cluster, and the results are returned to the user.

In parallel computing, the collaborative efforts of multiple processors necessitate the consideration of a number of factors, including the allocation of subtasks, feedback mechanisms following the completion of a task, and measures for abnormal processor behavior within a cluster. Researchers conducted a comprehensive analysis and developed an abstract model to address these challenges. Any programming implementation that conforms to this paradigm can effectively address big data computing issues through the use of parallel computing. MapReduce is a representative of the parallel computing model, which is a large-scale data processing method proposed by Google. The original design was intended to address the parallel processing challenges associated with large-scale web data using Google's search engine. MapReduce's inspiration comes from two core operations in the functional programming model: the "Map" procedure and the "Reduce" method. The Map procedure maps a set of data into another set of data in a one-to-one manner, with the incoming function specifying the mapping rules. The Reduce method reduces the mapping result, with the reduction rules are also passed by the incoming function. Therefore, Map and Reduce provide a framework for operations, while the incoming functions provide the rules for operations.

The Map and Reduce operations in the MapReduce programming model are analogous. The fundamental principle of the MapReduce programming model is to use an input key-value pair set to generate an output key-value pair set. The MapReduce model expresses the operational rules through the Map and Reduce functions. The user-defined Map function receives an input key-value pair and produces a set of mapped key-value pairs as an intermediate result. In the MapReduce framework, all the same key values are transferred to the user-defined Reduce function, where they are merged to form a smaller set of values. Generally, the size of the Value set produced by the Reduce function is either 0 or 1. Algorithm 2.1 MapReduce Algorithm is an illustrative example of a MapReduce algorithm. The following example illustrates the operation of the MapReduce algorithm: Task Goal: Count the number of occurrences of each word in "to be or not to be."

Implementation Steps:

Step 1: Define the original data.
{"to","be","or","not","to", "be"}
Step 2: Convert the original data into a set of input key-value pairs <number of rows, content>
{1, "to","be","or","not","to", "be"}
Step 3: Convert the set of input key-value pairs into an intermediate result through a Map operation.
{{"to",1}, {"be",1}, {"or",1}, {"not",1}, {"to",1}, {"be",1}}
Step 4: Reduce the intermediate result output by Map to the final result through Reduce operation.
{{"to",2}, {"or",1}, {"not",1}, {"be",2}}

Algorithm: MapReduce-WordCount

```
1: function MAP (key, value)
2: //key: Line Number
3: //value: Line Content
4:  for i←0 to value.length -1
5:  do
6:   EmitIntermediate(w←value,1)
7:  repeat
8: end function
9: function REDUCE (key, values)
10: //key: a Word
11: //values: a List of Counts
12:  result←0
13:  for i←0 to values.length -1
14:  do
15:   result←result+ParseInt(v←values)
16:  repeat
17:  Emit (AsString(result))
18: end function
```

The Map operation, as illustrated in the example, can be defined as follows: for each word w in the value, the EmitIntermediate function is called to convert it into an intermediate result $\{w, 1\}$. In fact, the order of traversing the array is often irrelevant because the traversal operation does not have any effect on the intermediate results of the Map output. This feature allows that the Map operation to be divided into multiple subtasks, which are then handed over to multiple processors for parallel calculation. This improves the efficiency of the calculation process.

A consideration of the implementation of parallel computing reveals that the MapReduce programming model divides processors into three categories: master, mapper, and reducer. These three types of processors are responsible for system scheduling, mapping operations, and reduction operations, respectively. In the context of MapReduce parallel computing, the simple expression for a dataset comprising N elements is as follows: the master node divides all elements into m parts and records each part as Ei. The Mapper Mj obtains Ei and calls the map function to obtain the intermediate result Ij. Subsequently, Tj transmits the associated Reducer Rk to reduce the output to Ok (the submission process is called shuffle). The final output, O, can be obtained by synthesizing all the output, Ok. This shows that the MapReduce programming model is consistent with the "Divide and Conquer" design concept. Figure 2.7 depicts the workflow of the MapReduce programming model. The user program is distributed to multiple mappers through the master control node for concurrent processing. The intermediate results of the output are organized by the relevant reducer to form an output file.

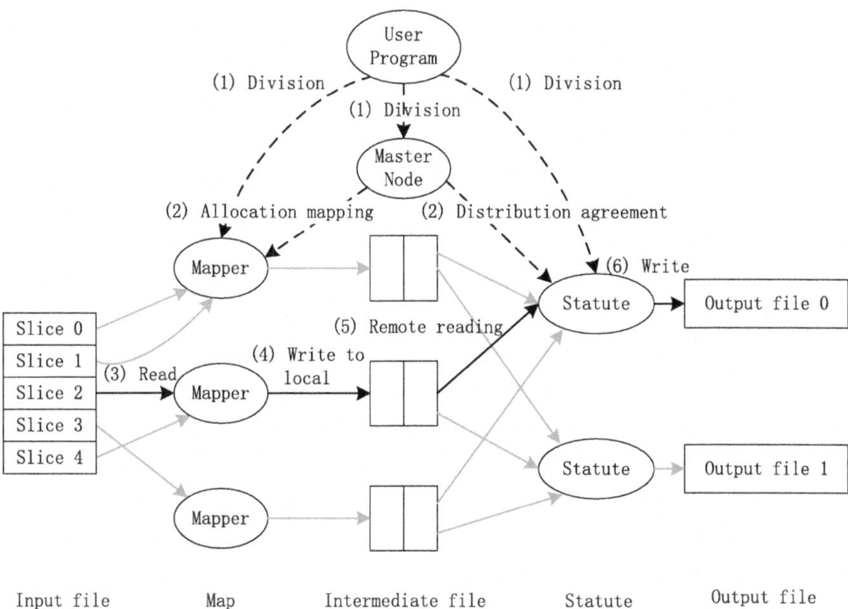

Fig. 2.7 MapReduce work process

2.2.4 Distributed Computing

The term "distributed computing" refers to the use of networks connecting thousands of computers to form a virtual supercomputer, thereby enabling the solution of extremely large-large-scale problems that cannot be accomplished by a single computer (see Fig. 2.8). The earliest form of distributed computing emerged at Intel Corporation in the late 1980s. Intel utilized the idle time of their space station to develop computing datasets for chips and employed a LAN to facilitate research adjustments. The field of distributed computing reached its zenith in the post-1990s era, with a notable surge in popularity on the Internet. The primary focus of distributed computing research is on distributed operating systems and distributed computing environments. Over the years, many distributed computing technologies have emerged, including middleware technology, grid technology, mobile agent technology, P2P technology, and web service technology. Each technology has been acknowledged to a certain extent and has been widely used in a specific area. The advantages of distributed computing are as follows:

1. Resource Sharing: Information resources, such as hardware, software, and data, can be shared within the system.
2. System Transparency: The user is presented with a unified, comprehensive view of the system.
3. High-cost Performance: The distributed system exhibits a high performance-price ratio.
4. The distribution of applications: Most applications are inherently distributed in nature and are therefore directly applicable to distributed computing.
5. High Reliability: Modern distributed systems are equipped with a robust fault tolerance mechanism.
6. Scalability: The performance of the system can be improved by the addition of processors.

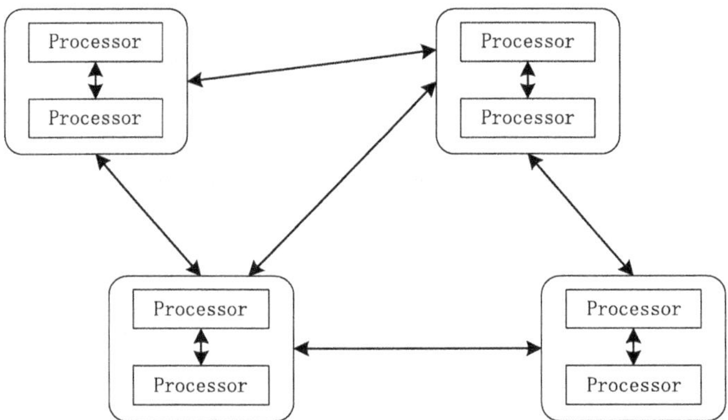

Fig. 2.8 Distributed computing

7. High Flexibility: The system is compatible with a range of hardware products, including low-configuration machines, enabling high-performance computing.

The working principle of distributed computing leverages the idle processing power of central processors across the Internet to solve large-scale computing problems. The proliferation of the Internet has made it possible to install and interconnect all these computer systems equipped with idle computing resources. In practical applications, intricate problems are often amenable to being decomposed into numerous smaller computational tasks. The server's role is to segment the complex problem into many manageable sub-problems, dispatch these to multiple computers engaged in the computation for concurrent processing, and ultimately aggregate the results to derive the final outcome. Hence, distributed computing represents an efficient embodiment of parallel computing.

The design of a distributed computing system is a complex undertaking, with the primary objective being to guarantee system transparency. Transparency reflects a particular attribute: when an entity possesses a characteristic that remains imperceptible from a certain vantage point, this quality is termed transparent. In the context of distributed computing system, transparency is manifested in various dimensions, mainly including the following points:

1. Transparent name: Use of unique global of object naming, so no matter where the object is accessed, the same used name is called.
2. Transparent access: Users do not need to distinguish between local resources and remote resources; access to both local resources and remote resources is the same.
3. Transparent location: The name of a resource in the distributed system does not contain any location information. In this way, when the resource moves in the system, the original program can normally run while the resource name remains unchanged.
4. Transparent migration: The user does not need to know whether a resource or job is being migrated to another location.
5. Transparent replication: Multiple copies of files or other objects are allowed to exist in the system at the same time, but this situation is transparent to the user, and modifications to one object should be effected on all copies of the object at the same time.
6. Transparent concurrent or parallel: Multiple processes may access the same resource concurrently or in parallel, or one process may use multiple resources simultaneously. In this case, it should be guaranteed that there is no mutual interference and destruction.
7. Transparent failure: If one part of the system fails, the entire system will not fail.

The challenges faced by a distributed computing system in the design process include error control, resource management, synchronization, protection, object representation, and coding conversion.

1. Error control: Error control requires all system levels to detect errors and restore the system from the error state to the correct state. The distributed computing

system may have errors in many links, such as errors caused by name reuse, packet delays, and damage to communication media. It is, in fact, challenging to resolve all layers through a singular error control mechanism.

2. Resource management: Resource management requires that each layer of the system has main memory, buffer space, channel access, communication bandwidth, and CPU time for object representation, and manages hardware/firmware access, address space, and other resources accordingly. The allocation and scheduling of resources are typically conducted locally due to the necessity for local autonomous management in addition to global scheduling and allocation. In order to achieve the dual objectives of low latency and high throughput of messages, it is necessary to pre-allocate certain state information and resources at each layer, as well as to retain them over time.
3. Synchronization: Synchronization problems typically pertain to the mechanism of shared resources or shared events among the cooperating processes. In a centralized system with shared main memory, all processes' system status is consistent, thus making synchronization relatively simple. However, in a distributed computing system, messages used to update and report status may experience unpredictable delays, resulting in the various cooperative processes not seeing the same system status, even in the absence of errors or node failures. Consequently, when an error occurs, the problem becomes significantly more complex.
4. Protection: The protection problem associated with distributed computing systems is inherently more complicated than that of standalone systems. The main reason for this is that the system is physically dispersed, its components are heterogeneous, and the control is multiple. In a single-processor system, all processes are assumed to be trustworthy and utilize a well-defined and reliable interface, such as the IPC mechanism, to access each other. In a distributed computing system, assuming that the core operating system and server running on each node are trusted, then encryption can be performed on all layers to protect the information and the entire system. Access control lists are usually used to control access permissions. An alternative approach is the use of capability technology, which provides the necessary permissions for access. In a standalone system, the kernel effectively manages system capabilities; however, in a distributed computing system, a reliable method for such management is less apparent. Nonetheless, it is feasible for intermediary agents, perhaps akin to "waiters" in this context, to employ encryption for capabilities or to implement password protection to safeguard against forgery.
5. Object representation and coding conversion: The issues of object representation, encoding, and conversion refer to the definition of specific objects at each layer of the system. This includes defining files, processes, and directories at the high level, as well as defining packages at the IPC layer. It also entails elucidating the representation and encoding methods employed for these objects. In the event that the representation and encoding of an object differ between two host machines, a conversion process is required to ensure compatibility. The conversion method

generally entails the transformation of the local representation into a certain standard format, or vice versa. When the automatic conversion is required, the type of information is often transmitted to indicate the required conversion format. The object used for communication may be data or control information.

2.2.5 Cloud Computing

As the scale and complexity of enterprise application continue to expand, the necessity for enhanced performance, particularly in terms of computing power and concurrent users, has reached an unprecedented level. In order to support these expanding demands, the conventional approach entails the acquisition of additional hardware and software resources, as well as the formation of a dedicated operational and maintenance team to ensure the optimal functioning of these systems. In this model, the cost of supporting applications increases exponentially with the number and size of applications. For small- and medium-sized enterprises, even individual entrepreneurs, the cost of creating software product operation and maintenance is nearly unaffordable. In this context, cloud computing technology emerged as a solution. The National Institute of Standards and Technology (NIST) defines cloud computing as a pay-per-use model that provides users with usable, convenient, on-demand network access, and enters configurable computing resource sharing pools (resources include network, server, storage, application software, and service) [2]. These resources can be supplied rapidly, with minimal management effort or little interaction with service providers. Cloud computing represents a transition from the traditional single-generator model to a centralized power supply model akin to a power plant. It means that computing power can be distributed as a commodity. An excellent cloud software service provider provides software services to every corner of the world, just like the accessibility and affordability of gas, water, and electricity. The defining characteristics of cloud computing can be summarized as follows:

1. Super large scale: The term "Clouds" is used to describe a phenomenon that occurs on a vast scale. Google Cloud Computing has a server infrastructure comprising over one million servers. Notable industry leaders such as Amazon, IBM, Microsoft, and Yahoo have hundreds of thousands of servers. Enterprise private clouds typically comprise hundreds or thousands of servers. The cloud computing model offers users a previously unattainable level of computational capability.
2. Virtualization: Cloud computing enables to access application services from any location and via a range of terminals. The requested resource is sourced from the cloud rather than a fixed physical entity. The application is situated within the cloud, and users are not required to be aware of or concerned with the precise location of the application. The requisite device is merely a laptop or a mobile phone, which can be used to accomplish any necessary task through network services, including those of a supercomputing nature.

3. High reliability: The Cloud uses a variety of techniques, including multiple data copy fault tolerance, computing node isomorphic interchangeability, and other measures, to ensure the high reliability of the service. The reliability of cloud computing is superior to that of local computing.
4. Versatility: The objective of cloud computing is not the development of applications tailored to specific purposes. The cloud provides the necessary infrastructure to construct applications that are continuously evolving. The same cloud can simultaneously support multiple applications.
5. High scalability: The scale of the cloud can be dynamically scaled to accommodate the growth of both applications and users.
6. On-demand service: Cloud represents a vast resource pool, following the on-demand purchase model.
7. Extremely low cost: The fault-tolerant measures of the cloud permit the use of inexpensive nodes in its formation. The automated centralized management of the cloud relieves a large number of enterprises from bearing the increasing cost associated with data center management. Moreover, the versatility of cloud computing compared with traditional systems has led to a notable improvement in resource utilization, enabling users to fully benefit from the cost-effective advantages of the cloud. In many cases, it now takes only a few hundred dollars and a few days to complete tasks that previously required tens of thousands of dollars and a few months.

The deployment of cloud computing can be achieved through three principal methods: public cloud deployment, private cloud deployment, and hybrid cloud deployment (see Fig. 2.9 Cloud computing deployment). Public cloud is the most common way to deploy cloud computing. In a public cloud deployment, third-party cloud service providers own and operate the resources, including servers and storage spaces, which are accessed via the Internet. Microsoft Azure [3] is an exemplar of public cloud. In a public cloud, all hardware, software, and other supporting infrastructure are owned and managed by the cloud provider. In a public cloud, the same hardware, storage, and network equipment are shared with other organizations or cloud "tenants." Users can access services and manage accounts via a web browser. The most common applications of public cloud deployments are the provision of web-based email, online office applications, storage, testing, and development environments.

The advantages of public cloud computing include lower cost, no maintenance, nearly unlimited scalability, and high reliability. A private cloud is a computing environment that is dedicated to the exclusive use of a particular enterprise or organization. The private cloud may be physically located in the organization's on-site data center, or hosted by a third-party service provider. In contrast, a private cloud is comprised of services and infrastructure that are consistently maintained on a private network. Additionally, the hardware and software utilized within this environment are exclusively dedicated to organizations. In this way, this setup facilitates the customization of resources to align with specific IT requirements. The private cloud is usually used by government agencies, financial institutions, and other media

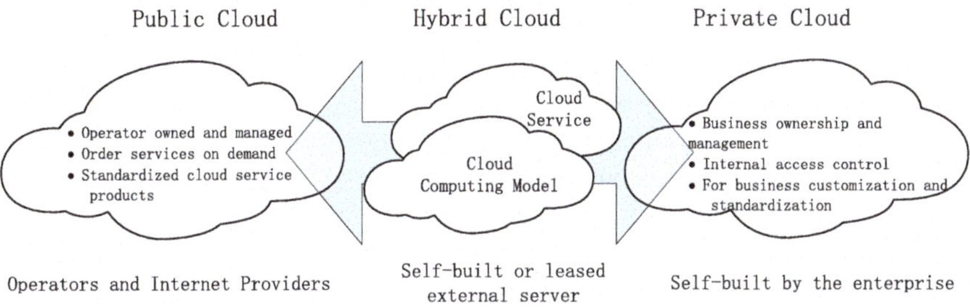

Fig. 2.9 Cloud computing deployment

to large-scale organizations with business-critical operations that seek greater control over their operational environment. The advantages of a private cloud include greater flexibility, enhanced security, and the scalability and efficiency of the public cloud [4].

A hybrid cloud is generally considered to be "the best of both public and private cloud models." A hybrid cloud combines local infrastructure or a private cloud with a public cloud. Organizations may thus benefit from the advantages of both. In a hybrid cloud, data and applications can be transferred between private and public clouds, thereby affording greater flexibility and a wider range of deployment options. For example, public clouds can be used for high-volume and low-security requirements such as web-based email. Private clouds (or other local infrastructure) can be used for the execution of sensitive and business-critical operations, such as financial statements. In a hybrid cloud, the option of "cloud burst" is also available. In the event of a surge in demand, such as during a seasonal event or tax reporting period, applications or resources running in a private cloud can choose to utilize the public cloud, allowing the organization to access additional computing resources. The advantages of hybrid cloud computing include flexibility, controllability, and high-cost efficiency. All three cloud computing deployment solutions share similar advantages, including cost-effectiveness, performance, reliability, and scalability. Nevertheless, the specific choice of deployment method is contingent upon the specific requirements of the enterprise.

The services provided by cloud computing to users include Infrastructure-as-a-Service (IaaS), Platform-as-a-Service (PaaS), and Software-as-a-Service (SaaS), as shown in Fig. 2.10. Among these services, IaaS provides users with complete computer infrastructure services, which include hardware-related services. In the traditional model, enterprises are required to procure hardware devices, such as servers and switches, to construct websites or implement File Transfer Protocol (FTP). In the context of cloud computing services, enterprises have the option of renting facilities and other service providers' facilities through IaaS services, which can help to reduce system deployment costs. PaaS provides users with access to software research, and development platforms, which are also referred to as middleware. These include virtual servers, operating systems, database systems, etc. Users are able to develop customized software applications based on middleware. In the

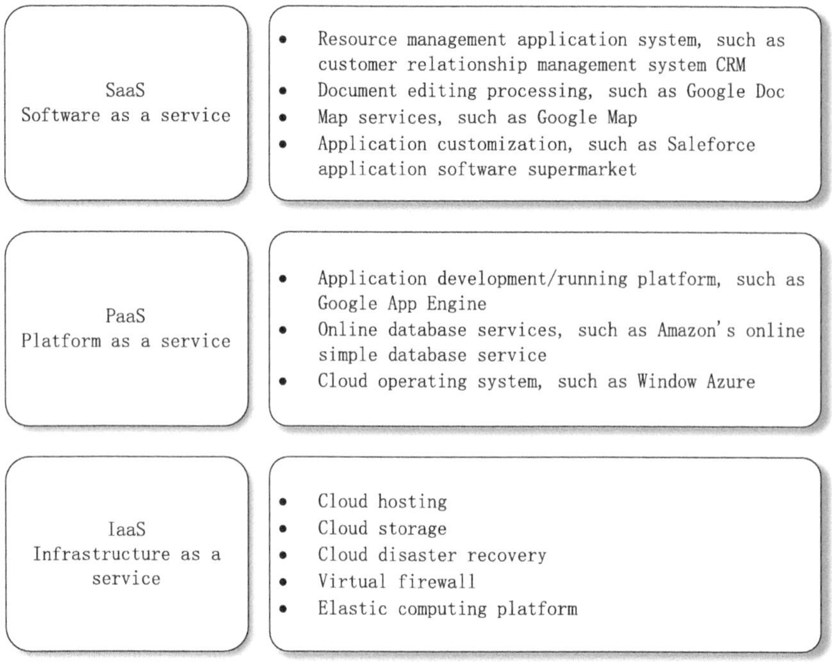

SaaS Software as a service	• Resource management application system, such as customer relationship management system CRM • Document editing processing, such as Google Doc • Map services, such as Google Map • Application customization, such as Saleforce application software supermarket
PaaS Platform as a service	• Application development/running platform, such as Google App Engine • Online database services, such as Amazon's online simple database service • Cloud operating system, such as Window Azure
IaaS Infrastructure as a service	• Cloud hosting • Cloud storage • Cloud disaster recovery • Virtual firewall • Elastic computing platform

Fig. 2.10 Cloud computing service types

SaaS model, users are able to access and utilize web-based service functions without downloading.

2.3 Hadoop Architecture

2.3.1 Hadoop Basic Knowledge

Apache Hadoop [5] is a reliable and scalable open-source framework for distributed computing development. It allows developers to use a simple programming model for distributed large datasets in computer clusters. The supported cluster size can range from a single server to thousands of machines, and each machine in the cluster provides local computing and storage capabilities. The Hadoop library can perform anomaly detection at the application layer instead of relying on hardware. It can still provide stable and high-quality services in a computer cluster where a single device may fail.

Hadoop is a text search library developed by Doug Cutting, the Founder of the Apache Lucene project [5]. Hadoop originated from the Apache Nutch project (an open-source web search engine) that commenced in 2002 and is part of the Lucene project. In 2004, the Nutch project also emulated the Google File System (GFS) to develop its distributed file system, the Nutch Distributed File System (NDFS), which served as the precursor to HDFS. In 2004, Google published another paper with far-reaching impact, expounding MapReduce distributed programming ideas. In 2005, Nutch open-source software implemented Google's MapReduce. In the following year, in February 2006, NDFS and MapReduce in Nutch began to become independent, becoming a subproject of the Lucene project, called Hadoop. Subsequently, Doug Cutting joined Yahoo. In April 2008, Hadoop broke the world record, becoming the fastest system for sorting 1 TB of data. The Hadoop system uses a cluster of 910 nodes to perform calculations, with the sorting time taking only 209 s. In May 2009, Hadoop shortened the sorting time for 1 TB data to 62 s. Since then, Hadoop has gained a great reputation and has rapidly developed into the most influential open-source distributed development platform in the era of big data. It has become the standard for big data processing.

The Hadoop architecture is centered upon HDFS and MapReduce, providing developers a low-level implementation of a transparent distributed infrastructure. The high fault tolerance and high scalability of HDFS allow developers to deploy Hadoop on inexpensive hardware to form a distributed system. The MapReduce programming model allows developers to develop parallel computing applications without necessarily having to understand the underlying details of the distributed system. Consequently, developers can use the Hadoop architecture to readily organize computer resources to construct their distributed computing platform. This enables them to make full use of the cluster's computing and storage capabilities, thus facilitating the storage of massive data. The Hadoop architecture offers the following advantages:

1. Open source: Hadoop is a free and open-source framework that allows users to implement custom modifications.
2. Scalable: Hadoop supports storage and computing expansion.
3. Cost-effective: Hadoop framework can be deployed on any standard processor.
4. Reliable: Guaranteed by the HDFS backup and recovery mechanism and the MapReduce task monitoring.
5. Efficient: It is possible to achieve a processing mode that combines local data with MapReduce.

Hadoop architecture is commonly applied in data mining and business intelligence, including log processing, clickstream analysis, similarity analysis, precision advertising, bioinformation technology, file processing, web indexing, traffic analysis, user segmentation feature modeling, user behavior analysis and trend analysis, etc. The main production release of the Hadoop 2.0 framework was developed based on JDK 1.7 version development, and JDK 1.7 stopped updating directly, forcing the Hadoop community to re-develop and release the latest version of the Hadoop framework, Hadoop 3.0 [6], based on JDK 1.8. Compared to Hadoop 2, Apache Hadoop 3 integrates a number of essential enhancements that provide stability and high-quality APIs for real-world product development. Some of the new features of Hadoop 3.0 are listed as follows:

1. Minimum Java version is required to be upgraded to Java 8: All Hadoop jars are compiled and executed based on the Java 8 runtime version. Users still using Java 7 or lower will need to upgrade to Java 8.
2. HDFS incorporates support for erasure coding: Erasure coding is an advanced data persistent storage technique that enhances storage efficiency by requiring less space compared to traditional copying methods. Take the Reed-Solomon (10, 4) scheme as an example: this standard coding technique necessitates only 1.4 times the storage space, whereas the conventional replication method employed by HDFS demands a tripling of the storage space. The primary trade-off with erasure codes is their increased computational overhead during data reconstruction and the processing of remote read and write operations. As a result, erasure coding is optimally suited for the storage of cold data—information that is infrequently accessed. Moreover, when adopting this feature, users should be mindful of the potential impact on network bandwidth and CPU usage.
3. Rewrite shell script: Hadoop's shell script has been rewritten to fix bugs and add some new features.
4. Shaded client jars: HADOOP-11804 adds a new Hadoop-client-API and Hadoop-client-runtime artifact to isolate Hadoop dependencies into a single jar package, preventing Hadoop dependencies from penetrating the application's environment variables.
5. MapReduce task-level local optimization: MapReduce adds support for the localized implementation of the map output collector. This can provide a 30% performance improvement for intensive shuffling operations.
6. Support more than two NameNodes: Allow users to run multiple standby NameNodes. For example, by configuring three NameNodes and five JournalNodes, the cluster can tolerate two node failures instead of one. However, there is only one active NameNode, and the rest are all standby. The standby NameNode will continue to synchronize with the JournalNode to ensure that it obtains the latest transaction log and synchronizes it with the mirror it maintains. In this way, hot standby can be achieved. In the event of a failure, the standby NameNode immediately switches to the active state and provides external services. At the same time, the JournalNode allows only one active NameNode to write.
7. YARN resource type: The YARN resource model has been generalized to support user-defined computable resource types, not just CPU and memory. For example, the cluster administrator can define resources such as the number of GPUs and software serial numbers. Then, YARN tasks can then be scheduled for these available resources.

2.3.2 Hadoop System Architecture

The basic components of the Hadoop framework include

1. Hadoop Common: Common components that support other Hadoop components.
2. HDFS: Distributed file system.

3. Hadoop MapReduce: A large dataset parallel computing system based on the Hadoop YARN framework.
4. Hadoop YARN [7]: Framework for scheduling user task and managing cluster resources.

The basic components of the Hadoop framework and other related tools together form the Hadoop ecosystem. The ecosystem provides many powerful and easy-to-use functions that enable developers to more quickly develop and deploy distributed projects using Hadoop. The common tools are as follows:

1. Hive (MapReduce-based data warehouse): Facebook open-source Hive was originally used to solve the problem of massive structured log data statistics. It is an Extraction-Transformation-Loading (ETL) tool. It is also based on Hadoop. On the data warehouse, Hive uses MapReduce and HDFS to compute and store data. It defines an SQL-like query language, HiveQL. Apart from not supporting updates, indexes, and physical objects, HiveQL supports almost all other SQL features and is usually used for offline data processing. The HiveQL language can be regarded as a MapReduce language translator, and the MapReduce program can be simplified to the HiveQL language. However, some complex MapReduce programs cannot be described by HiveQL. Hive provides Shell, JDBC, Thrift, Web, and other interfaces.
2. Pig (data warehouse): Yahoo open-sourced Pig. The design motivation is to provide an ad hoc data analysis tool based on MapReduce, which is usually used for offline analysis. Pig defines Pig Latin, a data flow language similar to SQL. Pig Latin can perform operations such as sorting, filtering, summing, associating, and supporting custom functions. Pig automatically maps Pig Latin to MapReduce jobs and uploads them to the cluster to run, reducing the need for users to write Java programs. There are three ways to run Pig: Grunt shell, script mode, and embedded.
3. Mahout (data mining library): Mahout is a distributed computing framework based on Hadoop machine learning and data mining. It implements three major algorithms: recommendation, clustering, and classification.
4. HBase (distributed database): HBase originated from the Bigtable paper published by Google in November 2006. In other words, HBase is a clone of Google Bigtable. It supports multiple access methods such as Shell, Web, and API, and is a typical representative product of NoSQL. HBase features high reliability, high performance, column oriented, and scalability.
5. Zookeeper (distributed collaboration service): Zookeeper originated from the Chubby paper published by Google in November 2006. In other words, Zookeeper is a clone of Chubby. Zookeeper is mainly used to solve data management problems in a distributed environment, such as unified naming, state synchronization, and cluster management.
6. Sqoop (data synchronization tool): Sqoop is a bridge between Hadoop and traditional databases. It supports multiple databases, including MySQL and DB2.

Figure 2.11 shows the basic architecture of the Hadoop 2.0 ecosystem.

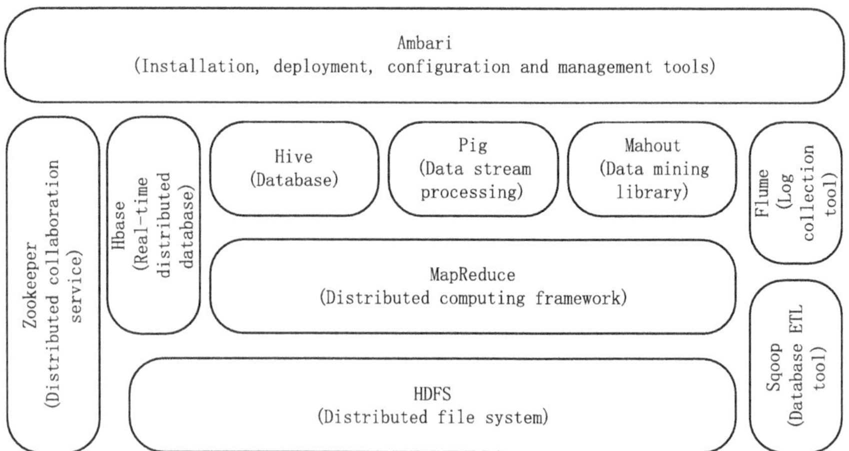

Fig. 2.11 Hadoop ecosystem

The development of the Hadoop framework has been divided into two distinct eras: Hadoop 1.0 and Hadoop 2.0. The Hadoop framework in the 1.0 era only had two core components: the distributed file system (HDFS) and Hadoop MapReduce. Its architectural simplicity and clarity, coupled with the rapid industry adoption observed during the early years of its introduction, contributed to its notable success. However, as distributed system clusters expanded and workloads grew, the Hadoop 1.0 framework began to exhibit shortcomings. These included single points of failure, centralized tasks, and difficulty in maintaining source code. The Hadoop team realized that repairs could not fully address these issues, prompting them to begin applying the refactored Hadoop MapReduce framework, which became the YARN framework after version 0.23.0. At this point, the Hadoop framework officially entered the 2.0 era. Figure 2.12 shows the evolution of Hadoop 1.0 to Hadoop 2.0.

YARN is a resource management and scheduling system in the Hadoop framework. It is mainly responsible for managing the resources in the cluster (in a similar role to the operating system) and allocating resources to upper-layer applications [6]. YARN is an important component added during the evolution of Hadoop 1.0 to

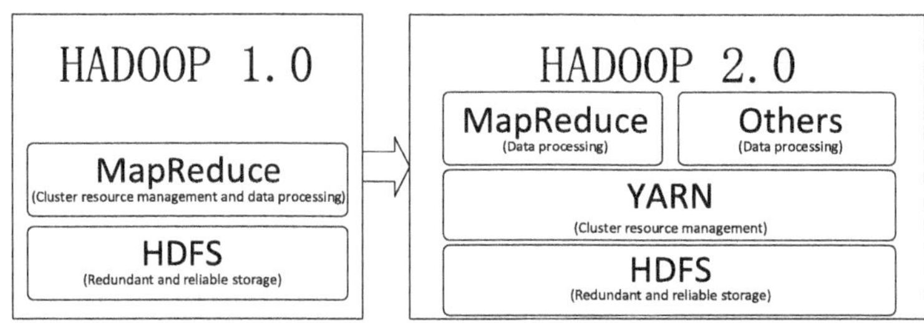

Fig. 2.12 Hadoop architecture evolution

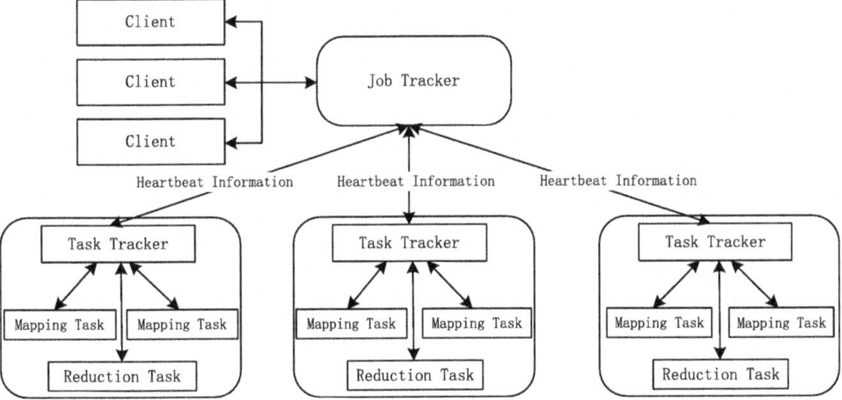

Fig. 2.13 Hadoop1.0 MapReduce framework workflow

Hadoop 2.0. To gain a deeper understanding of the role and significance of YARN in the Hadoop architecture, it is essential to first analyze the principles and shortcomings of the Hadoop 1.0 MapReduce (MR1) framework.

The Hadoop MapReduce framework is an offline processing computing framework based on the MapReduce programming model, as outlined in Sect. 2.2.3. Figure 2.13 shows the workflow of the MapReduce framework in the 1.0 era.

MR1's workflow includes 13 steps:

1. The client (JobClient) submits a job (Job) to the job tracker (JobTracker).
2. The job tracker requests the data to be analyzed through HDFS.
3. The job tracker retrieves the data.
4. The job tracker assigns the job and the corresponding data block to the task tracker (TaskTracker) node for calculation.
5. Each computing node loads the jobs and data blocks that need to be calculated.
6. The job tracker monitors the status of the task tracker. If the task node fails, the task is distributed to a new computing node. Once the task is completed, the scheduling is determined by the monitoring results.
7. Store the calculated intermediate data in HDFS.
8. The monitoring found the nodes with surplus calculation.
9. The job tracker reschedules the task block.
10. Schedule tasks between different computing nodes.
11. Once a computing node has completed its computing task, it notifies the job tracker.
12. The job tracker writes the results to HDFS.
13. The job tracker obtains the results from HDFS.

A challenge with Hadoop 1.0 is the single point of failure. In this scheduling mechanism, the JobTracker node serves as the central processing node of MapReduce. This topology often results in a hidden danger of a single point of failure within the system. Moreover, the JobTracker node is also under significant pressure from the master node. The scheduling and allocation of resources, as well as the monitoring and restarting of tasks, are the sole responsibility of the JobTracker node, increasing the likelihood of failure. Consequently, the industry generally concluded that Map-Reduce for Hadoop 1.0 could only support the management of 4000 node hosts at most. The bottleneck in system performance is the operation mode in which JobTracker has both resource management and job control is the bottleneck that restricts system performance.

The primary objective of refactoring is to divide the two main functions of JobTracker, namely, resource management and job control, into distinct components. YARN's design is based on this premise. Figure 2.14 shows the Hadoop YARN architecture.

The YARN framework is a robust system that provides extensive resource management and job control capabilities, facilitated by two key processes: the Resource Manager (ResourceManage, RM) and Application Controller (ApplicationMaster, AM). The RM's primary role is to monitor the resource management and scheduling across the entire cluster, ensuring an efficient distribution of computational resources. In contrast, the AM is tasked with the comprehensive lifecycle management of applications, including the handling of tasks, monitoring for failures, and orchestrating

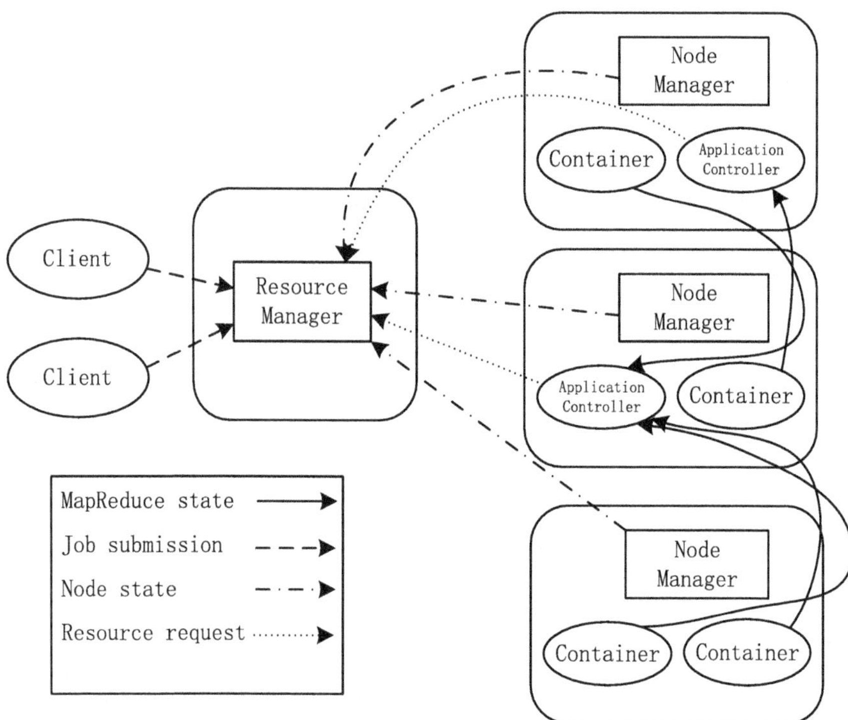

Fig. 2.14 Hadoop YARN architecture

restarts. It is worth noting that each individual job, rather than job types, is associated with a specific AM. The AM has the flexibility to operate on different machines, separate from the node hosting the RM. However, in the Hadoop 1.0 version, task tracking is fixed and cannot be relocated to alternative device. Besides, the YARN framework has various node types, including the Node Manager (NodeManager, NM), which serves as the local resource and task manager on each node.

The NM plays a dual role in the YARN ecosystem. Firstly, it periodically communicates with the RM, providing updates on the resource usage of the node and the operational status (Container) of all containers. Secondly, it will accept and execute various requests from the AM, which pertain to the start and stop of container processes. In YARN, a container is an abstraction of resources, encapsulating the multi-dimensional resources on a node, such as memory, CPU, disk, and network. When the AM applies requests resources from the RM, the RM acts as the application controller, assigning containers that represent the granted resources. YARN will allocate a container for each task, with the task being restricted to the use of the resources described in the container.

Upon submission of an application to YARN, the system will initiate two distinct stages of processing. The first stage involves the commencement of the application controller. The second stage is to create an application by the application controller and apply for resources to support its operation. Throughout this process, YARN will monitor the application's operation until it is completed. Figure 2.15 illustrates the workflow of the MapReduce framework based on YARN.

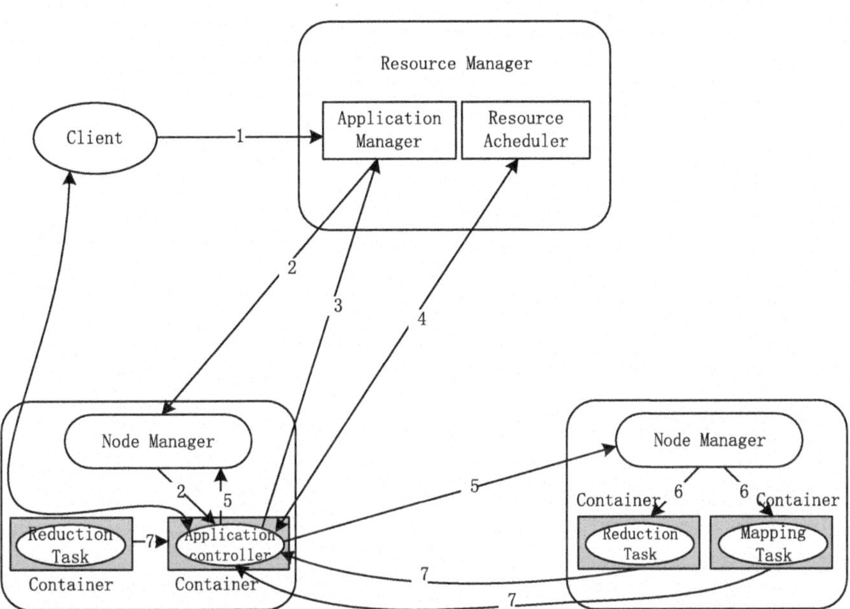

Fig. 2.15 Workflow of MapReduce framework based on YARN

The seven-phase workflow of the MapReduce framework, based on YARN resource manager, is described below:

1. The user submits the application program to YARN, including the application controller program, the command to start the application controller, and the user program.
2. The resource manager allocates the first container for the application and communicates with the corresponding node manager, requiring it to start the application controller of the application within this container.
3. The application controller first registers with the resource manager, thereby enabling the user to directly observe the running status of the application through the resource manager. Then it will apply for resources for each task and monitor its running status until the end of the run.
4. The application controller uses polling to request and obtain resources from the resource manager via the Remote Procedure Call Protocol (RPC) protocol.
5. Once the application controller has applied for resources, it communicates with the corresponding node manager, requesting that it initiates the task.
6. Once the requisite environment has been established (including the configuration of environment variables, JAR packages, and binary programs), the node manager initiates the task by writing it to a script and launching it.
7. Each task reports its status and progress to the application controller via an RPC protocol. This enables the application controller to monitor the status of each task at any given time, allowing for the task to be restarted in the event of failure.

In comparison to the operational mode of MapReduce 1, YARN dramatically reduces the resource consumption of the job tracker (i.e., the resource manager within the framework), thereby overcoming the limitations of the previous framework. Additionally, the Hadoop 1.0 architecture is only compatible with the MapReduce programming model. Concurrently, YARN enables the use of conventional communication modes, such as MPI (message passing interface), while facilitating the execution of diverse programming models, including those pertaining to graphics processing, iterative processing, machine learning, and general cluster computing. This represents a further development of Hadoop 2.0, with YARN exceeding the capabilities of its predecessor.

2.3.3 Hadoop Typical Case

Yahoo applies Hadoop in a variety of its products, including data analysis, content optimization, anti-spam systems, optimization of advertising selection, big data processing, and ETL. Additionally, Hadoop is also fully applied in user interest prediction, search ranking, and advertising positioning. In 2013, the real-time service system utilized Apache to read the mapping of a user to interest from the database in order to personalize the homepage of Yahoo. The Hadoop cluster in the production environment rearranges the content based on the latest data every five minutes, and will update the content on the page every seven minutes.

In terms of mailboxes, Yahoo uses Hadoop clusters to score mail based on the spam model and to improve the cluster's anti-spam model every few hours. The cluster system can also promote more than 5 billion mail deliveries per day. Yahoo's largest Hadoop production application is the Search Webmap application, which runs in a Linux system cluster comprising over 10,000 machines. Yahoo's web search queries utilize the data generated by this application. The construction of Webmap is comprised of several steps. Initially, web pages are crawled, concurrently with the generation of a database comprising all known web pages and Internet sites, as well as a massive dataset encompassing all pages and sites. Subsequently, these data are transferred to the Yahoo Search Center to perform the sorting algorithm. Throughout this process, the number of links between pages in the index reaches 1 TB, while the data output after compression reaches 300 TB. To run a MapReduce task, it is necessary to use more than 10,000 cores, and data stored within a production environment. Consequently, the requisite volume exceeds 5 PB.

As a world-renowned social networking site, Facebook boasts over 2.5 billion active users in 2019, with an average of 510,000 comments posted per minute, 239,000 statuses update per minute, and 136,000 photos uploaded per minute. Consequently, the quantity of data that Facebook is required to store, and the associated process, is considerable. It is therefore evident that a high-performance cloud platform is significant important to Facebook. The Hadoop platform is employed by Facebook for a number of functions, including log processing, recommendation systems, and data warehousing.

Facebook stores data in a data warehouse constructed with Hadoop/Hive. This data warehouse has 4800 cores and boasts a storage capacity of 5.5 PB. Each node is capable of storing 12 TB of data. Additionally, it is equipped with a two-layer network topology [7]. As shown in Fig. 2.16, the MapReduce cluster in Facebook is subject to dynamic changes, with the capacity to be relocated based on load conditions and configuration information between cluster nodes.

eBay stores information on hundreds of millions of commodities, with millions of new commodities added daily. This necessitates the use of a cloud system to

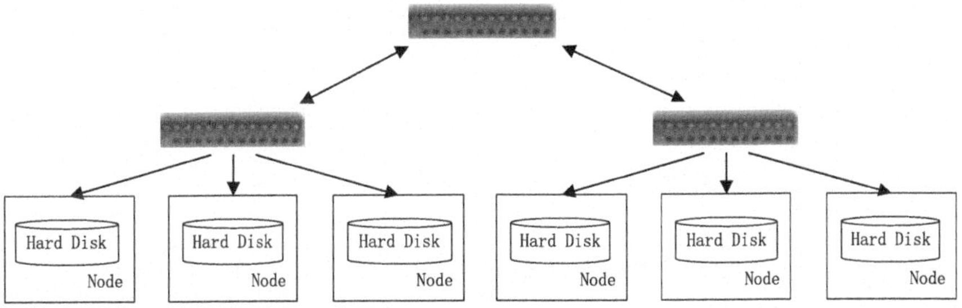

Fig. 2.16 Cluster network topology

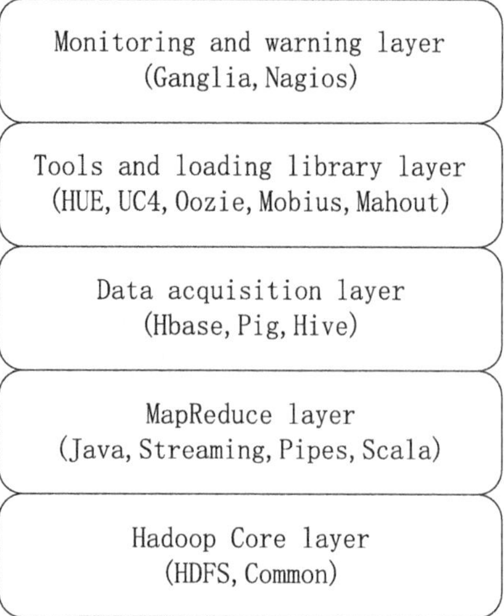

Fig. 2.17 Athena layer

store and process petabytes of data. Hadoop is an optimal choice for this purpose. Hadoop is a fault-tolerant, scalable, and distributed cloud computing framework built on commercial hardware. eBay has used Hadoop to construct a large-scale cluster system called Athena, comprising the following five layers (see Fig. 2.17):

1. Hadoop core layer includes the Hadoop runtime environment, a number of general facilities, and HDFS. The file system has been optimized to read and write large blocks of data, such as the alteration of the block size from 128 to 256M.
2. The MapReduce layer provides APIs and control mechanisms for developing and executing tasks.
3. The data acquisition layer: This layer is primarily comprised of HBase, Pig, and Hive, which collectively serve as the main framework for data acquisition.
4. The tool and loading library layer: UC4 is eBay's enterprise-level scheduler that automatically loads data from multiple data sources. The loading library includes a statistical library (R), a machine learning library (Mahout), a math-related library (Hama), and an eBay-specific library (Mobius) developed to parse weblogs.
5. The monitoring and warning layer: Ganglia is a distributed cluster monitoring system. Nagios is used to issue warnings of critical events, such as server unavailability and full hard disk capacity.

Baidu is the largest Chinese search engine company. The company offers a range of products that utilize search engines, including functional search based on the Internet search, community search based on Post Bar, vertical search for regions and industries, MP3 music search, and encyclopedia. It fulfills the majority of the search requirements within the Chinese online domain. Baidu's requirements for

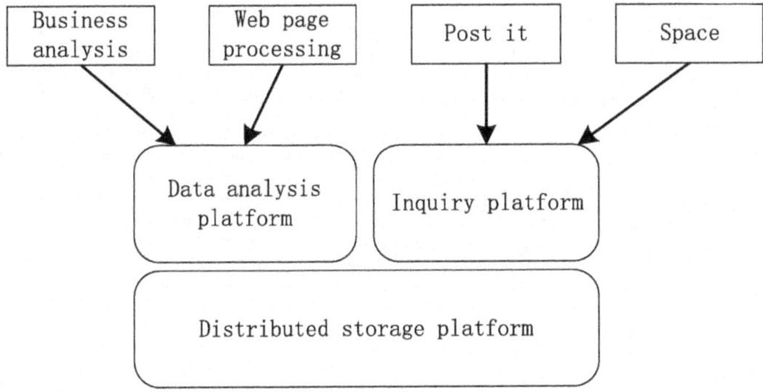

Fig. 2.18 Platform requirements in the Internet field

massive data processing are relatively high. To analyze the data offline, it must be processed within the specified time and the results must be fed back to the platform. Figure 2.18 illustrates the platform requirements of Baidu in the Internet field. To meet these requirements, a cloud platform with enhanced performance is necessary. Hadoop is a good option. In Baidu, Hadoop is mainly used in the following areas:

1. Storage and subsequent analysis of log data.
2. Analysis and mining of web data.
3. Business data analysis, such as user behavior and advertising attention.
4. Online data feedback to get timely online ad clicks.
5. Clustering of user web pages, analyze the user's recommendation degree, and examining the relationships between users.

MapReduce is not a universal solution to all computing problems. According to Baidu researchers, a good model should be as shown in Fig. 2.19. HDFS implements a shared storage system, with some calculations solved using MapReduce, and others solved using MPI. Nevertheless, there are still some calculations that require a combined approach. Given that MapReduce is suitable for processing massive and partitioned data, it can be employed to perform some filtering when processing this type of data to obtain a basic vector matrix. This can then be returned after further processing through MPI. Consequently, integration techniques are more effective in addressing the issue.

2.3.4 Hadoop Programming Interface

The Hadoop framework standalone environment can be constructed according to the following methodology:

1. Preliminary preparations: A Linux server, the downloading the JDK, and the installation of Hadoop packages.

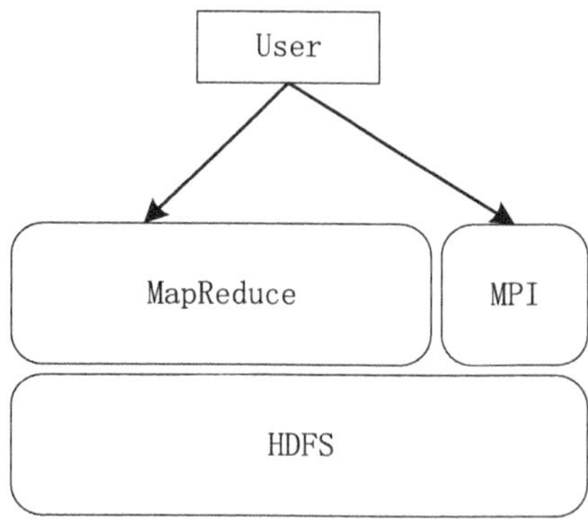

Fig. 2.19 Calculation model

2. Server configuration: Enter the vim/etc./sysconfig/network command to change the hostname and then reboot the system to implement the change. Enter the vim/etc./hosts command to add the mapping of the host IP address and hostname. Prior to configuration, it is necessary to close the firewall. Enter the service iptables stop command under CentOS 7 version. Otherwise, enter the systemctl stop firewalld.service command. Enter the date command to check whether the server time is consistent. If it is inconsistent, enter the command date —s'MMDDhhmmYYYY.ss' to modify.
3. JDK environment configuration: The JDK and Hadoop compressed packages should be placed in the home directory, and new Java and hadoop folders should be created. Enter the tar—xvf jdk-8uXXX-linux-x64.tar.gz command and the tar—xvf hadoop-2.8.X.tar.gz command to decompress the JDK and Hadoop (the file names are different in different versions). The files should then be moved to the Java and Hadoop files, respectively. Finally, the folder should be renamed to JDK 1.8 and Hadoop2.8. Enter the java-version command to check whether the JDK is installed. If the version is not suitable, uninstall it. Enter the vim/etc./profile command to edit the etc./profile file. The configuration is as follows:

```
export JAVA_HOME=/home/java/jdk1.8
export JRE_HOME=/home/java/jdk1.8/jre
export CLASSPATH=.:$JAVA_HOME/lib/dt.jar:$JAVA_HOME/lib/
tools.jar
export CLASSPATH=.:$JRE_HOME/lib:$CLASSPATH
export PATH=.:${JAVA_HOME}/bin:$PATH
```

The JAVA_HOME path is the JDK path. Enter source/etc./profile to make the configuration take effect.

4. Hadoop environment setup: edit etc./profile file as well. The configuration is as follows:

```
export HADOOP_HOME=/home/hadoop/hadoop2.8
export HADOOP_COMMON_LIB_NATIVE_DIR=$HADOOP_HOME/lib/native
export HADOOP_OPTS="-Djava.library.path=$HADOOP_HOME/lib"
```

```
export PATH=.:${JAVA_HOME}/bin:${HADOOP_HOME}/bin:$PATH
```

Create new folders in the root directory: hadoop, hadoop/tmp, hadoop/var, hadoop/dfs, hadoop/dfs/name and hadoop/dfs/data. Switch to the hadoop directory, enter the command vim core-site.xml, and add in <configuration>:

```
<configuration>
    <property>
      <name>hadoop.tmp.dir</name>
    <value>/root/hadoop/tmp</value>
        <description>Abase for other temporary directories.</
description>
  </property>
  <property>
    <name>fs.default.name</name>
    <value>hdfs://host_name:9000</value>
  </property>
</configuration>
```

Input the command vim hadoop-env.sh, will ${JAVA_HOME} modify to your own JDK path.

Enter the command vim hdfs-site.xml and add it in <configuration>:

```
<configuration>
  <property>
    <name>dfs.name.dir</name>
    <value>/root/hadoop/dfs/name</value>
    <description>Path on the local filesystem.</description>
  </property>
  <property>
    <name>dfs.data.dir</name>
    <value>/root/hadoop/dfs/data</value>
    <description>Comma separated list of paths.</description>
  </property>
  <property>
    <name>dfs.replication</name>
    <value>2</value>
  </property>
  <property>
    <name>dfs.permissions</name>
    <value>false</value>
    <description>need not permissions</description>
  </property>
</configuration>
Enter the command vim mapred-site.xml and add it in <configuration>:
<property>
    <name>mapred.job.tracker</name>
  <value>host_name:9001</value>
</property>
<property>
    <name>mapred.local.dir</name>
    <value>/root/hadoop/var</value>
</property>
```

```
<property>
     <name>mapreduce.framework.name</name>
     <value>yarn</value>
</property>
```

The Hadoop standalone environment is set up.

5. Hadoop startup: Upon initial startup of Hadoop, the configuration process is required. Switch to the directory /home/hadoop/hadoop2.8/bin and enter./hadoop namenode format to complete the initialization. Switch to the /home/hadoop/hadoop2.8/sbin directory and enter start -dfs.sh to start HDFS, and start -yarn.sh to start YARN. This will complete the startup of the Hadoop framework.

This section presents an introduction to the programming of Hadoop through the Java API, with the programming example of the Word Count.

The pseudocode implementation of the WordCount is given in Sect. 2.2.3. In short, the objective of this task is to count the frequency of occurrence of each word within the input data. The Hadoop framework is designed for straightforward development. It is sufficient for developers to focus on the implementation of the Map and Reduce logic, leaving the remaining transactions to the processing framework.

First, a brief introduction to the related Java API is provided:

1. MapReduceBase class serves as a base class that implements the Mapper and Reducer interfaces. The implementation content of the MapReduceBase class is empty, and is mainly used to extend the MapReduce function.
2. Mapper interface represents the interface of the Mapper node in the MapReduce programming model. The abstract method void map (K1 key, V1 value, OutputCollector <K2, V2> output, Reporter reporter) defined in this interface is the Map logic mentioned above. This method maps the input key-value pair <K1, V1> to the intermediate key-value pair <K2, V2>. There is no constraint relationship between the output key-value pair type and the input key-value pair type, and the input key-value pair can be mapped to 0 or more output pairs. The parameter OutputCollector interface is used to collect <k, v> pairs of Mapper and Reducer output, and its collect (k, v) method is used to add a (k, v) pair to the output.
3. Reducer interface represents the interface of the Reducer node in the MapReduce programming model. The abstract method void reduces (K2 key, V2 value, OutputCollector <K2, V2> output, Reporter reporter) defined in this interface is the Reduce logic mentioned above. This method merges input key-value pairs <K2, V2> into output. The output key-value pair type is the same as the input key-value pair type, and the input key-value pair is generally mapped to 0 or one output key-value pair. It is worth noting that value is usually passed in the form of an iterator.
4. Jobconf class: A class is representing the concept of Job in the MapReduce framework. The framework implements Job-related work; developers only need to call the relevant API according to the format to start the Job.

According to the above explanation, the process of programming with the Java API-based Hadoop can be summarized as follows: the implementation of MapReduce logic is achieved by rewriting the Map and Reduce interfaces and by creating tasks to execute this logic.

```
Reference source code:
public class WordCount
{
/**
* Map class
*/
   public static class Map extends MapReduceBase implements
         Mapper<LongWritable, Text, Text, IntWritable>{
      //LongWritable is a serializable (serilizable) data type
         encapsulated by Hadoop
      //Serialization facilitates data exchange in a
         distributed environment
      //The class obtains comparison through the Writable
         interface, and all types as keys should implement
         this interface
      private final static IntWritable one = new IntWritable(1);
         private Text word = new Text()
      /**
      * Map main
      */
         @Override
         public void map(LongWritable key, Text value,
      OutputCollector<Text, IntWritable> output, Reporter reporter)
         throws IOException{
      String line = value.toString();
      StringTokenizer tokenizer = new StringTokenizer(line);
         while (tokenizer.hasMoreTokens()){
         word.set(tokenizer.nextToken());
         output.collect(word, one);}}}
/**
* Reduce class
*/
   public static class Reduce extends MapReduceBase implements
         Reducer<Text, IntWritable, Text, IntWritable>{
      /**
      *Reduce main
      *@param key Key: Comparable
      *@param values Value: passed in as an iterator
      */
         @Override
         public void reduce(Text key, Iterator<IntWritable> values,
      OutputCollector<Text, IntWritable> output, Reporter reporter)
         throws IOException{
      int sum = 0;
            while (values.hasNext()){
         sum += values.next().get();}
         output.collect(key, new IntWritable(sum));}}
            public static void main(String args) throws Exception{
         //Configure Job
         JobConf conf = new JobConf(WordCount.class);
         conf.setJobName("wordcount");
```

```
conf.setOutputKeyClass(Text.class);
conf.setOutputValueClass(IntWritable.class);
conf.setMapperClass(Map.class);
//conf.setCombinerClass(Reduce.class);
   conf.setReducerClass(Reduce.class);
   conf.setInputFormat(TextInputFormat.class);
      conf.setOutputFormat(TextOutputFormat.class);
FileInputFormat.setInputPaths(conf, new Path(args));
    FileOutputFormat.setOutputPath(conf, new Path(args));
      JobClient.runJob(conf); }
```

2.4 Hadoop Distributed File System (HDFS)

2.4.1 HDFS Basics

HDFS [8] is a distributed file system that is easily scalable and is used by the Hadoop framework. MapReduce implements massive data processing by distributing tasks to multiple servers. During this process, HDFS is responsible for providing data access capabilities for each computing node. HDFS system can run on a large number of ordinary, inexpensive machines, has a robust fault tolerance mechanism, and provides high-throughput file access services for many users. The design goals of HDFS are as follows:

1. The system consists of numerous inexpensive, standardized components, with component failure considered an expected occurrence.
2. The system's read operations are dominated by two types: large-scale streaming read and small-scale random read.
3. The write workload of the system is characterized by a high volume of large-scale, sequential, data appending operations.
4. High-performance and stable network bandwidth are far more important than low latency.
5. The majority of target programs require high-speed and large-volume data processing capabilities.

In consequence, HDFS exhibits the following characteristics:

1. High fault tolerance:
   (1) Multiple copies of data are automatically saved, thereby improving fault tolerance by increasing the number of copies.
   (2) Following the loss of a copy, it can be recovered automatically through the implementation of HDFS internal mechanism.
2. Suitable for batch processing:
   (1) HDFS is based on mobile computing rather than mobile data.

(2) HDFS exposes data locations to the computing framework.

3. Suitable for big data processing:

   (1) Data scale: Capable of processing data with a data scale up to GB, TB, or even PB.
   (2) File size: It can handle more than one million files, which is quite large.
   (3) Node size: It can handle the size of 10 K nodes.

4. Streaming file access:

   (1) One write, multiple reads, cannot be modified, can only be appended.
   (2) Guarantees the consistency of data.

5. Can be built on cheap machines:

   (1) Improves reliability through a multi-copy mechanism.
   (2) Provides fault tolerance and recovery mechanism. For example, if a copy is lost, it can be recovered by other copies.

The integration of the HDFS and MapReduce programming models can provide high-throughput data access and data processing services for applications. In the event that the server within the Hadoop cluster encounters an error during the processing of large datasets, the entire computing process will not be terminated. Concurrently, HFDS can guarantee data redundancy in the event of faults and errors across the entire cluster. Upon completion of the calculation, the result is written into a node of HFDS. The HDFS does not impose strict formatting requirements on stored data, allowing for unstructured or other data types [9]. In contrast, relational databases necessitate the structuring of data and definition of an architectural framework prior to the storage of data.

The most significant shortcoming of the HDFS system is its poor scalability. The system's primary performance characteristics are as follows:

1. The namespace expansion problem: In the current HDFS architecture, the name node is responsible for the management of the namespace. It supports the basic operations of creating, modifying, and deleting file systems, similar to the directories, files, and blocks in HDFS. As the name node is stored in memory, the number of objects (files, blocks) is constrained. The throughput of the entire distributed file system is also limited by the name node's throughput, which is not conducive to the namespace Expand.
2. The expansion of blocks: An increase in the number of blocks results in an increase in the reporting scale and the management burden of data nodes. The necessity for scalability has prompted a shift in the underlying storage architecture of HDFS toward a distributed storage layer. In response to this shortcoming, the community has proposed the concept of object storage, and introduced a new key-value metadata storage model to implement object storage. The objective is to use this model to replace the existing HDFS metadata management method in the future. This metadata storage model is designated as the Hadoop Distributed Storage Layer (Hadoop Distributed Storage Layer, HDSL).

The prospective evolution of the HDFS system will focus on resolving the above issues. In addition, to migrate the existing model to the new model, the community has proposed two potential methodologies. One is to build a new name node directly on the HDSL. The other is to provide a key-value namespace file system, designated as the Hadoop Compatible File System (HCFS). While the system offers a compatible file system API, the HDFS structure remains rebuilt.

The HDFS distributed file system is built on three fundamental concepts: blocks, name nodes, and data nodes.

1. Block: In the file system, a block represents the fundamental unit of operation unit for reading and writing to the disk. A file system is responsible for managing the disk by processing data blocks, which are composed of integer blocks. The basic disk read and write operations are the reading of blocks from the disk into the memory and the writing of memory to the disk, respectively. The system is responsible for maintaining detailed information about each block, which is accessible to the user. Furthermore, the HDFS distributed file system adheres to the aforementioned principles governing file operations. In this system, the block serves as the fundamental unit of the HDFS file content organization. By default, the HDFS block size is 64 MB, and the system will not allow files smaller than 64MB to occupy the entire block space alone. The HDFS increases the block size to the greatest extent possible to reduce the overhead of addressing, thereby greatly improving the efficiency and disk transmission rate. In a distributed file system, dividing files into blocks confers numerous advantages. For instance, it enables files to be larger than any disk's capacity in the network, which is conducive to fault tolerance, replication operations, and streamlined storage subsystem design.
2. Name node (NameNode): The name node is the main control node within the HDFS system, serving to store the block's metadata information (specifically, the block index information). This node is a singular entity within the entire HDFS cluster. The name node is responsible for maintaining the entire file system tree and all files and index directories in this tree. It permanently saves files on the local disk in two forms: namespace mirroring and editing logs. Concurrently, the name node also records the location of the data node, which is where the constituent blocks of each file are stored. This location is not permanently saved, as the data node generates this information at startup.
3. Data node (DataNode): The data node is a slave node of the HDFS system used to store the block content data. The data node is responsible for storing blocks and providing services for their location. It periodically transmits a list of their stored blocks to the name node. An HDFS cluster may comprise numerous data nodes.

2.4.2 HDFS System Architecture

The development of HDFS has also been divided into two eras. Figure 2.20 presents the system architecture of HDFS 1.0, which only sets a unique name node. Although this approach greatly simplifies the design of the system, it also brings certain limitations.

1. Namespace limitation: Name nodes are stored in memory, and the number of objects (files, blocks) that the name node can hold is limited by the size of the memory space.
2. Performance bottleneck: The throughput of the entire distributed file system is limited to the throughput of a single name node.
3. Isolation problem: There is only one namespace in the cluster, so different applications cannot be isolated.
4. Single point of failure: The name node, as the master node, is susceptible to a single point of failure. Damage to the processor running the name node process will result in the loss of the entire file system data.

HDFS solves the above problems by copying the constituent files into persistent files and running secondary name nodes. Figure 2.21 shows the improved HDFS 2.0 system architecture. The architecture is divided into three layers: the client layer, the name node layer, and the data node layer. The client layer is primarily responsible for providing users with access to the HDFS system. The initial step in the workflow is to read the file data from the corresponding data node. The name node layer comprises

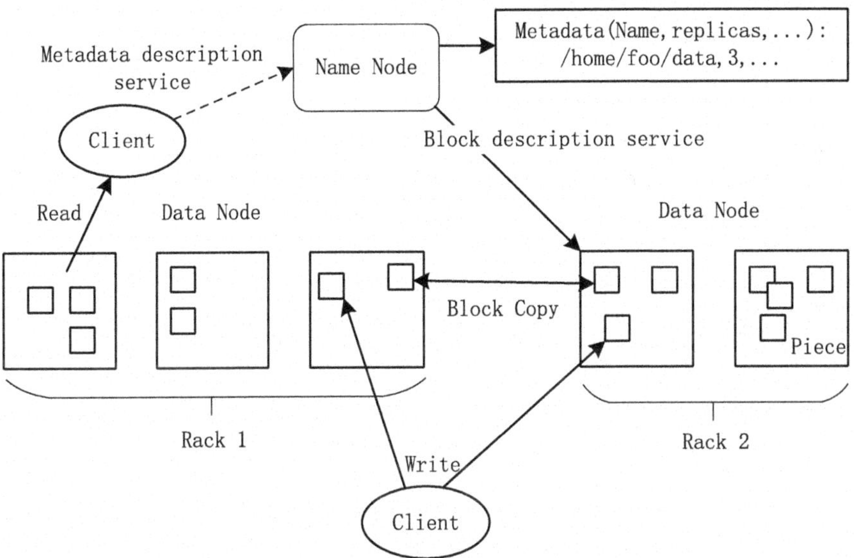

Fig. 2.20 HDFS1.0 system structure

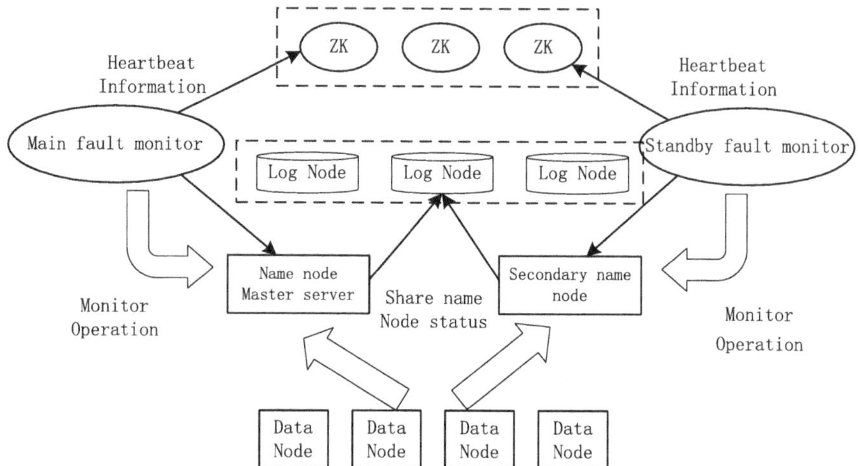

Fig. 2.21 HDFS 2.0 system structure

two components: a name node main server (NameNode Active) and a second-level name node (NameNode Standby). The name node master server is responsible for performing the namespace operation of the file system. The second-level name node facilitates the processing of the mirror file and transaction log by the master node, resulting in the creation of a new mirror file and log. This process is then completed by returning the master node. The data node layer is composed of many data nodes, each of which is tasked with processing the client's read and write requests.

HDFS implements streaming data reading and writing. The client and the name node interact through the instruction stream, while the data node interacts through the data stream. The operation mechanism of HDFS can be discussed separately according to the two different file writing operations and file reading operations. This mechanism is completely transparent to HDFS users.

The data writing operation includes seven steps:

1. The client calls the create method of the DistributedFileSystem object to create the specified file. The HDFS DistributedFileSystem object engages in an RPC with the name node to submit a request for the creation of a new file.
2. Upon receiving instructions to create a file from the RPC, the name node executes a series of verification steps to ensure the file's non-existence and the client's authorization to generate the file. In the event of a successful verification, the name node will generate a new file; otherwise, the file creation will fail, resulting in an exception being thrown to the client. The distributed file system will return a data output stream object, DFSOuputStream, to assist the client in managing the communication between the data node and the name node.
3. The client writes data through the write () method of the DFSOutputStream object. This method divides the data into packets and writes it to the internal data queue. The data queue flows in the form of a data stream, which requires nodes

recorded by a reasonable data node list to allocate new blocks to the replica. These data nodes will form a pipeline.
4. The data stream offloads the packet to the first data node in the pipeline. This node then stores and forwards the packet to the second data node in the pipeline. The behavior of subsequent nodes can be deduced by analogy.
5. The DFSOutputStream object maintains an internal packet queue called the confirmation queue to await confirmation from the data node. All nodes must confirm a packet in the pipeline before moving out of this queue.
6. Once the client has completed writing data, it should call the close method of the DFSOutpulStream object. This function will put all the remaining packets into the data node management and await confirmation before transmitting the name node's completed request.
7. The name node already knows which blocks the file is made of, and requests confirmation from each data node through the Data Streamer. Therefore, it is only necessary for it to await the confirmation result before returning a successful response.

The data reading operation can be divided into six steps:

1. The client calls the open method of the FileSystem object to delegate the opening of the specified file. The FileSystem object serves as the unified agent for file operations within the HDFS system. The FileSystem will then delegate the interaction with the name node to the DistributedFileSystem object via the RPC mechanism, with the objective of obtaining the location information for the target file block.
2. Upon receipt of the RPC instruction to open the file, the name node will return the address of the data node that stores a copy of each block of the specified file. The DistributedFileSystem object receives the information from the name node and returns an FSDatalnputStream object to the client, which allows to read the data.
3. The client calls the read method of the FSDatalnputStream.DFSInputStream object to read the address of the data node where the block is located from the returned information, and establishes a communication connection with the data node.
4. The data node returns the block data to the client. Once the block data has been read, the DFSInputStream object will close the connection with the data node and then establish a connection with the next block's data node to read the remaining data.
5. Repeat the previous step until all data blocks have been read.
6. Close the data input stream and complete a client read operation.

2.4.3 *Main Features of HDFS*

As a distributed file system, HDFS uses multiple copies to store data redundantly to ensure the system's fault tolerance and availability. It is typical for multiple copies of

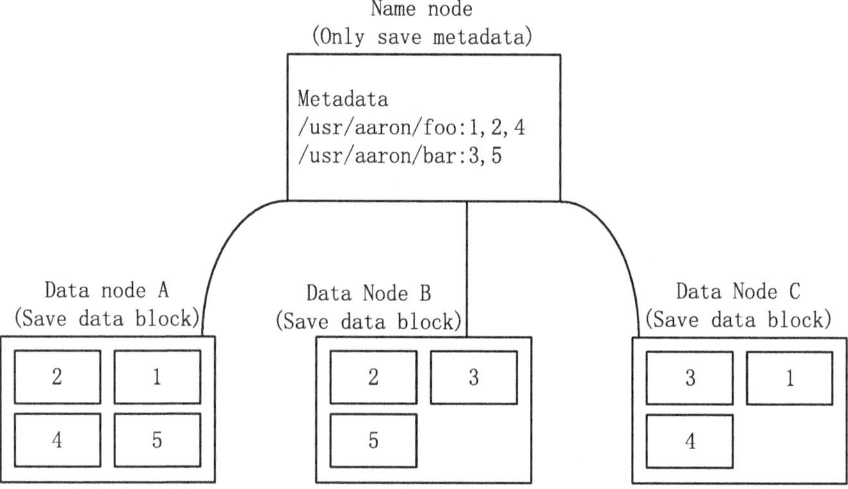

Fig. 2.22 HDFS multi-copy database storage

a data block to be distributed to different data nodes. Taking Fig. 2.22 as an example, data block 1 is stored in data nodes A and C, respectively, while data block 2 is stored in data nodes A and B, respectively. The advantages of multi-copy data block storage include accelerated data transmission speed, ease of detecting data errors, and enhanced data reliability.

Furthermore, HDFS has designed a series of mechanisms for detecting data errors and performing automatic recovery. Hardware errors are regarded as a standard occurrence in HDFS, with the following scenarios representing the most common instances:

1. Name node error: The name node is responsible for the storage of all metadata information. The most critical data structures are FsImage and Editlog. In the event of damage, the entire HDFS instance will fail. Therefore, HDFS implements a backup mechanism to synchronously copy these core files to the backup server SecondaryNameNode. In the event of a failure of the name node, the relevant data may be restored from the backup server's FsImage and Editlog data.
2. Data node error: Each data node will periodically send a "heartbeat" message to the name node to report its status. In the event of a data node failure or network disconnection, the name node is unable to receive heartbeat information from some data nodes. At this time, these data nodes will be marked as "down," the data on the nodes will be marked as unreadable, and the name node will not send any I/O requests to them. In this state, there may be a situation where the number of copies of some data blocks is less than the redundancy factor (the lower limit of the number of copies). This is due to the unavailability of some data nodes. The name node will regularly check this situation. Upon identification of a data block with a lower number of copies than the redundancy factor, the name node initiates data redundancy replication, resulting in the generation of a new copy.

3. Data error: Factors such as network transmission and disk error may cause data error. Upon reading the data, the client uses the MD5 and SHA1 algorithms to verify the data block, thereby ensuring that the correct data has been read. Upon the creation of the file, the client extracts information from each file block and writes it to a hidden file in the same path. When the file is read, the client first reads the information excerpt of the file, and then uses it to verify each read data block. In the event that the verification is unsuccessful, the client requests that the file block be read from another data node and report the file block to the name node for errors. The name node then performs a periodic check and updates the file block in error.

HDFS data access strategy will affect the read and write performance of the entire distributed file system, as it forms the core of the entire system. HDFS uses a Rack-based data storage strategy in terms of data storage to improve data reliability and system availability, while optimizing network bandwidth. By default, HDFS stores different data nodes on different physical racks.

The advantages of this strategy are as follows:

1. High data reliability: The failure of one rack will not affect the copy of data on other racks.
2. High reading speed: When reading data, it can be read from multiple racks in parallel, which improves the reading speed.
3. Easier to achieve load balancing and error handling within the system.

In terms of data reading, HDFS provides an API to determine the rack ID to which a data node belongs, and the client can also call the API to obtain the rack ID to which it belongs. Upon reading the data, the client obtains a list of storage locations for different copies of the data block from the name node, which contains the data node where the copy is located. Subsequently, the API is called to determine the rack ID to which the client and these data nodes belong. If the rack ID corresponding to a copy of a data block is the same as the rack ID corresponding to the client, the copy is preferentially selected to read data. If the copy is not found, a random copy will be selected to read the data.

In terms of data replication, HDFS uses a pipeline replication strategy, which greatly improves the efficiency of the data replication process. When a client writes a file to HDFS, the file is first written to the local system, divided into several blocks, and the HDFS parameters specify the size of each block. Subsequently, each block initiates a write request to the name node within the HDFS cluster. The name node selects a data node list and returns it to the client, indicating which data nodes are currently in use within the system. The client then writes the data to the first data node, while simultaneously passing the list to the first data node. Upon receipt of the 4 KB data, the first data node will proceed to write the data to the local storage, initiate a connection request to the second data node in the list, and transfer the 4 KB data and the list of data it has received to the second data node. When the second data node receives 4 KB of data, it will write the data locally, and initiate a connection request to the third data node in the list. The list of data nodes forms a data replication pipeline. Once the file writing has been written, the data replication process is also completed.

2.4.4 HDFS Programming Interface

The HDFS access interface is a programming rule exposed by the HDFS system to users. Developers can access distributed file system services via interfaces without knowing the internal details of the system. The Java API is one of the most commonly used application programming interfaces of HDFS. The reference routine is as follows:

1. Obtain the file system:

```
//FileSystemClass Represents File System.
public static FileSystem getFileSystem() {
   //Read the configuration file
   Configuration conf = new Configuration();
   // File system
   FileSystem fs = null;
   //SystemURI
   String hdfsUri = HDFSUri;
   if(StringUtils.isBlank(hdfsUri)){
      //Run under Hadoop cluster, we should get the default system
      try {
         fs = FileSystem.get(conf);
      } catch (IOException e) {
         logger.error("", e);
      }
   }else{
     // Testing locally should get the specified file system returned
      try {
         URI uri = new URI(hdfsUri.trim());
         fs = FileSystem.get(uri,conf);
      } catch (URISyntaxException | IOException e) {
         logger.error("", e);
      }
   }
   return fs;
}
```

2. Create a file directory:

```
public static void mkdir(String path) {
   try {
      // Get file system
      FileSystem fs = getFileSystem();
      String hdfsUri = HDFSUri;
      if(StringUtils.isNotBlank(hdfsUri)){
```

```
            path = hdfsUri + path;
        }
        // Create file directory
        fs.mkdirs(new Path(path));
        // Free up system resources
        fs.close();
    } catch (IllegalArgumentException | IOException e) {}
}
```

3. Delete files or directories:

```
public static void rmdir(String path) {
    try {
        FileSystem fs = getFileSystem();
        String hdfsUri = HDFSUri;
        if(StringUtils.isNotBlank(hdfsUri)){
            path = hdfsUri + path;
        }
        // Delete file or file directory
        fs.delete(new Path(path),true);
        fs.close();
    } catch (IllegalArgumentException | IOException e) {
        logger.error("", e);
    }
}
```

4. Get the specified file according to the filter:

```
public static String ListFile(String path,PathFilter pathFilter) {
    String files = new String;
    try {
        FileSystem fs = getFileSystem();
        String hdfsUri = HDFSUri;
        if(StringUtils.isNotBlank(hdfsUri)){
            path = hdfsUri + path;
        }
        FileStatus status;
        if(pathFilter != null){
            //List directory contents according to filter
            status = fs.listStatus(new Path(path),pathFilter);
        }else{
            // List directory contents directly
            status = fs.listStatus(new Path(path));
        }
        // Get all file paths in the directory
        Path listedPaths = FileUtil.stat2Paths(status);
        // Convert listedPaths to String
        if (listedPaths != null && listedPaths.length > 0){
            files = new String;
            for (int i = 0; i < files.length; i++){
                files = listedPaths.toString();
            }
```

```
        }
        fs.close();
    } catch (IllegalArgumentException | IOException e) {
        logger.error("", e);
    }
    return files;// Return the resulting file
}
```

5. File upload HDFS:

```
public static void copyFileToHDFS(boolean delSrc, boolean over-
write,
        String srcFile,String destPath) {
        // Source file path
        Path srcPath = new Path(srcFile);
    // Destination path
    String hdfsUri = HDFSUri;
    if(StringUtils.isNotBlank(hdfsUri)){
        destPath = hdfsUri + destPath;
    }
    Path dstPath = new Path(destPath);
    // File Upload
    try {
    FileSystem fs = getFileSystem();
        fs.copyFromLocalFile(srcPath, dstPath);
        fs.copyFromLocalFile(delSrc,overwrite,srcPath, dstPath);
        // Free up resources
        fs.close();
    } catch (IOException e) {
        logger.error("", e);
    }
}
```

6. Download files from HDFS:

```
public static void getFile(String srcFile,String destPath) {
    // Source file path
    String hdfsUri = HDFSUri;
    if(StringUtils.isNotBlank(hdfsUri)){
        srcFile = hdfsUri + srcFile;
    }
    Path srcPath = new Path(srcFile);
        // Destination path
        Path dstPath = new Path(destPath);
    try {
        FileSystem fs = getFileSystem();
        // Download from HDFS
        fs.copyToLocalFile(srcPath, dstPath);
        // Free up resources
        fs.close();
    } catch (IOException e) {
        logger.error("", e);
```

```
    }
}
```

7. File rename:

```
public boolean rename(String srcPath, String dstPath){
    boolean flag = false;
    try {
        FileSystem fs = getFileSystem();
        String hdfsUri = HDFSUri;
        if(StringUtils.isNotBlank(hdfsUri)){
            srcPath = hdfsUri + srcPath;
            dstPath = hdfsUri + dstPath;
        }
        flag = fs.rename(new Path(srcPath), new Path(dstPath));
    } catch (IOException e) {
        logger.error("{} rename to {} error.", srcPath, dstPath);
    }
    return flag;
}
```

HDFS also provides a variety of non-Java access interfaces to facilitate access to other non-Java applications. For example, Thrift, C, FUSE, WebDav, HTTP, FTP, etc.

2.5 Spark Architecture

2.5.1 Spark Basics

Apache Spark is a parallel computing framework, reminiscent of Hadoop's MapReduce, that was developed by UC Berkeley AMP lab (University of California, Berkeley). Spark not only inherits the benefits of Hadoop MapReduce but also excels in several key areas. A distinctive feature of Spark's architecture is its ability to cache intermediate job outputs in memory rather than relying solely on the HDFS system. This in-memory processing significantly enhances Spark's performance, particularly for iterative MapReduce algorithms used in data mining and machine learning [10]. Furthermore, Spark introduces the abstraction of Resilient Distributed Dataset (RDD), which represents a fault-tolerant collection of read-only objects distributed across a cluster of nodes. These collections are flexible. In the event of data loss, the "bloodline" allows for reconstruction of the affected subset, thereby providing Spark with a high degree of fault tolerance. Spark is implemented in the Scala language, and uses Scala as its application framework. Scala is a modern multi-paradigm programming language that runs on the Java platform and is compatible with existing Java programs. Scala has a robust concurrency model, supports functional programming, and can more effectively support distributed systems. Unlike Hadoop, Spark and

Scala can be tightly integrated, and Scala can operate distributed datasets as easily as local collection objects. However, Spark has been created to support iterative jobs on distributed datasets. It is actually a supplement to Hadoop that can run parallel in the Hadoop file system. This behavior can be supported through a third-party cluster framework called Mesos. In comparison to Hadoop MapReduce, Spark exhibits the following advantages:

1. Spark's calculation mode also belongs to MapReduce, but it is not limited to Map and Reduce operations. It also provides a variety of dataset operation types. The programming model is more flexible than Hadoop MapReduce.
2. Spark offers in-memory computing, which enables the storage of intermediate results in memory, thereby enhancing the efficiency of iterative operations.
3. Spark's DAG-based task scheduling and execution mechanism exhibits superior performance compared to MapReduce's iterative execution mechanism.

All in all, Apache Spark is a rapid and versatile computing engine designed for large-scale data processing, and has formed an ecosystem with accelerated development and wide application.

The inventor of Apache Spark, Matei Zaharia, introduced the key development direction of Apache Spark. In 2016, deep learning and big data were two of the most popular trends in the field of data science. As a result, it was crucial for Spark to provide support for deep learning and to develop a user-friendly API. In addition, although Spark Streaming exhibits considerable advantages in terms of throughput, the development of streaming technology has led to an increasing focus on not only throughput of real-time streaming, but also the impact of delays. In the future, Spark will reduce the stream delay to 1 ms while maintaining the user interface unchanged.

Furthermore, the development of Spark will be aligned with the development trend of hardware. First, the cost of memory will continue to decline, with machines capable of storing more than 256 GB of data becoming increasingly prevalent. Additionally, the adoption of SSD as the standard for server-grade hard drives will gradually become more widespread. As Spark is a memory-based big data processing platform, the data bottleneck will be caused by the hard disk during processing. As the machine's memory capacity increases, a distributed file system like HDFS stored on disk will be gradually replaced by a distributed storage system with shared memory, such as Tachyon. Furthermore, the AMPLab laboratory of Berkeley University provides superior performance compared to HDFS. Consequently, in the future, Spark will undergo major changes in internal storage interfaces to enhance support for SSDs and shared memory systems such as Tachyon.

In general, the enterprises that have started to apply Spark mainly focus on the Internet. The factors restricting the adoption of Spark by traditional enterprises mainly include three aspects. First, it depends on the maturity of the platform. Traditional enterprises tend to exhibit a greater degree of stability in their technology selection, although they are often perceived as being somewhat conservative in this regard. In the event that a technology pertains to the selection of major platforms, a heightened level of caution is typically observed. In the absence of multi-faceted verification and successful industry experience, it is unlikely that this option will be. The second factor is the availability of SQL support. The data processing of traditional enterprises is mainly conducted within relational databases, with numerous legacy systems in operation. In these legacy systems, most data processing is conducted through SQL or even stored procedures. If a big data platform does not support the SQL of the relational database well, the cost of migrating data analysis business logic will be significantly increased. The third factor is the learning curve of teams and technology. In the absence of familiarity with the platform and related technologies among team members, enterprises may experience concerns regarding development progress, cost, and potential risks. Consequently, the future development of Spark will aim to address these challenges.

2.5.2 Spark System Architecture

In practical applications, big data processing mainly includes the following three types (Table 2.1):

1. Complex batch data processing: The time span ranges from tens of minutes to several hours.
2. Interactive query based on historical data: The time span ranges from tens of seconds to minutes.
3. Data processing based on real-time data stream: The time span ranges from hundreds of milliseconds to seconds.

In the event that the aforementioned three conditions coexist, the deployment of three distinct software solutions is required concurrently, including MapReduce, Impala, and Storm. This approach inevitably introduces certain complications. Firstly, the input and output data between different scenarios cannot be seamlessly shared, and usually data format conversion is required. Secondly, different software requires different development and maintenance teams, which brings high use

Table 2.1 Application scenarios of Spark ecosystem components

Application scenario	Time span	Other frameworks	Spark ecosystem components
Complex batch data processing	Hourly	MapReduce, Hive	Spark
Interactive query based on historical data	Minute level, Millisecond level	Impala, Dremel, Drill	Spark SQL
Data processing based on real-time data stream	Millisecond level, Second level	Storm, S4	Spark Streaming
Data mining based on historical data	–	Mahout	MLlib
Processing of graph structure data	–	Pregel, Hama	GraphX

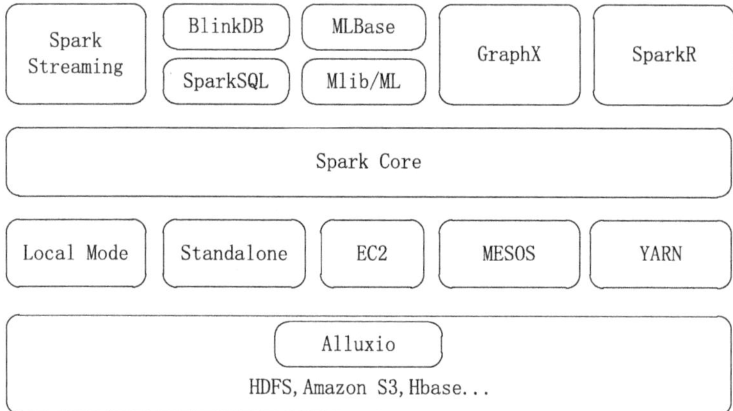

Fig. 2.23 Spark ecosystem

costs. Thirdly, it is challenging to coordinate and allocate resources uniformly to each system in the same cluster. Spark's design follows the concept of "one software stack meets different application scenarios," gradually evolving into a complete ecosystem. In addition to providing an in-memory computing framework, Spark also supports SQL ad hoc queries, real-time streaming computing, machine learning, and graph computing. Besides, Spark can be deployed on top of YARN. Consequently, the Spark ecosystem is capable of supporting batch processing, interactive query, and streaming data processing simultaneously.

The Spark ecosystem has become an integral component of the Berkeley Data Analytics Stack (BDAS), including Spark Core, Spark SQL, Spark Streaming, MLlib, GraphX, and other components (Fig. 2.23).

1. Spark Core represents the foundational element of the Spark ecosystem, encompassing the framework's fundamental functionalities. In particular, define the RDD API, operations, and actions on both. Other Spark libraries are built on RDD and Spark Core.
2. BlinkDB is a massively parallel query engine that runs interactive SQL queries on massive amounts of data. It allows users to optimize query response time by weighing data accuracy. The accuracy of the data is controlled within the allowable error range.
3. Spark SQL provides an API for interacting with Spark-based on HiveQL. Each database table is treated as an RDD, and Spark SQL queries are converted into Spark operations.
4. Spark Streaming is designed to process and manage real-time data streams. Spark Streaming allows programs to process real-time data like ordinary RDDs. Therefore, Spark is a mixed data type processing framework.
5. MLlib is a library of commonly used machine learning algorithms. The algorithms are implemented as Spark operations on RDDs. This library contains

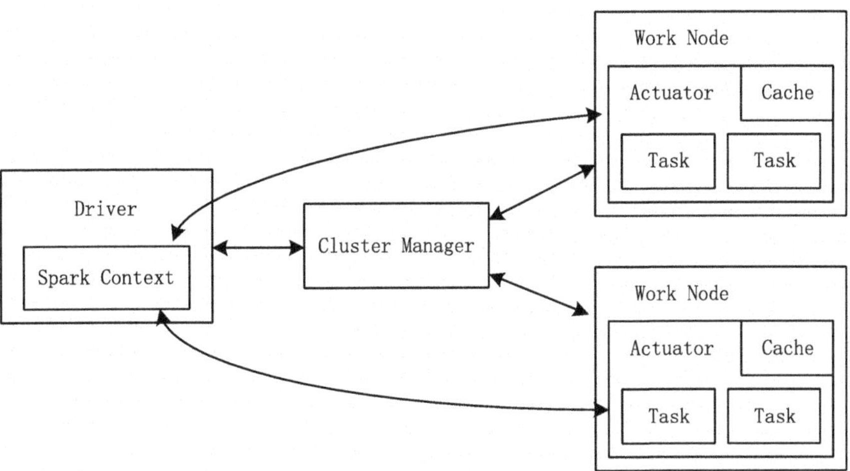

Fig. 2.24 Spark architecture

scalable learning algorithms, such as classification and regression, which require iterative operations on large datasets.

6. GraphX is a collection of algorithms and tools for control graphs, parallel graph operations, and calculations. GraphX extends the RDD API, including control graphs, creating subgraphs, and accessing all vertices on the path.

The primary distinction between the Spark and the Hadoop frameworks is that the former work executes all operations within the memory of the system. Only the data is initially read into memory, and the final result must interact with the storage layer. Consequently, the Spark framework itself does not provide a separate distributed file system. Figure 2.24 shows the fundamental basic system architecture of the Spark framework.

1. Cluster Manager: Spark has three deployment modes: standalone, spark one Mesos, and spark on YARN. In standalone mode, the cluster manager is the master node, and in YARN mode, the cluster manager is the resource manager.
2. Worker Node: The slave node is responsible for controlling the computing node and starting the executor or driver.
3. Driver: Execute the main function of the application and create a Spark context. The purpose of creating the Spark context is to prepare the running environment of the Spark application. In Spark, the Spark context is responsible for communicating with the cluster manager to apply for resources, allocate and monitor tasks, etc. Once the executor is partially finished, the driver is responsible for terminating the Spark context. The Spark context is usually used to represent the drive.
4. Actuator: It is a process that an application runs on the worker node.

In the Spark architecture, an application program written by a user is called Application. The driver function code and executor code are distributed across multiple nodes in the cluster. The driver is used to execute the Application's main function and is responsible for creating Spark context. The Spark context is responsible for communicating with the cluster manager to perform resource applications, task allocation, and monitoring. Once the executor code has completed its execution, the driver is still tasked with recycling the Spark context.

The Spark architecture uses the Master–Slave model in distributed computing. Master is the node that contains the Master process in the cluster, while Slave is the node that contains the Worker process in the cluster. The Master node serves as the centralized controller for the entire cluster, and is responsible for the normal operational functions of the entire cluster. The Worker is equivalent to a computing node, receiving commands from the Master node and providing status reports. The executor is responsible for the execution of tasks. The Client is responsible for submitting applications, while the Driver is responsible for controlling the execution of an application. Once the Spark cluster has been deployed, it is necessary to initiate the master and worker processes on the master node and slave node to control the entire cluster. During the execution of a Spark application, the roles of the Driver and Worker are of critical importance. The Driver program is the starting point of application logic execution. It is responsible for job scheduling, the distribution of Task, and the management of computing nodes through the utilization of multiple workers, which are employed to create executors for the parallel processing of tasks. In the execution phase, the Driver will serialize the Task and any dependent files and JARs to the corresponding Worker machine, after which the executor will process the tasks of the corresponding data partition.

The task division of the Spark architecture is similar to Hadoop MapReduce. An application consists of one or more jobs, each of which is divided into multiple task sets (TaskSet). A task set is the basic scheduling unit of a job. It consists of a group of related tasks that do not exhibit shuffle dependencies between them. The task is the basic unit of application's execution, which is the unit of work sent to the executor. The scheduling component is responsible for the task scheduler (TaskScheduler).

The workflow of the Spark framework includes the following five steps (Fig. 2.25):

1. The Spark application environment is set up, and the Spark context is initiated.
2. The Spark context is applied to the resource manager for running actuator resources.
3. The resource manager allocates executor resources and starts StandaloneExecutorBackend. The running status of the actuator is sent to the resource manager with the heartbeat.
4. The Spark context is constructed as a DAG graph. The DAG graph is decomposed into task sets, which are then sent to the task scheduler. The task scheduler issues the executor's task, and the Spark context issues the application code to the executor.
5. Upon task completion, the executor releases all resources.

The Spark running architecture has the following characteristics:

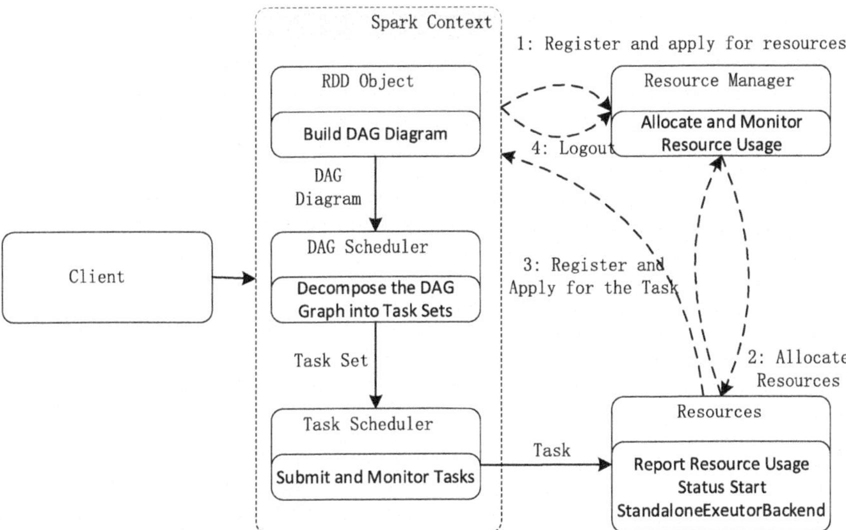

Fig. 2.25 The workflow of the Spark framework

1. Each Application gets its executor process, which has been resident during the execution and runs tasks in a multi-threaded manner. From a scheduling perspective, each driver schedules its task, and from a running perspective, tasks from different Applications run in different JVMs. This application isolation mechanism has several advantages. It is also important to note that this architecture precludes the sharing of the data across applications unless the data is written to an external storage system.
2. Spark is not dependent on the resource manager, provided that it can acquire the executor process and maintain mutual communication with it.
3. The client submitting the Spark context should be close to the working node, preferably in the same rack, due to the large amount of information exchange between the Spark context and the executor during the Spark Application operation. If running in a remote cluster, it is recommended that RPC be employed to submit the Spark context to the cluster, rather than running the Spark context away from the worker node.
4. The task adopts the optimization mechanism of data locality and speculative execution.
5. During the operation of Spark, the job scheduling module is independent of the specific deployment operation mode. In this sense, the logic is consistent across different modes of operation. The implementation of difference deployment and operating modes entails the incorporation of specific task scheduling modules. Different deployment and operation modes, according to different underlying resource scheduling methods, have implemented their specific task scheduling module, which is used to schedule tasks to corresponding computing resources.

2.5.3 Main Features of Spark

A significant number of iterative algorithms, such as machine learning, graph algorithms, as well as interactive data mining tools, require the reuse of intermediate results. The MapReduce framework uses HDFS for the storage of intermediate results, which are all written into HDFS. This approach introduces significant data replication, disk I/O, and serialization overhead. The Spark framework's core is that the distributed data architecture, based on Resilient Distributed Dataset (RDD) model, is designed to enhance the storage of intermediate results. The primary functionality of Spark lies in its ability to perform memory calculations. RDD is the most basic abstraction of the Spark framework. It provides an abstract use of distributed memory, enabling the implementation of operating distributed datasets through the manipulation of local collections. It represents a dataset that has been partitioned, is immutable, and can be operated in parallel. Different dataset formats correspond to different RDD implementations. RDD must be serializable and can be cached in memory. All results generated by RDD dataset operations are stored in memory. The next operation can be directly inputted from memory, thus eliminating a large number of MapReduce disk I/O operations.

RDD has the following characteristics:

1. Creation: RDD can only be created from data in stable storage or other RDD data sources through conversion.
2. Read-only: The RDD state is immutable and cannot be modified.
3. Partitioning: It supports partitioning of elements in RDD based on key and saving to multiple nodes. When restoring, only the data of the lost partition will be recalculated without affecting the entire system.
4. Path: It is called lineage in RDD, that is, RDD has sufficient information to determine how other RDDs generate it.
5. Persistence: It supports RDD cache that will be reused, such as in-memory or overflow to disk.
6. Delay calculation: Spark will delay the calculation of RDD to achieve pipeline transformation.
7. Operation: A multitude of transformations and actions. Despite the number of transformation operations are performed, the RDD will not actually operate; rather, it will merely record the lineage. The operation is only triggered when the action operation is executed.

The advantages of RDD are as follows:

1. RDDs can only be generated from persistent storage or through conversion operations. Compared with distributed shared memory (DSM), RDDs can achieve fault tolerance more efficiently, and they can be recalculated based on their lineage for the missing data partition.
2. RDDs can implement speculative execution like Hadoop MapReduce.
3. RDDs can improve performance through the locality of data.

4. RDDs can be serialized. When the memory is insufficient, the RDD can be automatically downgraded to disk storage, where it can be stored on the disk. Although there will be a notable decline in performance at this stage, the resulting performance will not fall below that of Hadoop MapReduce.
5. Batch operations: The distribution of tasks according to data locality can be employed to enhance performance.

In Spark applications, the entire execution flow is represented by a directed acyclic graph. Once the Action operator has been triggered, all accumulated operators will form a directed acyclic graph. Subsequently, the scheduler will schedule the tasks on the graph to perform operations. The scheduling methodology used by Spark differs from that of MapReduce. Spark is divided into different stages according to interdependencies between RDDs. A stage contains a series of functions for pipeline execution. In Fig. 2.26, A, B, C, D, E, F, and G represent different RDDs, and a box in the RDD represents a data block. The data is input into Spark from HDFS to form RDD A and RDD C. The map operation is performed on RDD C and converted to RDD D. RDD B and RDD F are joined and converted to G. Shuffle is performed during the process from B to G. Finally, RDD G is saved to HDFS through the function saveAsSequenceFile output.

RDD implements fault tolerance of distributed datasets through the use of data checkpoints and record updates. RDD uses a record update method that records all update points, which is very costly. Therefore, RDD is only capable of supporting coarse-grained transformation, whereby a single operation performed on a single block (partition) is recorded, and then stored in the form of an RDD transformation sequence (that is, lineage). The transformation sequence refers to each RDD, which

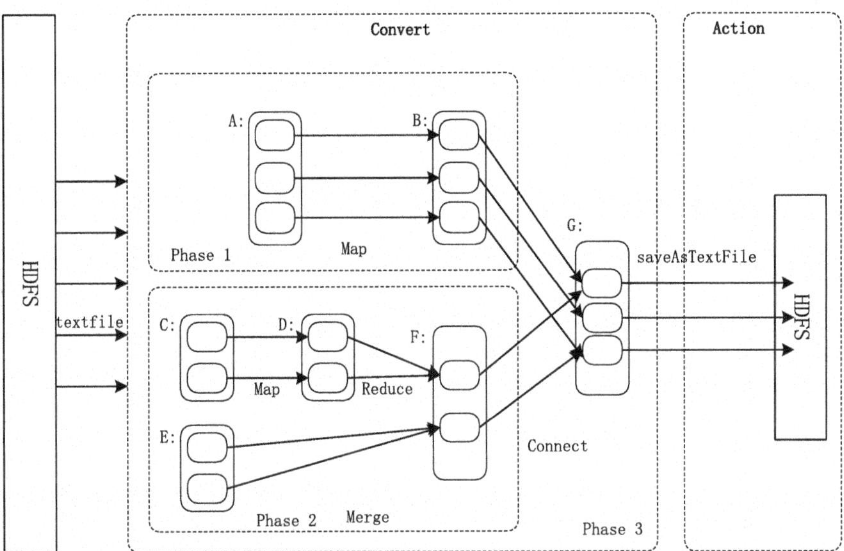

Fig. 2.26 Spark RDD work process

contains information about how other RDDs transform it and how to reconstruct a certain piece of data. In consequence, RDD's fault tolerance mechanism is also referred to as the "bloodline" fault tolerance mechanism. The primary challenge in implementing this "bloodline" fault tolerance mechanism, the biggest problem is the expression of the dependency between the parent RDD and the child RDD. In fact, the dependency relationship can be divided into two types: narrow dependency and wide dependency. In the context of narrow dependency, it can be stated that each data block in the child RDD is dependent on a limited number of fixed data blocks that correspond to the parent RDD.

In contrast, a wide dependency means that a single data block in the child RDD can depend on all data blocks in the parent RDD. For example, the child RDD's data block depends only on the corresponding data block in the parent RDD in the case of the map transformation. Conversely, the data block in the child RDD depends on the data blocks in the multiple parent RDDs in the case of the groupByKey transformation, because one of the keys may be distributed in any data block of the parent RDD.

The characteristics of narrow and wide dependencies differ. Firstly, in the case of narrow dependence, the dependency can be calculated on a computing node directly by calculating a piece of data of the parent RDD to obtain a piece of data corresponding to the child RDD. In contrast, a wide dependence necessitates the completion of the entire parent RDD before the calculation is completed. The outcomes of the parent RDD are then hashed and transmitted to the corresponding node, enabling the computation of the child RDD. Secondly, in the event of data loss, narrow dependencies require only the missing piece of data needs to be recalculated for narrow dependencies. However, for wide dependencies, a complete recalculation of all data blocks in the ancestor RDD is mandatory to restore the dataset. Therefore, when the "bloodline" chain is particularly dependent, it is necessary to set data checkpoints appropriately. These two characteristics require tailored task scheduling and fault-tolerant recovery mechanisms for various dependencies.

2.5.4 *Typical Case of Spark*

Case Name: Analysis of the recommendation system based on Spark.

The recommendation is a common application scenario in big data analysis and large-scale machine learning. Recommendation engines are employed across a range of digital platforms, including e-commerce, retail, social networking, and video sites. A good recommendation engine can provide users with a positive user experience while simultaneously generating great economic benefits for businesses. The specific recommendation system is a widely used tool. Notable examples of recommendation engines in use abroad include Amazon and Netflix, while in China, Taobao and Douban are two prominent platforms using recommendation engines to facilitate product recommendations to users. A variety of recommendation algorithms have been proposed, ranging from relatively basic content-based algorithms to more sophisticated collaborative filtering algorithms, which are suitable for real-world applications. Furthermore, a number of more mature recommendation systems have also been adopted in a variety of fields. The distributed recommendation algorithm that is widely used in the big data processing platform is the collaborative filtering algorithm based on ALS, which was proposed in the Netflix competition. The distributed implementation of this algorithm is encapsulated in MLlib and can be used to construct a big data search engine. It is currently a popular area of application.

Based on Spark's recommendation system analysis, HDFS is used to store movie rating datasets. A distributed collaborative filtering algorithm is used to build a recommendation engine. The best recommendation model is obtained through multiple training. The recommendation model is then used to recommend movies to new and old users and Spark Streaming is used to simulate real-time recommendations, providing a feasible solution for deploying the recommendation engine on the Spark platform. Spark not only reads data locally, but also supports the HDFS distributed file system, HBase, and Hive data warehouses, and Amazon S3 (simple storage service). Besides, AMPLab has developed the Tachyon distributed memory file system, which separates the memory storage function from Spark and enables clusters to access data stored in Tachyon at the speed of accessing memory. The Tachyon architecture is between the underlying storage platform and the upper computing platform. Storing data that does not need to be stored in the underlying layer in memory enables memory sharing and improve efficiency.

The primary objectives of Spark-based recommendation system analysis are as follows:

1. Construct a movie recommendation engine based on the collaborative filtering algorithm of ALS on the Spark platform.
2. Use the recommendation engine to train the Netflix competition dataset to achieve movie recommendations for both new and old users.
3. The real-time streaming capabilities of Spark Streaming can be combined with the recommendation engine to simulate real-time recommendations.
4. Study the principle and parallel implementation of a collaborative filtering algorithm based on ALS.

As shown in Fig. 2.27, the architecture of Spark-based recommendation system analysis is divided into four layers: The HDFS distributed storage layer, the Spark distributed processing layer, the Spark component layer, and the recommended applications. The HDFS distributed storage layer is used to store the movie dataset of the user's rating data. The Spark distributed processing layer is used to provide distributed processing capabilities, data operations, resource allocation, and task execution capabilities for the entire system. This is the application of the upper layer, the Executor. The Spark component layer is the Spark component for the upper recommended applications, including Spark Streaming and MLlib. Spark Streaming offers the system the real-time processing capabilities, while MLlib provides the system with algorithm interfaces. The recommended applications are the codes of

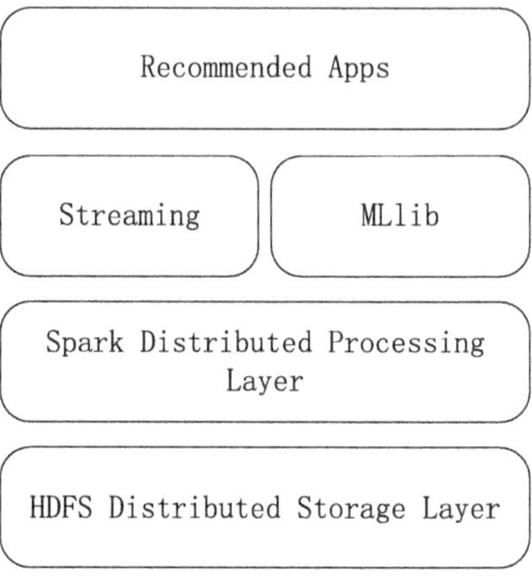

Fig. 2.27 Architecture of movie recommendation system based on Spark

the recommended system that have been written and that are capable of realizing specific functions, including the realization of various modules of the system.

The system workflow is shown in Fig. 2.28. The system first reads the user's rating data and movie information data from the HDFS storage layer and converts the data into a format called MLlib. Subsequently, the collaborative filtering algorithm of ALS is invoked, the data is trained, various parameters are set, the process is repeated multiple times, the optimal recommendation model is established, and movies are recommended to existing users by invoking the model's recommendation interface. Similarly, for the purpose of simulated real-time recommendation, real-time data is read through Spark Streaming, which comprises the user's rating of the movie. The data is then mixed with the original data, and the most optimal model is employed to recommend the user in real time.

The advantages of the case are as follows. The conventional methodology for data processing entails the initial collection of data and its subsequent storage within a database. In response to a request for data services, a series of queries and calculations are performed on the database's data. From a macro-perspective, this is a passive service method and is not real time. Flow computing, however, can thoroughly analyze large-scale flow data in real time during the changing motion process, capture potentially useful information, and feedback the results to the next calculation node. The data, the calculations, and the service are provided in a continuous, uninterrupted stream. The whole process is continuous, and the response is also real time, reaching seconds. The advantages of Spark Streaming are as follows:

1. Capable of running on 100+ nodes and can reach second-level delay.
2. Uses memory-based Spark as the execution engine, with high efficiency and fault tolerance.
3. Can integrate Spark batch processing and interactive queries.
4. Provides a simple interface similar to batch processing for implementing complex algorithms.

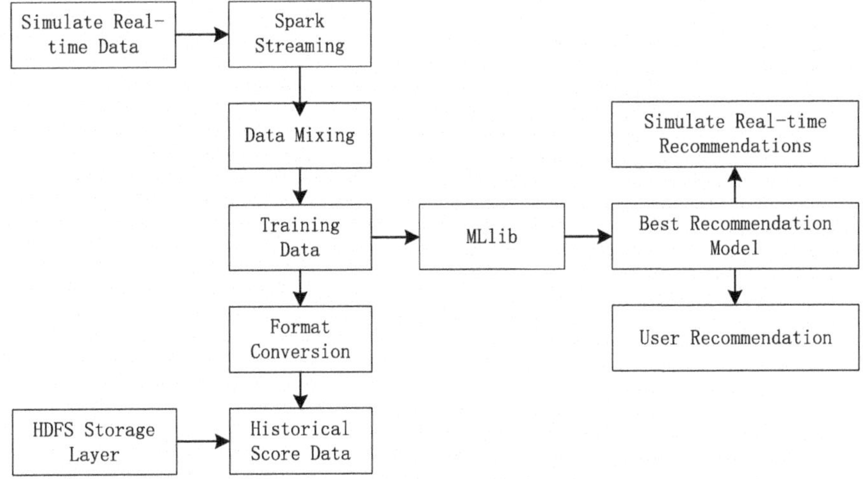

Fig. 2.28 Workflow of movie recommendation system based on Spark

Spark has the following four advantages:

1. The speed of Spark is rapid. Spark extends the MapReduce computing model, which is an iterative computing framework based on memory. It uses an advanced DAG (directed acyclic graph) execution engine to convert intermediate results. The data is stored in memory, eliminating the need for repeated reading and writing to the disk. In terms of inner-loop operations, Spark is 100 times faster than Hadoop MapReduce. Even when the calculations are performed on disk, Spark remains 10 times faster than MapReduce.
2. Spark is a convenient and user-friendly platform that offers more than 80 advanced APIs (application program interface). These APIs are designed to provide a robust set of interfaces, allowing users to rapidly construct parallel applications. Additionally, Spark provides support for four languages, including Java, Scala, Python, and R, enabling developers to select the most suitable programming language according to their needs.
3. Spark is a general-purpose engine for processing big data. It seamlessly combines SQL interactive query, streaming processing, graph processing, and machine learning. It can be used in a variety of application scenarios and complete a variety of computing models. Users are not required to use a different engine to address various needs.
4. Spark has multiple operational modes. It can be deployed in a standalone mode or on Hadoop Yarn or Apache Mesos clusters. Additionally, Spark can be deployed on Amazon EC2 (Elastic Compute Cloud). Furthermore, Spark supports multiple data sources, including HDFS, Cassandra, HBase, Hive, Tachyon, and other Hadoop data sources.

2.5.5 Spark Programming Interface

The Spark framework implements various calculations, which are executed on RDDs. The main objective of Spark programming is to perform a range of operations on RDDs. It is possible for program developers to treat RDD as an "array." The operation of an RDD is very similar to that of an array. Prior to initiating any operations on the RDD, it is first necessary to construct the RDD. The method of constructing an RDD can be divided into two categories from the perspective of data sources: first, read data from memory and build RDD; second, read data from the file system and build RDD. The file system here can be HDFS or a local file system. The reference code for the first method is as follows:

```
/* Create RDD using makeRDD */
/* List */
val rdd01 = sc.makeRDD(List(1,2,3,4,5,6))
val r01 = rdd01.map { x => x * x }
println(r01.collect().mkString(","))
/* Array */
val rdd02 = sc.makeRDD(Array(1,2,3,4,5,6))
val r02 = rdd02.filter { x => x < 5}
println(r02.collect().mkString(","))
val rdd03 = sc.parallelize(List(1,2,3,4,5,6), 1)
val r03 = rdd03.map { x => x + 1 }
println(r03.collect().mkString(","))
/* Array */
val rdd04 = sc.parallelize(List(1,2,3,4,5,6), 1)
val r04 = rdd04.filter { x => x > 3 }
println(r04.collect().mkString(","))
```

The code is written in Scala language, and the API (also called Spark operator) used is the makeRDD method and the parallelize method. It can be seen that an RDD can be regarded as an array which uses linked lists and array types during the construction of data. The second method to construct an RDD is through the file system. In this example, the local file system (protocol is file://) is used to construct the RDD object. The reference code is as follows:

```
val rdd:RDD = sc.textFile("file:///D:/sparkdata.txt", 1)
val r:RDD = rdd.flatMap { x => x.split(",") }
println(r.collect().mkString(","))
```

After constructing the RDD object using memory or local system data, we can perform operations on the RDD object. RDD operations are divided into transformation operations and action operations. The rationale behind the division of operations into these two categories is related to the concept of RDD lazy operations. In the event that an RDD performs a conversion operation, the actual calculation is not performed. It is only when an action operation is performed by the RDD that the corresponding calculation task is submitted and the requisite calculation is performed. Furthermore, it is also straightforward to distinguish between conversion operations and action operations. A conversion operation generates a new RDD operation from an RDD, whereas an action operation is used to perform the actual calculations. Table 2.2 presents a list of the standard RDD operators.

Examples of conversion actions:

```
/*Create RDD*/
val rddInt:RDD = sc.makeRDD(List(1,2,3,4,5,6,2,5,1))
val rddStr:RDD = sc.parallelize(Array("a","b","c","d","b","a"), 1)
val rddFile:RDD = sc.textFile(path, 1)
val rdd01:RDD = sc.makeRDD(List(1,3,5,3))
val rdd02:RDD = sc.makeRDD(List(2,4,5,1))


/* map operating */
println(rddInt.map(x => x + 1).collect().mkString(","))
/* filter operating */
println(rddInt.filter(x => x > 4).collect().mkString(","))
/* flatMap operating */
```

Table 2.2 List of RDD operators

Operation type	Function name	Effect
Conversion actions	map()	The parameter is a function, the function is applied to each element of the RDD, and the return value is the new RDD
	flatMap()	The parameter is a function, the function is applied to each element of the RDD, the element data is split into an iterator, and the return value is the new RDD
	filter()	The parameter is a function, the function will filter out the elements that do not meet the conditions, and the return value is the new RDD
	distinct()	No parameters, deduplication of RDD elements
	union()	The parameter is RDD, and a new RDD containing all elements of two RDDs is generated
	intersection()	The parameter is RDD, find the common elements of the two RDDs
	subtract()	The parameter is RDD, filter the same elements in the original RDD as the parameter RDD
	cartesian()	The parameter is RDD, find the Cartesian product of two RDDs
Action Operation	collect()	Return all elements in RDD
	count()	Calculate the number of RDD elements
	countByValue()	Count the number of occurrences of each element in RDD
	reduce()	Consolidate all RDD data in parallel, such as sum operation
	fold(0)(func)	The function is the same as reduce(), but the operator has initial values
	aggregate(0)(s,c)	The function is the same as reduce(), but the RDD data type returned by this operator is different from the original RDD
	foreach(func)	Apply the specified function to each element of RDD

```
println(rddFile.flatMap { x => x.split(",") }.first())
/* distinct Deduplication */
println(rddInt.distinct().collect().mkString(","))
println(rddStr.distinct().collect().mkString(","))
/* union operating */
println(rdd01.union(rdd02).collect().mkString(","))
/* intersection operating */
println(rdd01.intersection(rdd02).collect().mkString(","))
/* subtract operating */
println(rdd01.subtract(rdd02).collect().mkString(","))
/* cartesian operating */
println(rdd01.cartesian(rdd02).collect().mkString(","))
Examples of actions:
/*Create RDD*/
val rddInt:RDD = sc.makeRDD(List(1,2,3,4,5,6,2,5,1))
val rddStr:RDD = sc.parallelize(Array("a","b","c","d","b","a"), 1)
/* count operating */
println(rddInt.count())
/* countByValue operating */
println(rddInt.countByValue())
/* reduce operating */
println(rddInt.reduce((x ,y) => x + y))
/* fold operating */
println(rddInt.fold(0)((x ,y) => x + y))
/* aggregate operating */
val res:(Int,Int) = rddInt.aggregate((0,0))((x,y) => (x._1 + x._
2,y),(x,y)
=> (x._1 + x._2,y._1 + y._2))
println(res._1 + "," + res._2)
/* foeach operating */
println(rddStr.foreach { x => println(x) })
```

2.6 HBase: Distributed Database

2.6.1 HBase Basics

HBase is a distributed column-oriented database system built on HDFS. Most relational databases prioritize data production, without considering the characteristics of large-scale data and distribution [11]. Many relational database systems expand the database use replication and partitioning to break through the limitation of a single node. This method is not without its drawbacks, such as complicated installation and maintenance, and low-level code to be refactored to accommodate changes. HBase addresses scalability from alternative angles. It expands the scale by adding nodes linearly. HBase is not a relational database, rather, it is a non-relational database (NoSQL). Its characteristic is that the structure of the database table can be dynamically expanded, and the structure of the same database table at different times may be different.

HBase's database file operations are inherited from the file system-related operations of HDFS. Its database files are also stored in the form of data blocks. HBase not only exhibits all the characteristics of the HDFS file system, but also has its special features. For example, the content of HBase files has a unified structure, which in turn determines the rules that govern HBase's read and write operations. Additionally, the design of HBase has the following objectives:

1. HBase necessitates a unified structure, no matter how many externally represented structural fields. Its internal storage structure must remain unchanged and allow the same table's structure to change dynamically over time.
2. HBase should support dynamic database table access. The user specifies the table name, table column, and table row to be accessed, and then organizes access to the database table according to user requirements. In contrast, traditional relational databases often use fixed language structures (such as SQL) to enforce database table access operations.
3. HBase should consider the increase in the amount of data in the database and avoid a decline in its query performance.
4. HBase should also allow rapid and dynamic insertion of physical nodes as the number of data increases.
5. HBase should provide different access interfaces for different clients.

As a distributed database, HBase introduced many new concepts because of its distributed architecture:

1. Regions: HBase divides the table horizontally into different regions, and each region contains a subset of all rows in the table. The initial configuration of the table comprises a single region. Upon reaching the set boundary, the table is divided into two regions of similar size via the behavior boundary, and the formed regions will be expanded separately. A region is a discrete unit of the HBase cluster oversees the entirety of the table by managing the nodes within a part of the entire region. Consequently, regardless of the table's size, it can be processed by the server cluster.
2. Basic Unit: The basic unit of HBase distributed database refers to tables, row keys, column families, and regions.
3. Regional Servers: All horizontal partitions in a table will be distributed on different regional servers, and the data in a region will only be stored on one server. All data is physically stored on the machine by calling the HDFS file system interface, and the regional server provides data services. Commonly, a Region process runs on a server, and this process is responsible for managing multiple Region instances. The regional server provides disaster backup services through HLog. The HLog log will be written to the HDFS distributed system instead of being stored locally. In the event of a failure of the regional server, data will not be lost.
4. Master Server: The master server is responsible for allocating regions to regional servers, coordinating the regional service region's load, and maintaining the status of the cluster. There is only one running HMaster in a certain period of time in

the HBase cluster. It is worth noting that HMaster does not provide any data services (but can provide management services) to HBase Client. It is designed to serve as an internal manager, which cooperates with ZooKeeper to manage Region and Region servers.

5. .META. metadata table: The metadata of multiple regions corresponding to a table (such as the table name, the start row, and end row of the table in the region) are saved in the. META. metadata table created by HBase. The size of the .META. metadata table will increase as the area increases. Therefore, the .META. metadata table will also be divided into multiple regions. So, how does HBase locate the region related to the .META. metadata table? The answer is through the -Root- metadata table.
6. -ROOT-metadata table: -ROOT-metadata table is the unified agent of .META.metadata table, which saves the metadata of .META.metadata table. The user access data must go through the -ROOT-metadata table, because only the information of the .META.metadata table can be obtained from the table to understand the target data is allocated to those regions. -ROOT-metadata table only exists in one region (located on a regional server), and has the characteristics of inseparability. Thus, the capacity of the -ROOT-metadata table determines the maximum number of regions that can be managed by the HBase cluster, which can reach more than 16.67 million.

2.6.2 HBase System Architecture

HBase is a distributed cluster database system that can be deployed independently. However, in practice, it is often necessary to consider the actual application, which may include importing existing database systems or massive textbooks. It frequently interacts with other systems, such as HDFS, ZooKeeper. The HBase system has taken into account the integration with the existing system in the design, thus ensuring the availability of an interface. A user need to only perform some simple configurations in HBase to achieve a multi-system integration application.

Fig. 2.29 shows the basic architecture of the HBase distributed database system. The core functional modules depicted in the architecture diagram are the Client, ZooKeeper (coordination service component), HMaster (master server), and HRegionServer (region server).

1. The client is the entrance of HBase system and is the interface through which users interact with the HBase database. The client uses HBase's RPC mechanism to facilitate communication with both the HMaster and HRegionServer. This enables the completion of management operations via communication with the HMaster, as well as data read and write operations through communication with the HRegionServer.
2. The ZooKeeper Quorum queue is responsible for managing the election of multiple HMasters in HBase, the state synchronization between servers, etc., for avoiding the single point problem of the HMaster. The coordination functions of

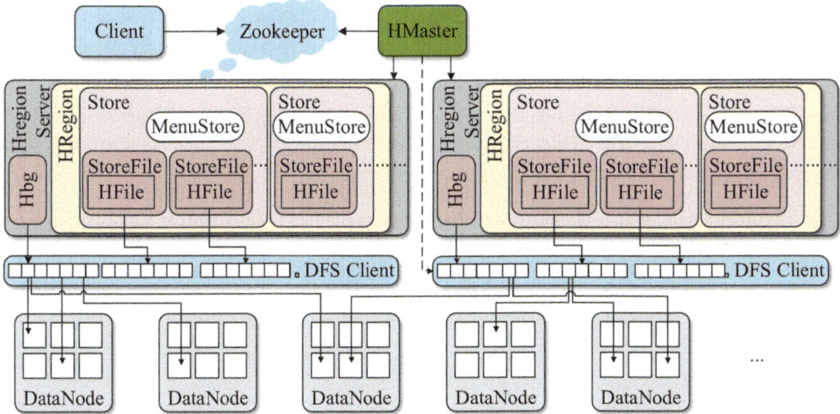

Fig. 2.29 HBase system architecture

the ZooKeeper instance in HBase include the storage of HBase metadata information, the real-time monitoring of RegionServer (perceive the health status of each HRegionServer), and the storage of addressing entries of all regions. Additionally, Zookeeper also ensures that there is only one HMaster node in the HBase cluster.

Section 2.6.1 provides an overview of the HMaster and HRegionServer. This section provides further details on the HRegionServer architecture. The HRegionServer is composed of two main components: HLog and HRegion. HLog is used to provide disaster backup, while HRegion is used to save data. The component unit of HRegion is HStore, which is the real data storage structure of HBase. HStore can be further divided into MemStore and StoreFile. MemStore is a sorted memory buffer (sorted memory buffer). The data written by the user first enters MemStore. After filling MemStore, it will be refreshed as a StoreFile. Once the number of StoreFile files reaches a certain threshold, the Compact operation will be triggered. This operation will merge multiple StoreFiles into a single StoreFile. During the merge process, version consolidation and data deletion will be performed. It is clear that the basic operation of HBase is to increase data. All the update and delete operations are performed during the subsequent Compact process, which allows the user's write operations to be returned as soon as they are buffered in memory, thereby ensuring the IO performance. Following the triggering of a Compact operation, StoreFiles will gradually form a larger StoreFile. Once a single StoreFile reaches a certain threshold, a Split operation will be triggered. This operation divides the current region into two, decommissioning the parent and activating the two new child regions. The region is then distributed by the HMaster to the corresponding HRegionServer, thereby distributing the load pressure of the original region across two regions. This process mirrors the regional split mechanism detailed in Sect. 2.6.1.

The client mainly initiates the operation request of the distributed database system. During the operation process, the HMaster does not actively engage in data reading and writing. Instead, the client interacts with the HRegionServer. The client can initiate requests to multiple HRegionServers according to the size of the database. The HRegionServer processes requests in a distributed and parallel manner. The operation process of HBase includes seven steps:

1. Read and write request: HBase Client submits the target database table name, column family, row key, and other information to the HBase system through the client interface.
2. Cache query: HBase Client first queries the host node of HRegionServer in the local cache. If it cannot find it, it sends a request to ZooKeeper to query the location information of -ROOT-.
3. Obtain -ROOT-: After HBase Client connects to ZooKeeper, first obtain the address of -ROOT-, and then query the address of the .META. region according to -ROOT-. The .META. table stores all the row information to be queried, and the client queries the address of the user space area HRegionServer corresponding to the row in the .META. table based on the row information. Then the client can interact with the relevant HRegionServer.
4. Submit request: The client establishes a connection with HRegionServer and submits a request to it.
5. Status check: If the client sends a write request, HRegionServer will write the operation request to the log, add it to the internal cache for processing, and issue a write row lock request to Zookeeper to add the row to be written lock. If the client sends a read request, HRegionServer first queries the distributed cache, and returns if it contains the query content, otherwise it reads the Region data to memory and queries it.
6. Return data: After receiving and processing the request, HRegionServer will return the client's data. At this point, a user's submission request has been processed. HRegionServer will keep the loaded cache file to speed up the next operation request.
7. Log submission: The log data is initially stored in the memory of each HRegion-Server. Once a certain threshold is reached, the data be written to disk and written into the background thread's HDFS file system. Once HMaster monitor has identified a failure in an HRegionServer, it can retrieve the region's log file from HDFS and assign it to a new HRegionServer node.

The workflow of the HBase client and the HBase system is shown in Fig. 2.30. This mechanism is transparent to users.

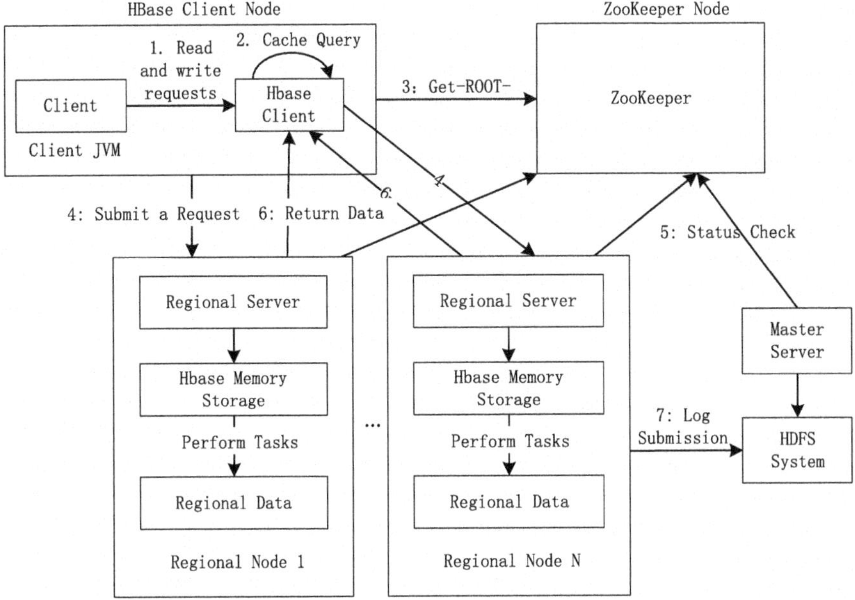

Fig. 2.30 HBase system workflow

2.6.3 Main Features of HBase

The basic concepts of distributed databases are different from those of relational databases. Attention should be paid to this when identifying HBase. It is important to be aware of the related concepts of distributed database tables, including:

1. The logical model refers to the logical model of a multi-table collection table. A traditional database table only stores all the data of a structure (such as a score table), while a distributed database table is a collection of multiple tables related to a particular subject. HBase's database table also stores multiple rows of data in a logical manner. Each row consists of three elements: the row key, the data column, and the time stamp. Table 2.3 shows examples of the logical storage of HBase class tables.

The database table records the teacher and student information for each class. The row key is the primary key of the index, which is automatically generated by the HBase database. HBase offers three methods for accessing row data: specifying

Table 2.3 HBase logical model example

Row key	Timestamp	Column teacher	Column student
class_1	T1	name: teacher_1	
class_1	T2		name: stu_1
class_1	T3	name: teacher_2	name: stu_2

a single row key, specifying a row key range, and performing a full table scan. The timestamps, T1, T2, T3, are used to indicate the version number of the data. The user can specify access to the non-latest version of the data. Column family Teacher and Student have the same concept as tables in traditional databases. Therefore, the distributed database table can be regarded as a "table consisting of multiple tables." The following example illustrates HBase's dynamic access to database tables.

Suppose that a database table is designed according to requirements. The table name is user_info and it has three fields: id, name, and tel. When dealing with a relational database table, it is essential to specify the table name and field name of the table at the time of its construction.

Create table user_info (id type, name type, tel type)

id	name	tel

To insert records into a table, the table name and the value of each field should be specified.

Insert into user_info values (…)

id	name	tel
id_1	name_1	tel_1
id_2	name_2	tel_2

If the requirements change and the address information are required to be recorded, this can be achieved by adding fields or adding extended tables.

id	name	tel	addr
id_1	name_1	tel_1	addr_1
id_2	name_2	tel_2	addr_2

HBase adopts a distinct methodology. When creating a database table, users are required to specify the table name and column family.

In this example, when creating a new distributed database table named user_info, it contains two column families: base_info and ext_info. The row key is automatically generated by the database and does not need to be specified by the user.

Create 'user_info', 'base_info', 'ext_info'

row_key	base_info	ext_info

Basic operation of inserting records: Add name and tel data to the base_info column family with row key row1 in the user_info table. It is important to note that the name and tel fields are not required to be pre-defined by the user.

put 'user_info', 'row1', 'base_info: name', 'name_1'
put 'user_info', 'row1', 'base_info: tel', 'tel_1'

row_key	base_info	ext_info
row1	name:name_1,tel:tel_1	

Insert another record with address information into the table:
put 'user_info', 'row2', 'base_info: name', 'name_2'
put 'user_info', 'row2', 'ext_info: addr', 'addr_2'

row_key	base_info	ext_info
row1	name:name_1,tel:tel_1	
row2	name:name_2	addr:addr_2

Compared with the relational database, HBase distributed databases offer greater flexibility. Table 2.4 provides a summary and comparison of the characteristics of relational databases and distributed databases.

2. Physical model: The physical model refers to the physical storage method of HBase. A row of a distributed database is logically composed of N column families, and its physical storage is N rows composed of three columns: row + column family + time. Table 2.5 is the physical model of the HBase class table (Table 2.5).

Please note that null values in the logical model will not be stored. As a result, the Teacher.name of the query timestamp T2 will return empty. If the query is to specify a specific timestamp, the latest data in the table is returned, i.e., teacher_2.

Table 2.4 Comparison between relational database and distributed database

Attributes	Database type	
	Relational database	Distributed database
Table structure	Two-dimensional structure	Multidimensional structure
Build table	Define the table name and specific fields	Define table names and column families
Insert	Insert multiple fields at once	Insert one field at a time
Expand	Pre-defined fields are required	No pre-defined fields

Table 2.5 HBase physical model example

Row key	Timestamp	Columns
class_1	T1	teacher_1
class_1	T3	teacher_2
Row key	Timestamp	Columns
class_1	T2	stu_1
class_1	T3	stu_2

2.6.4 HBase Programming Interface

The HBase distributed database standalone environment can be built according to the following steps:

1. Preliminary preparations: A Linux server, the downloading the JDK, and the installation of Hadoop and HBase packages.
2. Configuration of the Hadoop environment: See Sect. 2.3.4.
3. Configuration of the HBase environment: Enter tar -xvf hbase-1.2.X-bin.tar.gz to decompress the HBase installation package, and enter the command mv hbase-1.2.X /home/hbase to move it to the /opt/hbase path. Edit the /etc./profile file. The configuration is as follows:

```
export HBASE_HOME=/home/hbase/hbase-1.2.X
export PATH=.:${JAVA_HOME}/bin:${HADOOP_HOME}/bin:$PATH
export PATH=.:${HBASE_HOME}/bin:$PATH
```

Enter source /etc./profile to make the configuration take effect.
Create a new folder in the root directory

```
mkdir /root/hbase
mkdir /root/hbase/tmp
mkdir /root/hbase/pids
```

Switch to the /home/hbase/hbase-1.2.X/conf directory and edit the hbase-env.sh file:

```
export JAVA_HOME=/home/java/jdk1.8
export HADOOP_HOME=/home/hadoop/hadoop2.8
export HBASE_HOME=/home/hbase/hbase-1.2.X
export HBASE_CLASSPATH=/home/hadoop/hadoop2.8/etc/hadoop
export HBASE_PID_DIR=/root/hbase/pidsexport HBASE_MANAGES_
ZK=false
```

Edit hbase-site.xml file, add configuration:

```
<property>
     <name>hbase.rootdir</name>
     <value>hdfs://host_name:9000/hbase</value>
          <description>The directory shared byregion servers.</
description> </property>
<property>
     <name>hbase.zookeeper.property.clientPort</name>
     <value>2181</value>
        <description>Property from ZooKeeper'sconfig zoo.cfg. </
description>
</property>
<property>
     <name>zookeeper.session.timeout</name>
     <value>120000</value>
</property>
<property>
```

```
        <name>hbase.zookeeper.quorum</name>
        <value>host_name</value>
    </property>
    <property>
        <name>hbase.tmp.dir</name>
        <value>/root/hbase/tmp</value>
    </property>
    <property>
        <name>hbase.cluster.distributed</name>
        <value>false</value>
    </property>
```

The HBase standalone environment is built.

4. HBase startup: First start Hadoop as described in section then switch to the HBase bin directory, enter./start-hbase.sh to start the HBase distributed database.

HBase provides multiple access interfaces for clients. Common Hbase access methods are

1. Native Java API: The most conventional and efficient access method.
2. HBase Shell: HBase command-line tool, the simplest interface, suitable for HBase management.
3. REST Gateway: It supports REST-style Http API to access HBase, removing language restrictions.
4. MapReduce: It uses MapReduce jobs to process HBase data directly.
5. Use Pig/Hive to process HBase data.

The HBase Shell is a command-line operation interface for users provided by HBase. Users can complete operations such as creating tables, scanning tables, and obtaining row records through Shell commands. Taking Table 2.4 as an example, the relevant commands are

1. Start shell: hbase shell.
2. Create a table: Create 'class','Teacher'.
3. Insert data: put'class','class_1','Teacher:name','teacher_1'.
4. Specify line query: get'class','class_1'.
5. Delete the specified unit: delete'class','class_1','Teacher:name'.
6. Disable table: disable'class'.
7. Delete the table (provided that the table has been disabled): drop'class'.
8. Full table scan: scan'class'.

2.7 Hive: Data Warehousing

2.7.1 Hive Basics

Data Warehouse is a subject-oriented, integrated, non-volatile, time-variant data collection system that is used to support management decisions. Traditional data warehouses face many challenges, including the inability to meet the rapidly growing demand for massive data storage, the inability to process different data types effectively, and insufficient computing and processing capabilities [12].

Hive is a data warehouse tool based on Hadoop. It does not store or process data itself, rather, it relies on the distributed file system HDFS to store data. Hive defines a simple SQL-like query language, namely, HiveQL, which provides full SQL query functions and supports SQL statements' conversion into MapReduce task execution [13]. Hive supports large-scale data storage and analysis, has good scalability, and its characteristics are very suitable for data warehouses. Firstly, Hive uses batch processing to process massive data. The data warehouse is designed to store static data. The analysis of static data is suitable for batch processing. It does not require a rapid response to deliver results, and the data itself does not change frequently. Secondly, Hive provides a series of data warehouse operation tools for extracting, transforming, and loading (ETL) data. It can store, query, and analyze large-scale data stored in Hadoop. These tools can meet the needs of data warehouses in a variety of application scenarios.

As a component of the Hadoop ecosystem, Hive is associated with other components in the ecosystem. For example, Hive is dependent on the HDFS distributed file system for data storage, relies on MapReduce for data processing, and Pig can be used as an alternative tool for Hive in specific scenarios. Hive is similar to traditional relational databases in many respects, but its bottom layer is based on HDFS and MapReduce, which differentiates it from conventional databases in many respects (Table 2.6).

Hive has three operational modes, namely, embedded mode, independent mode, and remote mode. The embedded mode is integrated with HDFS to store metadata in the Derby database's local mode, which is provided with Hive. This mode can only have one session connection at a time, so it can only be used as a development and testing environment. In contrast, the independent mode stores metadata in the MySQL server. This mode is not limited by the number of session connections and is one of the modes in which Hive actually runs. The remote mode is a way to provide clients with remote access to Hive. Users can obtain Hive services through Thrift, the Hive command line, and the Web browser.

Hive is undergoing a phase of rapid development. In order to improve the performance of Hive, the Stinger plan, led by Hortonworks, has proposed a series of improvements to Hive. The most significant improvements are

Table 2.6 Comparison of Hive and traditional databases

Compare items	Hive	Traditional database
Data insertion	Support batch import	Support single and batch import
Data update	Not support	Support
Index	Support	Support
Partition	Support	Support
Execution delay	High	Low
Scalability	High	Limited

1. Vectorization: Changed Hive from single-line data processing to batch processing, which significantly improved the instruction pipeline and cache utilization rate.
2. Hive on Tez: Replace the MapReduce computing framework at the bottom of the Hive with the Tez computing framework. Tez can not only support multi-Reduce task MRR, but also submit an execution plan at a time, which allows for more efficient resource allocation.
3. Cost-based optimizer: Enable Hive to select the optimal Join sequence and accelerate query processing.

2.7.2 *Hive System Architecture*

The Hive system contains the following five basic components:

1. User interface: There are three main interfaces provided by Hive to users: CLI, Client, and WUI. The most commonly used interface is CLI. The CLI starts a copy of Hive copy at startup. The client is a Hive client, and users connect to Hive Server. In order to start the Client mode, it is necessary to specify the location of the Hive server and to initiate the Hive server on that node. WUI visits Hive through a browser.
2. Thrift server: Hive can be used as a Thrift server for client connection when running in server mode.
3. Metastore: Hive usually stores metadata in relational databases (such as MySQL). Hive metadata includes a multitude of elements, including table names, table columns, partitions and their attributes, table attributes (such as whether the table is external), and the directory where the table's data resides.
4. Parser (driver): It consists of the interpreter (Parser), compiler (Compiler), optimizer (Optimizer), and executor (Executor). These components are used to complete HQL query statements, which originate from lexical analysis, grammatical analysis, compilation, optimization, and the generation of query plans. The generated query plan is stored in HDFS and then executed by MapReduce calls.
5. Hadoop: The data warehouse and query plan are stored in HDFS, which is calculated by the MapReduce engine.

The Hive system architecture is shown in Fig. 2.31.

The Hive workflow includes eight steps, as shown in Fig. 2.32:

1. The user submits queries and other tasks to the parser.
2. The compiler obtains the user's task execution plan.
3. The compiler obtains the required Hive metadata information from the metadata database according to user tasks.
4. The compiler obtains metadata information, compiles the task, converts HiveQL into an abstract syntax tree, converts the abstract syntax tree into a query block, converts it into a logical query plan, and rewrites the logical query plan. The

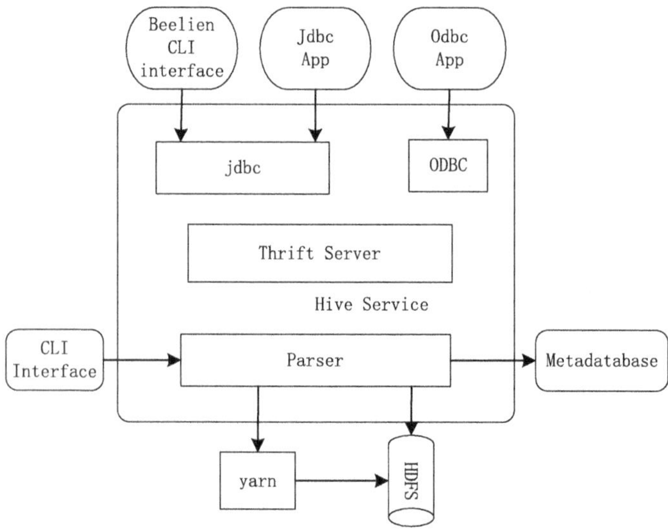

Fig. 2.31 Hive system architecture

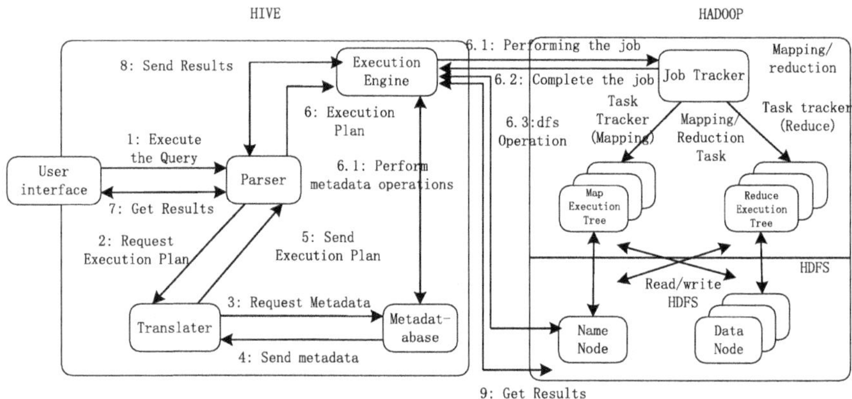

Fig. 2.32 Hive workflow

logical plan is transformed into a physical plan (MapReduce), and finally the optimal strategy is selected.

5. Submit the final plan to the parser.
6. The parser transfers the execution plan to the execution engine (ExecutionEngine) for execution, obtains metadata information, and submits it to the Hadoop job tracker or resource manager to execute the task. The task will directly read the files in HDFS for the corresponding operation.
7. Obtain the results of the execution.
8. Obtain and return the execution result.

2.7.3 Hive Main Features

Hive can be regarded as a user programming interface, which itself does not store and process data. It can map structured data files into a database table, while providing complete SQL query functions. Additionally, it facilitates the conversion of SQL statements into MapReduce task execution. The following example illustrates the principle of the MapReduce framework to achieve basic SQL operations.

1. The realization principle of Join: Mark the data of different tables in the output value of Map, and judge the data source according to the Tag in the Reduce stage. The MapReduce process of the statement "select u.name, o.orderid from order o join user u on o.uid = u.uid" is shown in Fig. 2.33.
2. The realization principle of Group By: Combine the fields of GroupBy into the output Key value of Map, use MapReduce sorting, and save the LastKey in the Reduce stage to distinguish different Keys. The MapReduce process of the statement "select rank, isonline, count (*) from city group by rank, isonline" is shown in Fig. 2.34.

Hive compiles SQL into MapReduce task mainly includes six steps:

1. Antlr defines SQL grammar rules, completes SQL lexical, grammatical analysis, and transforms SQL into an abstract syntax tree.
2. Traverse the AST Tree to abstract the basic query unit QueryBlock.
3. Traverse the QueryBlock and translate it into an operator tree.

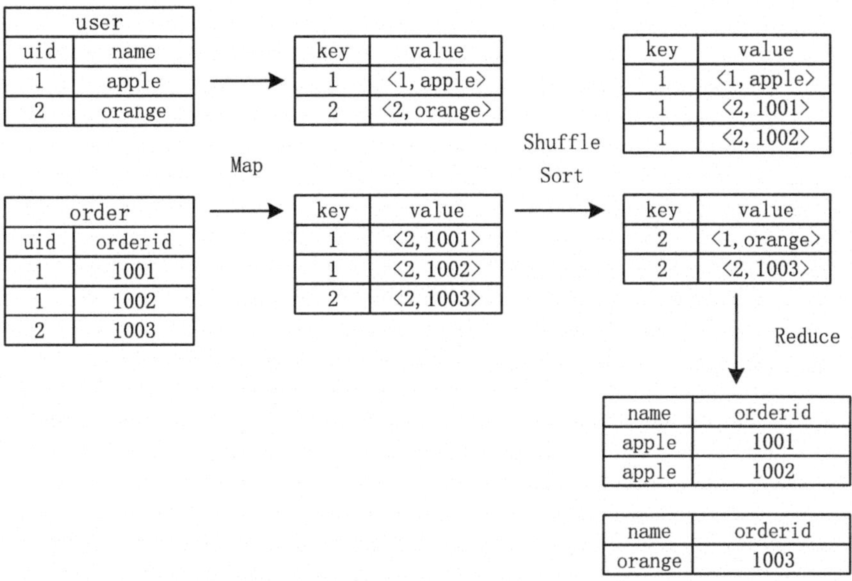

Fig. 2.33 Join's MapReduce implementation

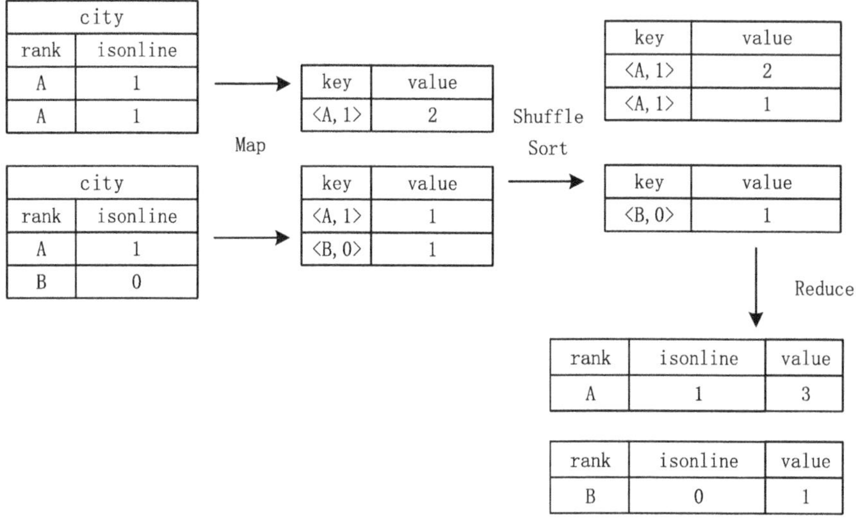

Fig. 2.34 Group by's MapReduce implementation

4. The logical layer optimizer performs OperatorTree transformation to reduce the amount of shuffle data.
5. Traverse OperatorTree and translate into MapReduce tasks.
6. The physical layer optimizer transforms MapReduce tasks to generate the final execution plan.

2.7.4 Hive Programming Interface

The installation and construction of the Hive data warehouse are usually divided into two stages. The first stage is to install and confirm the embedded mode. The second stage is to achieve the migration to the independent mode through the configuration mode (the independent mode already meets the business needs of most developers).

Hive embedded installation is divided into the following four steps:

1. Unzip and install: Enter the command tar -zxvf hive-0.7.X.tar.gz to decompress. Different versions of Hive have different file names.
2. Environment configuration: Edit the /etc./profile configuration file and add the following:

```
export HIVE_HOME = /home/hive-0.7.X
export PATH = $PATH:$HIVE_HOME/bin
export HADOOP_HOME=/home/hadoop-0.20.X
```

3. Directory creation: Create tmp, /usr/hive/warehousr directories on HDFS, and give group users write permission. These directories are storage directories for Hive data file and are used for configuration in hive-site.xml.

```
mkdir /tmp
mkdir /usr/hive/warehousr
chmod g+w /tmp
chmod g+w /usr/hive/warehousr
```

4. Test verification: Enter the command "hive"; If the Hive Shell interface is normally entered, this indicates that the build has been completed successfully.

Hive migration from embedded to independent mode is divided into the following six steps:

1. Install MySQL for this machine.
2. Create a Hive account (MySQL account) and grant permissions to the Hive account.
3. Create a meta-database: Create database hive.
4. Configure hive-site-xml:

```
<property>
     <name>hive.metastore.local</name>
     <value>true</value>
</property>
<property>
     <name>javax.jdo.option.ConnectionURL</name>
                          <value>jdbc:mysql://192.168.x.x:3306/
hive?createDatabaseIfNotExist=true</value>
</property>
<property>
     <name>javax.jdo.option.ConnectionDriverName</name>
     <value>com.mysql.jdbc.Driver</value>
</property>
<property>
     <name>javax.jdo.option.ConnectionUserName</name>
     <value>hive</value>
</property>
<property>
     <name>javax.jdo.option.ConnectionPassword</name>
     <value>12345678</value>
</property>
```

5. Copy the MySQL driver: Copy the JDBC driver package to the lib directory of Hive.
6. Test verification.

Hive uses HiveQL, a language similar to SQL, to manipulate data. It introduces many new features such as multi-table queries and MapReduce scripts based on the SQL language. HiveQL data types can be classified into basic types and complex types. The basic types include numeric types, Boolean types, and string types. The complex types include the ARRAY type, the MAP type, and the STRUCT type. The HiveQL language is based on Java, and thus there is a corresponding relationship between its data type and Java type. For example, the HiveQL BIGINT type (8-byte integer) is the Java long type. Table 2.7 represents data types for HiveQL.

HiveQL mainly contains four data models: database, table, partition, and bucket. They are listed below:

1. Database: Equivalents to the namespace in a relational database. The role is to isolate database applications to different database modes. Hive provides the ability to create, use, and drop database dbname, as well as other related statements to operate on the database.
2. Table: A table is composed of stored data and associated metadata describing the table. The user data is stored in a distributed file system, while the metadata is stored in a relational database. A table that fails to load data after creation corresponds to a single directory on HDFS. Upon loading the data, the data file is then copied to the HDFS directory. Hive tables can be classified into two categories: managed tables and external tables. A managed table has the characteristic of being loaded with its data file into a data warehouse directory set by Hive. In contrast, an external table can be stored in HDFS, not in the Hive data warehouse directory.
3. Partition: A Hive partition is a collection of some columns within a table. Each partition corresponds to a directory on HDFS. The significance of designing partitions is that it reduces the retrieval range and improves the query efficiency on Hive tables.
4. Bucket: The bucket is a more efficient way of data organization than the partition query, which organizes specific fields by row. Tables and partitions can be organized into buckets, with each bucket corresponding to a Reduce operation.

Table 2.7 HiveQL data type

Classification	Type of data	Storage space (bytes)	Type description
Basic type	TINYINT	1	Signed integer
	SMALLINT	2	
	INT	4	
	BIGINT	8	
	FLOAT	4	Single precision floating point
	DOUBLE	8	Double precision floating point
	BOOLEAN	1	Logical type
	STRING	≤ 2G	String type
Complex type	ARRAY	Unlimited	Ordered data
	MAP	Unlimited	Unordered key-value pairs
	STRUCT	Unlimited	Structure

HiveQL has similar functions to SQL, but the syntax is different. The following is a basic operation example of Hive:

1. Create a table: Create table student (id int, name string, score double) row format delimited fields terminated by '\t'; This statement is used to create a student table, the fields are number, name, and score. Row format delimited fields terminated by '\t' means that the data is separated by '\t'. It should be noted that HiveQL allows the user to define the data separator.
2. View the table: Show tables.
3. Add fields: Alter table student add columns (school string).
4. Delete the table: Drop table student.
5. Import data from HDFS: Load data local inpath '/home/hive-0.7.X/reservedata.txt' overwrite into table stu.
6. Create a partition: Create table person (id int, age int) partitioned by (name string) row format delimited fields terminated by ',';. This statement is used to create a person table, the fields are id, age, and name. Among them, the name column is partitioned separately, and the data is separated by ','.
7. Load data into the partition: Load data local inpath '/home/hive-/reservedata.txt' into table person partition (name = 'reservename');. This statement loads the data in reservedata.txt into the reservename partition table entry.
8. Query by partition: Select * from person where name = 'reservename'.
9. Display data table partition information: Show partition person.
10. Start the bucket service: Set hive.enforce.bucketing = true.
11. Create a bucket table: Create table bucket_test (id int, name string, score double) clustered by (id) into 3 buckets row format delimited fields terminated by ','.
12. Add data to the bucket: Insert overwrite table bucket_test select * from student.
13. Query data based on bucket: Select * from bucket_test tablesample (bucket 1 out of 3 on id).

2.8 Summary

Software architecture is defined as the abstract description and modeling of the overall software structure and components. The objective of the big data analysis core architecture is to reduce the overall structure of the big data analysis system and the association between its components [14].

Hadoop addresses the issue of single point of failure by separating resource management and job control, thereby improving overall stability and performance. The Hadoop framework is based on the Java programming language. Developers only need to implement the Mapper interface, Reducer interface, and configure Job to implement distributed computing [15, 16]. The details of Hadoop MapReduce are transparent to developers, which means that developers only need to focus on business logic and obtain distributed capabilities through the Hadoop API.

The HDFS distributed file system represents a foundational element of big data technology, which has given rise to the development of numerous [17, 18] data analysis architectures. HDFS addresses the single point issue by copying the constituent files into persistent files and running second-level name nodes. It provides data redundancy and self-test repair mechanisms, thereby enhancing the system's overall availability.

Spark's ecosystem supports batch processing, interactive query, and streaming data processing simultaneously, thereby improving computational efficiency by using memory instead of HDFS to save intermediate results. This is primarily achieved by elastic distributed dataset (RDD) technology [19, 20].

HBase distributed database allows users to store substantial quantities of data and provides them with high-throughput reliable services based on cost-effective processors. HBase's database tables can support dynamic access and offer greater flexibility and scalability than traditional relational databases.

The Hive data warehouse is implemented based on HDFS. It does not store and process data itself, but relies on the distributed file system HDFS to store data, and supports the compilation of SQL into MapReduce jobs.

The distinction between Apache Spark and Hadoop MapReduce lies in their data processing approaches. Spark operates by processing data in RAM, while Hadoop MapReduce writes data back to disk after completing map or reduce tasks. Theoretically, Spark's in-memory processing offers superior performance over Hadoop MapReduce. However, this efficiency comes at the cost of higher memory requirements. Hive, a SQL-styled interface, facilitates the execution of MapReduce jobs, while HBase serves as a NoSQL database, providing a key/value storage solution within the Hadoop. Just as Google is the go-to platform for search and Facebook for social networking, Hive can be used for analytical queries, and HBase is ideal for handling real-time querying.

2.9 Exercises

1. Briefly describe the concepts and connections of data, databases, database management systems, and database systems.
2. Briefly describe the main work of the MapReduce shuffle phase.
3. Briefly describe the definition and main features of cloud computing.
4. In which modes can the Hadoop cluster run?
5. How does Hadoop handle Job Tracker downtime?
6. Briefly describe the steps to install and configure the Apache open-source version of Hadoop.
7. List the process names of Hadoop and explain their roles.
8. List common Hadoop schedulers and briefly explain their working methods.
9. What are the main components of HDFS?
10. Briefly describe the process of HDFS file reading and writing.
11. What are the characteristics of the Scala language?

12. What are the advantages of the Spark framework over Hadoop?
13. List the Spark technology stack's common components and briefly explain their functions and application scenarios.
14. Briefly describe the concept and characteristics of Spark RDD.
15. What is data skew in Spark? What are the causes?
16. How many deployment modes does Spark have? What are the characteristics of each deployment model?
17. Briefly describe the steps for setting up a Spark distributed cluster.
18. List common Spark operators and explain their function.
19. Briefly summarize the characteristics of HBase.
20. Is the more HBase Region, the better?
21. What is the difference between HBase and Hive?
22. What is the difference between Hive internal and external tables?

References

1. Yifeng R. Introduction to software architecture;2019 April 12. https://www.ruanyifeng.com/blog/2016/09/software-architecture.html.
2. NIST. NIST special publication 500-299 NIST cloud computing security reference architecture;2015 June 16. https://www.nist.gov/news-events/news/2013/06/nist-publishes-draft-cloud-computing-security-document-comment.
3. Horizon Cloud. Ten trends forecast for cloud computing in 2019;2019 April 12. https://xueqiu.com/9632927105/122952189.
4. Microsoft Azure. What are private public hybrid clouds;2019 April 20. https://azure.microsoft.com/zh-cn/overview/what-are-private-public-hybrid-clouds/.
5. Apache Hadoop. Apache Hadoop documentation;2019 April 20. https://hadoop.apache.org/.
6. Vavilapalli V K, Murthy A C, Douglas C et al. Apache Hadoop YARN: yet another resource negotiator. In: Proceedings of the 4th annual symposium on cloud computing. ACM;2013.
7. Jianshu. Current application of Hadoop;2019 April 12. https://www.jianshu.com/p/97b4373b506a.
8. Shvachko K, Kuang H, Radia S et al. The Hadoop distributed file system. In: 2010 IEEE 26th symposium on mass storage systems and technologies (MSST). IEEE;2010.
9. The top 20 valuable facebook statistics;2020 March 11. https://www.manny.pro/post/the-top-20-valuable-facebook-statistics-updated-january-2020.
10. Zaharia M, Xin RS, Wendell P, et al. Apache spark: a unified engine for big data processing. Commun ACM. 2016;59(11):56–65.
11. Apache Hbase. Apache HBase documentation;2019 April 20. http://hbase.apache.org/.
12. Apache Hive. Apache Hive documentation;2019 April 20. http://hive.apache.org/.
13. CSDN. The principle of MapReduce to implement SQL operations;2019 April 20. https://blog.csdn.net/sn_zzy/article/details/43446027.
14. Julong Y. The full solution of big data technology: basics, design, development and practice. China Inf Technol. 2014;(6).
15. Patrick B, Chernozhukov V, Horta C et al. The impact of big data on firm performance: an empirical investigation. AEA Pap Proc. 2019;33–7.
16. Revathy P. HadoopSec: sensitivity-aware secure data placement strategy for big data/Hadoop platform using prescriptive analytics. GSTF J Comput (JoC). 2020;6(1).
17. Dharminder Y, Maheshwari D, Chandra D et al. Big data Hadoop: security and privacy. In: Umesh, editor. Big data Hadoop: security and privacy;2019.

18. Deshai N, Sekhar B V D S, Venkataramana S et al. Big data Hadoop MapReduce job scheduling: a short survey. In: Information systems design and intelligent applications;2019. p. 349–65.
19. Lunga D, Jonathan G, Lexie Y et al. Apache Spark accelerated deep learning inference for large scale satellite image analytics. IEEE J Sel Top Appl Earth Obs Remote Sens. 2020.
20. Shaik-Abdul-Khalandar B, Basha S, Vincent D et al. Challenges in storing and processing big data using Hadoop and Spark. In: Deep learning and parallel computing environment for bioengineering systems;2019. p. 179–87.

Chapter 3
Big Data Analysis and Computational Models

3.1 Introduction

Big data computing mode refers to various high-level abstractions or models that have been refined and established to address the inherent complexities and requirements of big data computing. This model is based on the unique characteristics of big data itself and its computing characteristics. According to the requirements of big data processing tasks and the diversity of application scenarios, various typical big data calculation modes and corresponding tools have emerged, each equipped with its suite of specialized tools. One notable example is the static batch processing model, which exemplifies the adaptability and innovation within the field of big data analytics. While the batch processing mode can execute large-scale data processing tasks in parallel, big data processing problems are complex and diverse. No single computing mode can address all computing needs. The batch mode is not well suited to situations requiring low latency, complex data relationships, or complex calculations. Consequently, stream and graph calculation modes were developed. Stream computing is a calculation mode for real-time analysis of data streams, which has the characteristics of low latency and high time efficiency. Graph calculation is a calculation mode for analyzing large-scale graph data structures, which can efficiently process graph structure data. At the same time, data visualization technology can clearly and effectively convey information with the help of graphical methods, in which the results of big data analysis can be better expressed and utilized. The current mainstream batch computing engine is MapReduce, a programming model and an associated implementation proposed by Google for processing and generating large datasets with a parallel, distributed algorithm on a cluster. Apache Storm, a stream computing distributed computing framework proposed by Twitter, is also widely used in real-time tasks. Apache Storm is written in Clojure and allows for the processing of both batch and distributed streaming data. It enables the customization of information sources and data operations. For distributed graph computing, Pregel

S. Luo and L. Pan, *Big Data Analysis*, Big Data Management,
https://doi.org/10.1007/978-981-95-5725-7_3

provides a scalable and fault-tolerant platform that provides a set of very flexible APIs that can describe various graph calculations and solve large-scale graph distributed computing problems.

This chapter addresses the following key topics: mastering typical big data computing models, understanding the basic concepts and programming models of batch processing, stream computing, graph computing, and architecture design, workflow, fault tolerance mechanisms, and application scenarios of its classic framework. It also covers the related concepts, tools, and development trends of data visualization.

3.2 Foundations of Data Mining and Analysis

3.2.1 Cognitive Basis of Pattern Recognition

Pattern recognition is the process of determining the category attribute (pattern class) of a sample. This process involves assigning a sample to a particular type among multiple types.

1. **Basic Concepts**

According to Bishop [1], in the context of a recognition task, the input is a sample of an unknown type, the classifier determines the sample type, and the output is the sample type. Additionally, there are numerous significant concepts in pattern recognition, including samples, patterns, features, and classifiers, as shown in Fig. 3.1.

A sample is defined as a specific research object, which may be a person, a Chinese character, or an image. Each instance is categorized, but some sample categories are unknown. The task of pattern recognition is to classify samples of unknown types. Mode refers to the description of the research object's characteristics, which is a

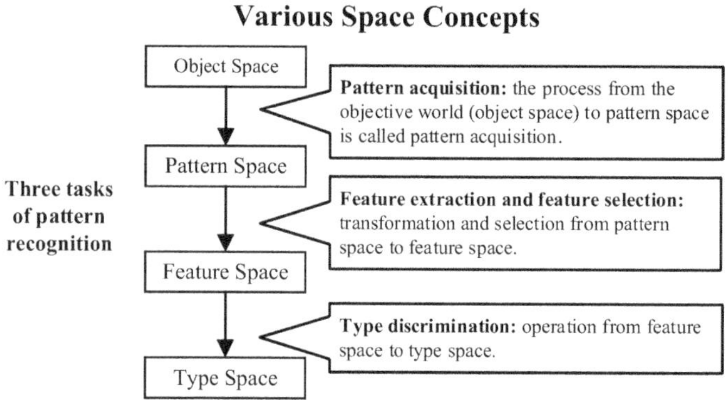

Fig. 3.1 Three tasks of pattern recognition

collection of measured values of a certain type of sample. Feature is used to describe the quantity that can be used to describe the characteristics of a given model. For example, in the gender classification task, the voice and appearance can be considered as the characteristics of the "gender" model. In statistical pattern recognition, a vector is usually used to represent features. Classifier refers to a system that can classify unknown samples by learning the characteristics of known samples in a pattern recognition task. The process of pattern recognition is the process of constructing a classifier with classification ability through the learning of samples with known categories, with the objective of using the classifier to classify samples of unknown types. The specific process and steps of pattern recognition will be explained in detail in Sect. 3.2.2, "Pattern Recognition Process."

There are three primary tasks in pattern recognition: pattern acquisition, feature extraction, feature selection, and type discrimination, [2] as shown in Fig. 3.1.

The term "object space" is used to describe the objective world, which is composed of various entities in the objective world. The term "model space" is used to describe a conceptual space composed of things with similar characteristics in the objective world. This concept is often referred to as "clustering by objects." Samples exhibiting specific similarities in the objective world are positioned in proximity in the model space to form a model group. "Pattern acquisition," one of the three major pattern recognition tasks, refers to the abstract process from the objective world to the pattern space. The formation of a specific pattern space is achieved through the discovery of similarities between objects in the objective world and the assignment of specific category labels. The feature space refers to a further abstraction of objects within the pattern space. It maps objects into representative metric values, removes redundant information of the objects, and combines a series of metric values. A pattern may be represented by a point or a feature vector in the feature space. This mapping process not only reduces the amount of information but also facilitates classification. This mapping process is also called "feature extraction and feature selection." Type space, also referred to as decision space, is defined as the collection of samples' classification results based on features. The value of the decision function corresponding to each category is calculated from the feature vectors in the feature space by introducing a decision function, as shown in Fig. 3.2. The samples are classified according to the value of the decision function. This process is referred to as "type discrimination."

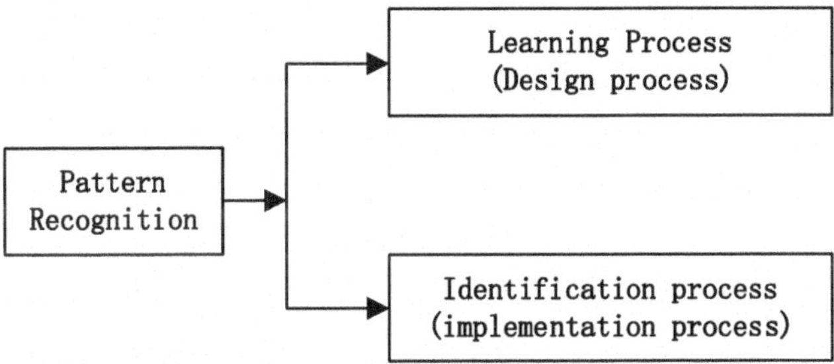

Fig. 3.2 Training and application process

2. **Pattern Recognition Process**

As shown in Fig. 3.3, the pattern recognition process includes both the processes of "learning" and "recognition." The term "learning process" is used to describe the training process of the original classifier. The classifier learns and analyzes the characteristics of a batch of known samples and can classify unknown samples. The recognition process is defined as the utilization of the trained classifier to identify and categorize the unknown type samples according to the classification criteria[3].

The statistical pattern recognition process generally includes the following steps, as shown in Fig. 3.4: data acquisition, data pre-processing, feature extraction, statistical analysis, and recognition classification.

Data acquisition: Through the use of sensors (such as cameras or microphones), the raw data, which may take the form of one-dimensional waveforms or two-dimensional images, is obtained following the processes of measurement, sampling, and quantization. During data acquisition, the data quality depends on the characteristics and limitations of the sensor, including bandwidth, resolution, sensitivity, and signal-to-noise ratio.

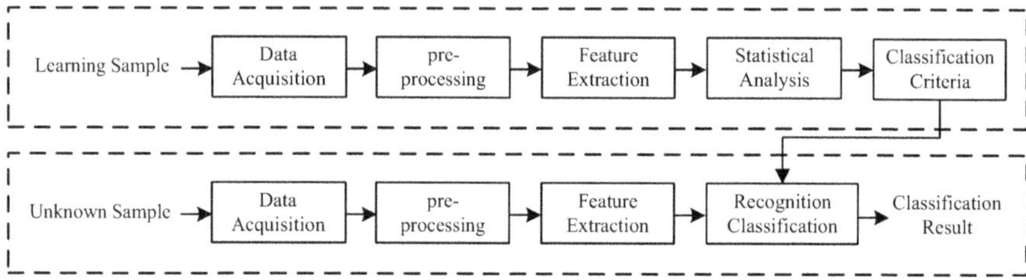

Fig. 3.3 General process of pattern recognition

Fig. 3.4 Classifier design process

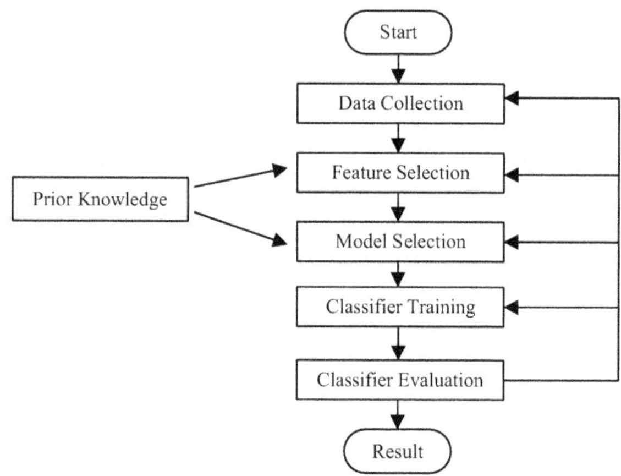

Data pre-processing: A specific method to segment the sample data, filter out noise, enhance the edge, and perform a series of processing operations that can highlight the characteristics of the sample data. The pre-processing method is closely related to the field to which the sample object belongs. For example, common pre-processing methods used in the field of voice analysis include emphasis and windowing, whereas those utilized in the image field include image enhancement.

Feature extraction: A specific method to extract the features that reflect the essential characteristics of the sample, thereby removing redundant information and compressing the original data.

Statistical analysis: The distribution of sample features is learned through specific statistical methods to establish a recognition model, form a classifier, and infer the classification criteria.

Recognition classification: According to the classification criteria, the samples to be identified are classified.

3. **Classifier Design**

The classifier's design, as shown in Fig. 3.4, mainly includes the following steps: data collection, feature selection, model selection, classifier training, and classifier evaluation.

Data collection is the basis of classifier design. In the development cost of a pattern recognition system, the cost of data collection occupies the main position. The number of training samples is an essential factor affecting the performance of the classifier. To ensure the performance of the classifier, there must be an adequate number of training samples. Insufficient training samples may result in the classifier being too close to the training samples' characteristics, leading to a lack of recognition ability for other samples of the same category in the objective world.

The objective of feature selection is to integrate prior knowledge and training data to remove redundant information in samples and explore efficient and accurate features. The selection of features depends on the specific application problem, which may pertain to areas such as speech recognition and image recognition. The objective of feature selection is to maximize the distance between sample classes and minimize the intra-class distance by extracting the quantity that most accurately reflects the sample's characteristics as a feature.

In model selection, the optimal model should be selected according to the specific characteristics of the actual application problem. For example, in the absence of known category labels, a clustering model should be selected for cluster analysis. In the event that the sample category labels are known, the classification model should be selected for classification training.

Classifier training refers to the learning process of classifiers on existing samples and their features. The characteristics of the learned samples may result in the trained classifier exhibiting different characteristics, such as under-fitting and over-fitting. Under-fitting means that the classifier cannot capture the data features adequately, resulting in an inability to properly fit the data. This can be effectively addressed by incorporating polynomial features and reducing regularization parameters. Over-fitting means that the classifier learns excessive information from the sample data,

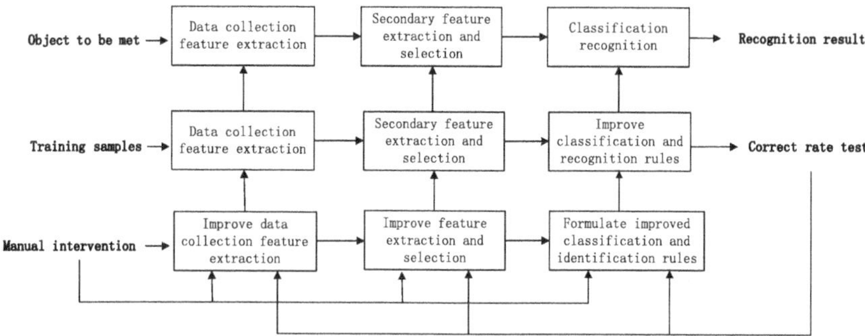

Fig. 3.5 Typical pattern recognition system

including noise information, making it difficult for the model to classify the data outside the sample set accurately. This can be mitigated by increasing the training dataset or incorporating regular items.

Classifier evaluation is the process of assessing the performance of a classifier. Common evaluation criteria include accuracy rate, detection rate, ROC curve, misrecognition rate, real-time performance, and computational complexity. Generally, the pattern recognition system adjusts the classifier according to the classifier evaluation index to optimize the classifier performance.

4. **Pattern Recognition System**

The computer system that realizes the learning process is called a pattern recognition system. The design principle of a typical pattern recognition system is displayed in Fig. 3.5.

The pattern recognition system can identify input samples that are unknown in category and subsequently output the corresponding categories. In the learning process, the following steps are undertaken: firstly, input training samples are provided and features are extracted; secondly, the classifier is trained and its performance evaluated; thirdly, the feature extraction method and classification rules are adjusted according to the evaluation results; and, finally the classifier is trained once more to form the optimal classifier after multiple iterations. In the recognition process, the input samples are initially subjected to preprocessing and feature extraction. The optimal classifier receives the sample features as input and produces the classification results according to the discrimination rules.

3.2.2 Cognitive Foundation of Data Mining

Data mining refers to the process of searching for information hidden in a large amount of data through the application of sophisticated algorithms. The objective is to analyze the data of the enterprise with a high degree of automation, make inductive reasoning, and identify potential patterns from the data to help decision-makers adjust market strategies, reduce risks, and make accurate decisions.

1. **Basic Concepts**

Data mining can be understood as the process of extracting or mining knowledge from a large amount of data, that is, "mining knowledge from data." "Data" here generally refers to a database that stores a large amount of data, so data mining can also be understood as "knowledge mining in the database"[4].

Data mining is a combination of "model" and "algorithm." The term "model" is used to refer to a description, summary, or representation of an overarching, global characteristic of the dataset. This characteristic is applicable to all points within a given space. The "algorithm" is a pre-defined, complete process that takes data as input and produces an output in the form of a model or pattern.

Data mining tasks can usually be divided into descriptive mining and predictive mining. Descriptive mining tasks summarize the general characteristics of the data in the database and present the data features in a convenient form. Predictive mining tasks are based on current data, observe object feature values, and make inferences to predict other feature values. Therefore, "description" and "prediction" are also the two main functions of data mining.

2. **Data Mining Process**

Data mining is the core link of Knowledge Discovery in Database (KDD). The process of knowledge discovery is presented in Fig. 3.6.

Generally, a knowledge discovery process consists of the following steps [5]:

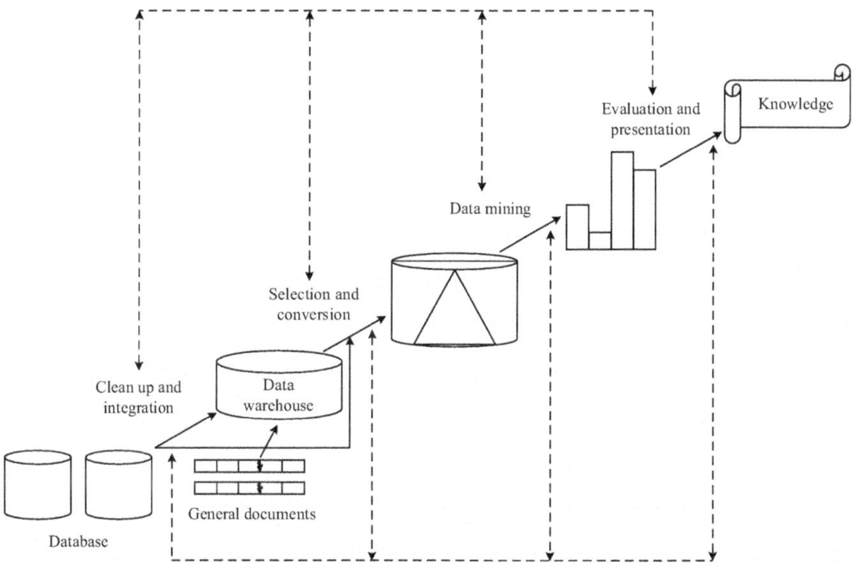

Fig. 3.6 Knowledge discovery process

(1) Data cleansing: Eliminate noisy data and inconsistent data.

The data in the database is usually incomplete, where some attribute values are missing, noise interference (attribute value error), etc., the data needs to be cleaned up, complete, correct, and consistent. The data information is stored in the warehouse for subsequent links to deal with.

(2) Data integration: Integrate data from different data sources.

For some enterprises, each branch may have its own database. In the process of knowledge discovery, to maximize the sample, it is necessary to integrate data from different sources, formats, and characteristics logically or physically through data integration to provide comprehensive data sharing for enterprises.

(3) Data selection: Extract data related to analysis tasks.

Data mining is based on a large amount of data in which the data mining process is often very time-consuming. To shorten the data mining cycle, organizations usually use data selection technology to reduce the dataset and filter out the data related to the analysis task while maintaining the data integrity, thereby shortening the data mining cycle.

(4) Data transformation: Summary or aggregation into a form suitable for mining.

The data is converted into a form suitable for data mining through smooth aggregation, data generalization, and normalization. For some real-number data, the data form can be effectively transformed through concept stratification and data discretization.

(5) Data mining: Use intelligent methods to extract data patterns.

According to the characteristics of the data in the data warehouse, apply appropriate statistical analysis methods, such as decision trees, rule reasoning, neural networks, genetic algorithms, and other methods, to process data and extract valuable information from it.

(6) Mode evaluation: Determine the accuracy of data mining results based on specific evaluation indicators.
(7) Knowledge representation: Visualize knowledge.

The data mining results are presented in a visual form, or stored as new knowledge in the knowledge base for use by other applications.

Additionally, the data mining process is iterative. If any step does not reach the expected goal, the past actions need to be adjusted and re-executed to optimize the data mining results.

3. **Common Methods of Data Mining**

Common data mining methods include classification, regression analysis, clustering, association analysis, variation deviation analysis, web information mining, etc. [6]

(1) Classification

Classification is a process to find the typical characteristics of a group of data objects in the database and divide them into different categories according to the classification mode. Its purpose is to map the data elements in the database to a given category through the classification model. They are usually used in customer classification, customer attribute analysis, customer satisfaction analysis, customer purchase trend prediction, etc. For example, auto retailers categorize customers according to their cars' preferences to conduct targeted marketing and increase sales.

(2) Regression analysis

The regression analysis method reflects the characteristics of the attribute value in the transaction database. It expresses the relationship of the data mapping through the function to find the dependency relationship between the attribute values. It is usually used for the trend characteristics of data series, the prediction of data series, and the study of the correlation between data. Typical applications include product lifecycle analysis, sales trend prediction, and targeted marketing activities in marketing.

(3) Clustering

According to the similarities and differences of data, clustering divides a group of data into multiple categories. The purpose is to maximize the similarity of data in the same category and minimize the similarity of data in different categories, that is, the cross-class data correlation is low. Typical applications include customer group classification, customer background analysis, and customer purchase trend prediction.

(4) Correlation analysis

Association analysis refers to mining the associations or mutual relationships that are hidden among data items. The appearance of other data elements can be derived based on the appearance of one data item. The mining association rules mainly include two stages: the first stage is to find all high-frequency combinations from massive raw data. The second stage is mining association rules based on the high-frequency combination. Correlation analysis has been widely used in financial industry enterprises to predict customer needs. Banks on their ATMs can improve marketing by bundling information that customers may be interested in. Besides, the typical "beer diaper" case also uses correlation analysis to find the correlation between commodities, thereby significantly increasing sales.

(5) Feature analysis

Feature analysis is the process of extracting the characteristic formulas of these data from a set of data in the database to express the overall characteristics of the dataset. For example, marketers can obtain a series of causes and main characteristics of customer churn by extracting the characteristics of customer churn factors. By using these characteristics, customer churn can be effectively prevented.

(6) Variation analysis

The deviation contains a lot of hidden knowledge, such as anomalous examples in the classification and the deviation of the observation results from expectations. The purpose of deviation analysis is to find meaningful differences between observations and reference quantities. For example, change bias analysis can mine unexpected rules. In enterprise crisis management and early warning, managers can realize the discovery, analysis, identification, evaluation, and early warning of various abnormal information types by paying attention to unexpected rules.

(7) Web data mining

Web information mining aims to analyze the massive information on the Internet and to collect information from a variety of sources, including political, economic, policy-related, scientific, technological, and financial data. This information can then be used to perform various data mining tasks. For example, in enterprise management, web information mining can facilitate the acquisition of internal or external information that has a significant impact on the enterprise. This information can then be analyzed to identify, analyze, evaluate, and manage potential crises.

4. The Relationship Between Big Data Analysis and Data Mining

Big data analysis includes the following six areas: visual analysis, data mining algorithms, predictive analysis capabilities, semantic engine, data quality and data management, and data warehouse[7]. Data mining is a crucial component of big data analysis. Using data mining technology and big data analysis allows for the exploration of the internal connections of massive data and the mining of hidden data. Additionally, data mining is based on the massive data of big data, and relies on the support of distributed computing of big data analysis platforms to accelerate data processing and shorten the data mining cycle.

In addition, data mining algorithms are an essential part of business computing logic within the big data technology stack. The big data analysis technology stack is presented in Fig. 3.7.

In the business computing layer, machine learning is the main implementation form of business computing logic. This logic is dependent upon a range of data mining algorithms to analyze the data. Therefore, data mining is one of the important functions of big data analysis. Concurrently, the business calculation logic of big data analysis depends on various data mining algorithms.

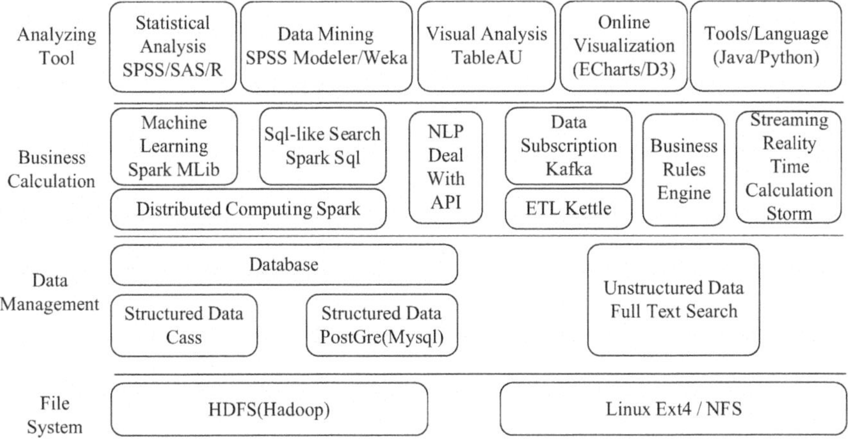

Fig. 3.7 Big data analysis technology stack

3.3 Batch Processing with MapReduce

3.3.1 Basic Knowledge

1. **Distributed Parallel Programming**

According to "Moore's law," the performance of a computer doubles approximately every 18 months. Thus, it has been reasonable to conclude that considerations of computer performance were not a significant concern in the past. However, as transistor circuits have gradually approached their physical performance limits, Moore's law has not been applicable since 2005[8]. Distributed parallel programming has become a major way to improve software performance.

The term "distributed parallel programming" is used to describe the execution of distributed programs on a large-scale computer cluster. This cluster includes a large number of inexpensive servers that can execute large-scale data processing tasks in parallel, thereby obtaining the computing power to process massive data [8]. The most significant advantage of this programming model is its robust scalability. The number of computing nodes can be directly increased by augmenting the computer and improving the cluster's computing power. Together, distributed parallel programming has strong fault tolerance, whereby the failure of some computing nodes will not affect the normal calculation and the correctness of the results.

Distributed computing is a computing model based on computer clusters, in which all hosts are divided into central nodes or child nodes. The central node divides large computing tasks into multiple subtasks, delivers them to the sub-nodes, and monitors the sub-nodes' operational status in real time. Multiple child nodes operate in parallel. Each child node completes the calculation of one or more subtasks, and finally merges and reduces the results to the central node to obtain the result. In comparison to the conventional computing paradigm, distributed computing offers a number of advantages, including.

(1) Rare resource sharing.
(2) Multiple computers work together to complete a computing task, balancing the load and improving computing efficiency and speed.

Distributed computing represents a standard solution for solving load problems. Based on distributed computing, researchers have established numerous concurrent programming models based on multi-core, multi-threading, and distributed parallel programming models by using large-scale computer clusters, such as MapReduce. The MapReduce programming model can effectively solve the calculation problem of big data.

2. **Introduction to MapReduce**

MapReduce is a core computing model of Google. As a distributed programming model, MapReduce represents the most outstanding big data batch computing model developed in the era of big data. For computing tasks with large amounts of data, distributed parallel computing is usually the optimal choice. However, implementing distributed computing is challenging, and many developers still face difficulties. MapReduce simplifies the distributed parallel computing programming model and provides an interface for users. The model shields numerous parallel computing details, especially those pertaining to distributed processing, thereby facilitating the broader adoption of distributed parallel computing.

The MapReduce algorithm highly abstracts the complex parallel computing process on large-scale clusters into two fundamental functions: map() and reduce() [9]. As the name implies, the map() function performs a one-to-one mapping of dataset to another, with the user defining the mapping rules. The reduce() function is the reduction, and the mapping results are reduced and merged, where the user also specifies the rules. In other words, MapReduce provides a framework for programming, and user-defined map() and reduce() functions provide the rules of operation.

The MapReduce's calculation process can be summarized as follows: the big data is decomposed into multiple sub-tasks. Each sub-task is then processed by a node within the cluster, resulting in the generation of intermediate results. Finally, the intermediate results are combined to obtain the final result and achieve an extensive calculation of data tasks. Consequently, the operation of MapReduce depends on the distributed operating system. The distributed job system consists of a job node and multiple task nodes. The job node is responsible for task scheduling and distribution, while the task node performs specific tasks. Among these components, the task node includes a mapping node and a specification node.

3. **MapReduce Design Ideas**

The MapReduce programming model follows the design philosophy of "divide and conquer, move logic, shield the bottom layer, and process customization" [10]. To clearly illustrate the aforementioned concepts, we introduce the story of Cao Chong weighing elephants. Cao Cao is attempting to ascertain the weight of elephants, but lacks the requisite apparatus for this purpose. His youngest son, Cao Chong, led the elephants to the boat and recorded the hull's water surface after the hull had reached a stable position. Once the elephant had been removed, the stones were placed on the boat until the water level reached the previously recorded position, after which the weight of the stone was separately measured. The weight of the elephant was obtained by adding together the individual measurements. The essential point to be derived from this narrative is that elephants can be both large and small; the quantity of data is considerable, yet the processing capacity is limited. Consequently, the elephant is divided into multiple stones, thereby the issue is solved. This revelation is also applicable to the processing of big data. In general, the big data processing task is divided into three steps:

Step 1: Divide the big data task into smaller sub-tasks.
Step 2: Each mapping node processes a split part. To increase the processing efficiency, multiple mapping nodes are processed in parallel.
Step 3: The processing results of the mapping node are merged by the reduction node to obtain the final result.

The following section provides a detailed description of MapReduce's "divide and conquer, move logic, shield the bottom layer, and process customization" design concepts.

"Divide and conquer" refers to the process of splitting large-scale data processing tasks into multiple sub-tasks. A master node assigns tasks, each task node processes the tasks, and then merges the intermediate results to obtain the final result. The above-mentioned process is highly abstracted into two functions, map() and reduce(). The former is responsible for analyzing the data, while the latter is tasked with summarizing the results.

Mobile logic means that MapReduce is not executed locally, but rather delivered by the master node to the task node for execution. MapReduce can be understood as a data collector that stores code data, which is subsequently handed over to the back-end for processing.

The bottom layer is shielded, which means that the code transmission and operation, data in and out, operation timing, and task allocation are all done by the distributed operating system. Developers only need to focus on the two functions, map() and reduce(), which significantly simplifies the implementation of distributed computing.

The customization of processing functions enables users to define map() and reduce() functions according to their needs. This implies that a single distributed data processing task can be customized to accommodate the diverse needs of users through the implementation of different map() and reduce() functions. The difference in processing logic does not affect the distributed operating system itself, which merely changes the calculation method without changing the data structure.

3.3.2 Programming Model

This section introduces the MapReduce programming model from four perspectives: MapReduce itself, distributed operating system, mobile code, and intermediate results.

1. **MapReduce Itself**

MapReduce itself has three key concepts: the map function (map()), the reduce function (reduce()), and the key-value pair (<key, value>). The map function's input and output are both key-value pairs. These key-value pairs may be converted into other key-value pairs or into a different batch of key-value pairs. The data flow model of the map function is shown in Fig. 3.8.

The task node previously divides the data into data blocks, also known as splits, with a default size of 64 MB. The task node then converts this data block into a set of <key, value> key-value pairs, which are referred to as the source key-value pairs. The intermediate result, <key1, value1> , is obtained by the user-defined map function.

The combination is then sorted through a shuffle process and input to the reduction node for the intermediate result. In the process of obfuscation, the intermediate results are typically handled by the combine function (combine()). The combine function combines the same values of the keys into an array. For a certain key, key1, the combine() function combines all key-value pairs of key1 into an intermediate set of key-value pairs, namely, <key1,> .

The input to the reduce function is a collection of intermediate key-value pairs. The output is the target key-value pair <key2, value2> , which is saved to HDFS. The data flow model for the reduce function is presented in Fig. 3.9.

The obfuscation process is the term used to describe the integration of the mapping task's output. The operational process of obfuscation is described in detail in Sect. 3.3.4.

The key-value pairs <key, value> can be understood as generalized arrays. The key is the array's subscript, which is unique, and the value is the element corresponding

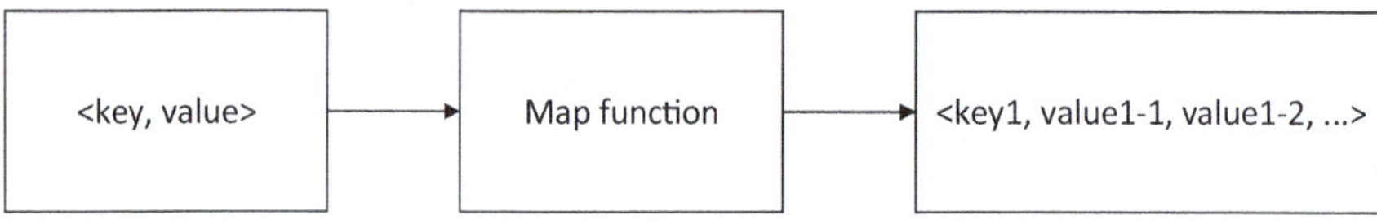

Fig. 3.8 Data flow model of map() function

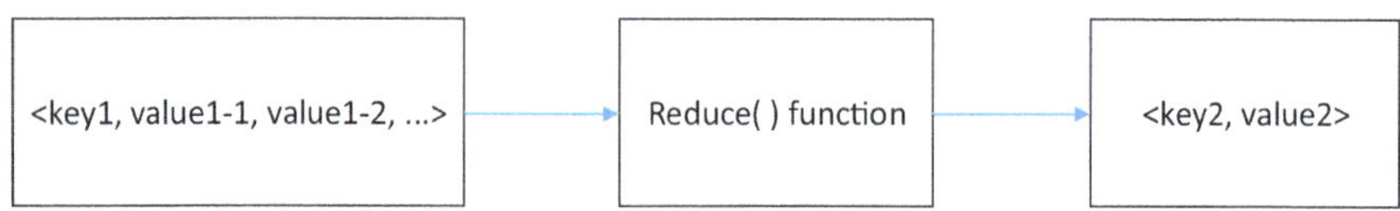

Fig. 3.9 Reduce function data flow model

to the subscript. In MapReduce, key-value pairs are used to describe the original data, intermediate results, and target data. In contrast to the array, the key is to remove the array name and element type, thus making the MapReduce programming model suitable for developing various data structures without being constrained by data types. In order to facilitate a more comprehensive understanding of the MapReduce process, it is necessary to divide key-value pairs into four categories, according to the processing stage: source key-value pairs, intermediate key-value pairs, intermediate key-value pairs sets, and target key-value pairs.

(1) The source key-value pairs are generated by distributed operating systems based on data blocks on data nodes.
(2) The intermediate key-value pair is processed by the map function, which defines the rules for forming the key and value.
(3) The intermediate key-value pair set is formed by the combine function. According to the same key-value principle, the combine function classifies the set of key-value pairs to obtain the set of intermediate key-value pairs.
(4) The target key-value pair is the result of the reduce function, which is formed according to the rules defined by the reduce function.

2. **Distributed Operating System**

The distributed operating system is a MapReduce hardware framework, which is mainly composed of a job node and multiple task nodes. The job node is responsible for scheduling and managing tasks, as well as assigning mapping tasks and specification tasks to idle task nodes. In the event of a failure of a task node, the job node will assign tasks to other idle task nodes for execution, thereby increasing fault tolerance.

3. **Move Code**

When a job request is submitted, the job node dynamically allocates computing power according to the idle state of the current task node, so the code cannot be pre-installed on a particular node, which uses code move in Hadoop. The code move is divided into the following four steps:

(1) The application specifies the job configuration, including the input and output format of the data, the source and flow of the data, the location of jar packet, and then sends a job request to the job node.
(2) The client program will then upload the jar package to the specified HDFS directory.
(3) The job node proceeds to obtain the jar package from each task node, ultimately reaching the specified HDFS directory.
(4) The task node calls the map() and reduce() functions to complete the corresponding work.

The user-provided application should indicate the input and output paths and provide the map() and reduce() functions by implementing the appropriate interface or abstract class to form a job configuration.

4. **Intermediate Results**

Upon completion of the obfuscation process on the map side, the task node writes the result to the task node's local disk and informs the job node of the location of the intermediate result file. Subsequently, the job node receives the information and transmits the requisite information to an idle task node, which is then responsible for collecting the location results. All intermediate results are then hashed according to the key value and modulo. Subsequently, the data is then divided into N parts, with each node N responsible for the statute of a specific key-value interval dataset. The reduce function is then called to form the final result, which is finally written to HDFS. The intermediate result is written to the local disk rather than HDFS because the intermediate result will be deleted after the completion of the task. Additionally, there will be additional data read and write operations on HDFS, which will result in a decline in the performance of the cluster system.

3.3.3 Architecture

The MapReduce architecture consists of four components: the user node, the job node, the task node, and the task, [11] as displayed in Fig. 3.10.

1. The specific tasks of the user node include
   (1) The MapReduce program written by the user is submitted to the job node by the user node.

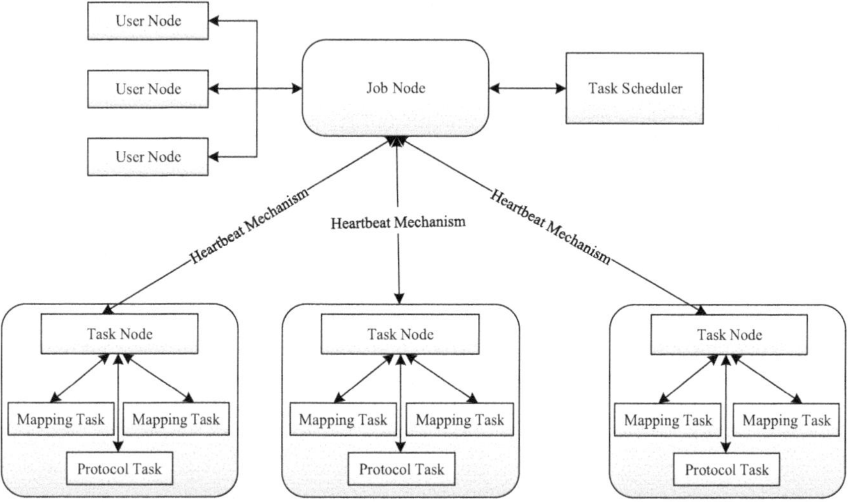

Fig. 3.10 MapReduce architecture

(2) The user can view the status of the job through the interfaces provided by the user node.

2. The tasks of the job node include

   (1) Monitoring and scheduling of resources and jobs.
   (2) Monitoring the survival status of all task nodes and tasks. In the event of a task failure or node failure, the corresponding task will be transferred to other nodes.
   (3) Tracking of information such as task execution progress, resource usage, and notification of this information to the task scheduler. The scheduler will search for free resources and assign tasks.

3. Task nodes can be divided into mapping nodes and protocol nodes according to the different tasks performed. The duties of the task node include

   (1) Periodically report the resource usage on the node and the task's running progress to the job node through the heartbeat mechanism. Concurrently, the task node receives the commands sent by the job node and performs the corresponding operations (such as starting a new task or terminating the task).
   (2) "slot" is used to divide the node resources (CPU, memory, etc.). A task only has the chance to run when it gets a slot, and the role of the Hadoop Scheduler is to assign the free slot on each task node to the task.

4. Tasks are divided into two categories: mapping tasks and specification tasks. Mapping tasks correspond to mapping nodes, while specification tasks correspond to specification nodes.
5. The execution process of the mapping task is presented in Fig. 3.11. The mapping node first parses the corresponding split iteration into a key-value pair and then calls the user-defined map() function for processing. Finally, the local disk's temporary results are stored. The provisional data is divided into several partitions, each of which will be processed by a specification task.

The following figure illustrates the implementation process of the protocol task. The process is divided into three phases (Fig. 3.12). The "obfuscation stage" is

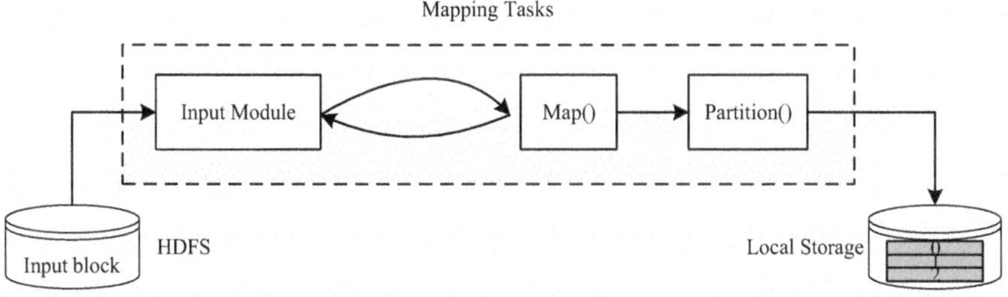

Fig. 3.11 Mapping task execution process

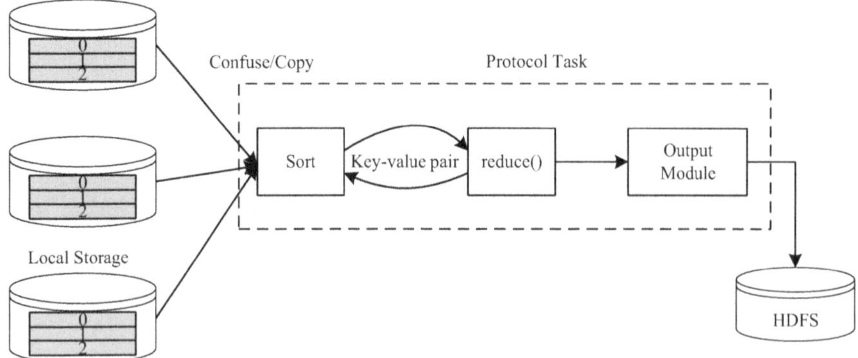

Fig. 3.12 Protocol task execution process

responsible for reading the mapping task results from the remote node. The "sorting stage" is responsible for sorting key-value pairs by key. The "specification stage" allows the specification node to read <key, value> list in turn, call the user-defined reduce() function to process it, and write the final result to HDFS, as indicated in Fig. 3.13.

3.3.4 Workflow

The essence of MapReduce is the segmentation of the input data and its subsequent distribution to different nodes for processing, which is then subsequently merged. The MapReduce processing can be divided into five stages: input = > map = > obfuscation = > specification = > output. Figure 3.14 briefly illustrates the MapReduce processing of big data.

1. **Input Stage**

   (1) The user creates a task and submits it to the job node.
   (2) Input format: The input module, InputFormat, loads files from the distributed file system and preprocesses them, including verifying the input and output formats. Concurrently, it divides them logically into input blocks InputSplit.
   (3) Record Reader (RR) processes the input block to get the source key-value pair.

2. **Mapping Stage**

Take the source key-value pair as the input of the mapping node, then execute the map() function. The output intermediate key-value pair will then be stored in the temporary file.

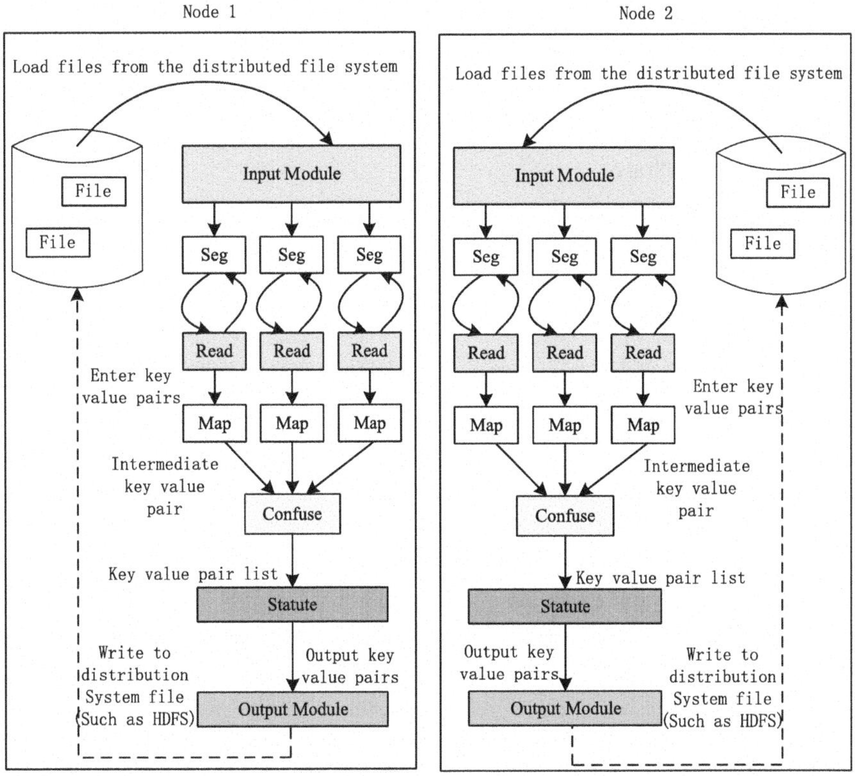

Fig. 3.13 MapReduce workflow

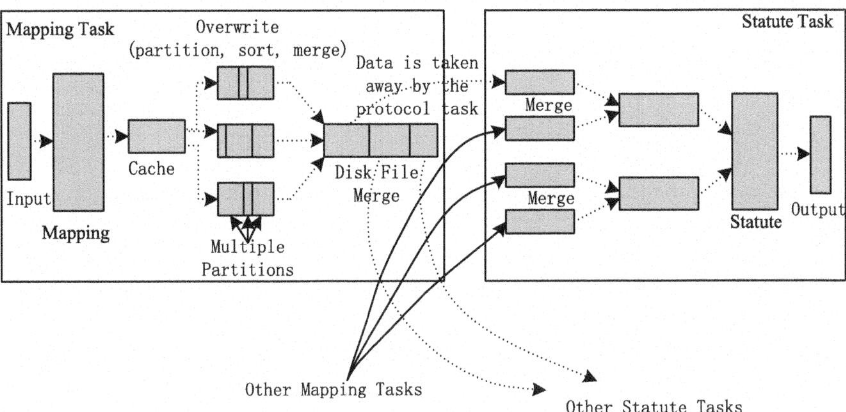

Fig. 3.14 The obfuscation process

3. **Obfuscation Stage**

In the obfuscation stage, the mapping node's output is sorted and segmented. The intermediate key-value pair set is obtained and handed over to the corresponding specification node.

4. **Specification Phase**

The obfuscated results are processed to get the target key-value pair.

5. **Output Phase**

The output module OutputFormat verifies the existence of the output directory exists. Once the output results match the configuration type, they are written to HDFS.

In the MapReduce process, to enable the specification node to process the mapping node results in parallel, it is necessary to sort and split the output of the mapping node before delivering it to the corresponding specification node. The sorting and splitting process is complex and is also the core of the MapReduce workflow. As displayed in Fig. 3.14, obfuscation occurs in both the mapping and specification tasks. The following section provides the obfuscation process in detail, beginning with the mapping task and then moving on to the specification task.

1. The obfuscation process of the mapping task:

The obfuscation process of the mapping task involves partitioning, sorting, and overwriting the results of mapping nodes, followed by the merging of the output belonging to the same specification task and its subsequent writing to disk. As presented in Fig. 3.15, the obfuscation of mapping tasks is typically divided into four processes: input, partition, overflow, and merge.

(1) The input process inputs the source key-value pair into the mapping node, which then performs the mapping task. The output of these mapping nodes is not directly written to the disk, as the high frequency of read and write operations will significantly impair performance. To mitigate this, each mapping task has a buffer serves as a temporary repository for intermediate results. Given that the output from multiple mapping nodes may be delivered to the same protocol node, it is crucial to understand the mapping relationships among them. The subsequent partitioning process plays a critical role in establishing these relationships.

(2) Partitioning refers to the process of assigning the result to the specified protocol node after the mapping task outputs key-value pairs. The MapReduce framework provides a partition interface that enables the determination of the protocol node to which the current key value should be assigned for processing according to the key, value, and the number of protocol nodes. The default rule involves hashing the key, which is then modulus-operated on by the number of protocol nodes that is the hash (key) mod R, where R is the number of protocol nodes. Additionally, users can also define their partition rules. Subsequently, the output is written to the buffer. Due to the limited size of the buffer, there is a possibility of a buffer overflow occurring. The process of handling a buffer overflow consists of the following overflow stages.

(3) Overflow is the process of writing the contents of a buffer to a temporary file. The buffer's default size is 100MB, and the mapping task's output is likely to be larger than 100MB. Therefore, under certain conditions, the buffer contents need to be written to the temporary file. This process is called "overflow." Overflow is executed by a separate thread and does not affect the mapper from writing the buffer's result. In the process of overwriting, the mapping task continues to write data to the buffer. Therefore, it is necessary to set an overflow threshold for the buffer, which is 0.8 by default. When the buffer data size exceeds 100MB $\times$ 0.8 = 80 MB, the overflow will begin. Once the overflow thread has been initiated, it is necessary to sort the data in the 80 MB space to improve the follow-up process's efficiency. The default sorting rule is the ascending order of the key. Additionally, to reduce the amount of data transferred from the mapping node to the protocol node, the user can set the combine function to add the values of the key-value pairs with the same key. This reduces the amount of data overwritten to the disk and subsequently reduces the amount of data sent by the mapping node. The process of processing the overflow file is the following merge phase.

(4) Merging refers to the process of merging multiple overflow files into a single data file. In the overwriting process, when the buffer data reaches the threshold, an overflow occurs, resulting in a temporary overflow file. Therefore, multiple overflow files may be generated following the completion of a mapping task. At the end of the mapping task, these overflow files are merged into one file, and this is known as the merging process. Since the merge is to merge multiple files, and that the above-mentioned merge procedure is executed when an overflow file is written into the buffer, it is possible that the merged file may have the same key. Furthermore, users can also merge the same key by setting the combine function.

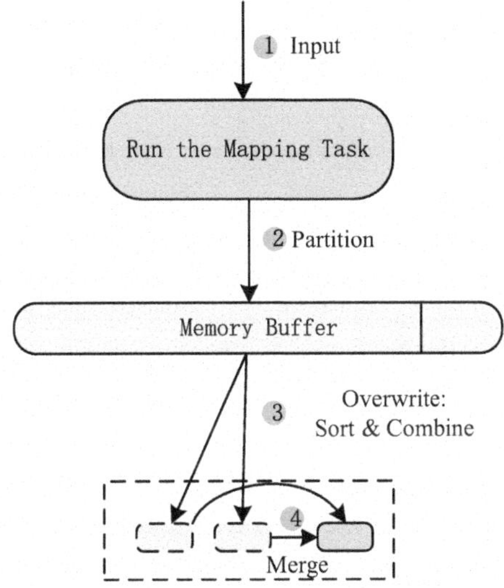

Fig. 3.15 The obfuscation process of the mapping task

So far, the obfuscation process of the mapping node is finished, and the final generated file is written to the local disk of the mapping node. The job node is then informed of the directory of task completion and result storage. Each specification node periodically queries the job node about the implementation of the mapping task. Upon completion, the job node will send the mapping node's result storage directory to the protocol node. The latter will then retrieve the mapping task result and start the second half of the confusion process. Additionally, all temporary overflow files will be deleted at the end of the job.

2. The obfuscation process of the protocol task:

The obfuscation of specification tasks is a process whereby the results of each mapping node are continuously read before the specification task is executed. The obfuscation of specification tasks is typically divided into three phases: copy, merge, and input, as shown in Fig. 3.16.

(1) Copy is the process of collecting input data from the specification task. The specification task's input data is distributed in the output of multiple mapping tasks in the cluster, with each mapping task completing its task at different times. The specification node queries the job node to determine when a particular mapping task is complete and to obtain the directory to save the results. Once a mapping task is completed, the protocol node copies the results of the mapping task to the specified directory, completing the copy phase of the process. The protocol node has multiple copy threads, which can obtain the mapping node's output in parallel. The number of threads can be modified by setting mapred.reduce.parallel.copies.
(2) Once the result file has been obtained, the output files of various mapping tasks must be merged into a single file, which is the merge phase.
(3) Finally, the merged file is used as the input file of the specification task, which is the input phase.

3.3.5 Fault Tolerance Mechanism

In distributed programming, it is inevitable that some errors will occur in a distributed operating system. Generally, errors are classified into four categories according to their location within the system: task errors, task node errors, job node errors, and HDFS errors.[12] The MapReduce computing framework provides a relatively perfect fault-tolerant mechanism. The following section describes various types of errors and the corresponding fault-tolerant mechanisms.

1. **Task Error**

One of the most common error tasks is the wrong one. The causes of such task errors can be attributed to factors like low code quality, data corruption, or node failure. Typically, if the task node detects an error, the task node will report the error to

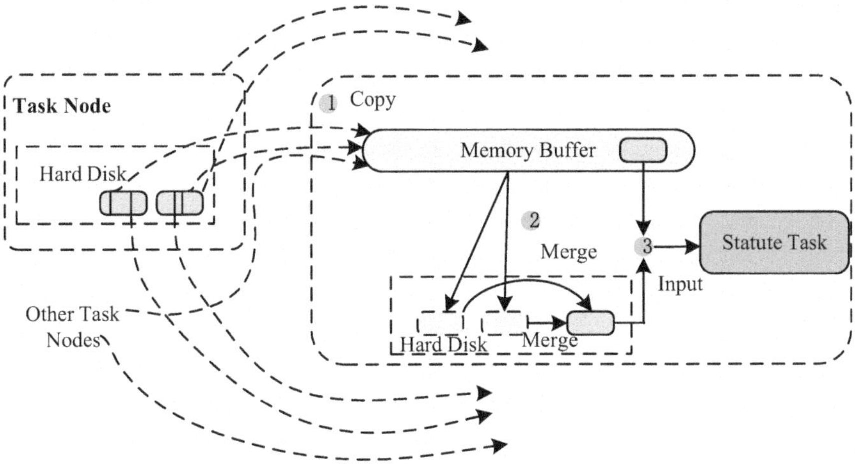

Fig. 3.16 The obfuscation process of the specification task

the next heartbeat's job node. Upon receiving the reported error, the job node will determine whether it is necessary to try again. Should a retry be deemed necessary, the task is rescheduled accordingly. By default, tasks are assigned four attempts, but this threshold can be adjusted through the configuration settings in the mapred-site.xml file.

If multiple tasks of the same job repeatedly fail in the same task node, the job node will put the task node on the job level blocklist to prevent the assignment of further tasks to that node. If multiple tasks of multiple jobs repeatedly fail in the same task node, the job node will put the task node on a global blocklist for 24 h to prevent the assignment of tasks to the task node.

2. **Task Node Error**

In the event of a task node process crash or failure of the node where the task node process is located, the task node will cease to transmit heartbeat information to the job node. In the event of a failure of the task node, the job node will consider that the task node is invalid. Consequently, the tasks that have been run on the task node will be considered as failed, and they will be rescheduled to other task nodes for execution. From the user's perspective, the reschedule of tasks due to errors in task nodes will result in a prolonged job execution time, but will not lead to errors or crashes.

3. **Job Node Error**

According to the MapReduce architecture and the distributed job system architecture, the job node is tasked with the scheduling and monitoring of all tasks and jobs. Therefore, the job node error represents a significant challenge. In the event of a failure of the job node, the status information pertaining to all currently executing jobs will be lost. To ensure the reliability of task completion, even if the job node returns to normal in a relatively short time, all jobs will be marked as failed and need to be executed again.

4. **HDFS Error**

The MapReduce framework relies on the underlying HDFS for file reading and writing. In the event of a failure of the HDFS, there is a possibility of exceptions being generated in the reading and writing of data, which may result in the failure of tasks and jobs. In the event of a DataNode failure, MapReduce will read the required data from other DataNodes. Generally, the data stored on each DataNode will be backed up on other DataNodes. Therefore, the only instance in which a read–write exception may occur is when the related DataNodes are damaged simultaneously. In the event of a failure of the NameNode, the system will generate an error message upon its subsequent attempt to access the NameNode. The MapReduce computing framework will attempt to access four times, with the default maximum number of attempts of four. If the access fails, it is considered that the NameNode is in a failed state, resulting in the failure of the job execution.

3.3.6 Programming Examples

This section compares various programming ideas with wordcount programming examples. It then lists the code and analyzes it so that readers can better understand MapReduce and distributed parallel programming [13]. In order to count the most frequently used words in computer papers in the past decade, there are usually four methods as follows:

Method 1: Traverse all papers in order and count the number of each word. This method is suitable for small-scale data, efficient, and easy to implement. However, this approach is not ideal for processing large amounts of data.

Method 2: Use a multi-threaded program to traverse the paper concurrently. This method is suitable for multi-core or multi-processor machines. This method is both efficient and challenging. In order to avoid the creation of duplicate files, the data must be shared in a synchronous manner between the two threads.

Method 3: Hand over to multiple computers to complete the task. The paper is assigned to several computers to traverse by using the program of the train one. This method is the most efficient but the most difficult. It requires manual allocation of data and integrates the results.

Method 4: Use the MapReduce framework. In essence, it is method 3. However, the method will autonomously distribute and integrate data. The users are only required to develop the distribution rules, with MapReduce then completing the specific tasks.

An analysis of the four methods reveals that method 4 is the most efficient and straightforward to implement. The WordCount method begins by dividing the file content into individual words. Subsequently, the same words are combined. Finally, the number of times is counted. In the MapReduce model, each node that receives the data must segment the data to complete the word segmentation task through the mapping phase. In order to improve the efficiency, the frequency calculation of the identical words can also be processed in parallel. This entails transferring the same word to the same machine for statistical analysis, thereby enabling the specification stage to count the frequency. The middle word grouping process is completed by obfuscation. In the design of a key-value pair, the word is used as the key and frequency serving as the value.

1. **Code Implementation**

```
public class WordCount
{
public static class TokenizerMapper extends Mapper<Object, Text,
Text, IntWritable>
{
private final static IntWritable one = new IntWritable(1);
private Text word = new Text();
public void map(Object key, Text value, Context context) throws
IOException,, InterruptedException
{
StringTokenizer itr = new StringTokenizer(value.toString());
while (itr.hasMoreTokens())
{
word.set(itr.nextToken());
context.write(word, one);
}
}
}
public static class IntSumReducer extends Reducer<Text,
IntWritable, Text,
IntWritable>
{
private IntWritable result = new IntWritable();
public void reduce(Text key, Iterable<IntWritable> values,Context
context)
throws IOException, InterruptedException
{
int sum = 0;
for (IntWritable val: values)
{
sum += val.get();
}
result.set(sum);
context.write(key, result);
}
}
public static void main(String args) throws Exception
{
Configuration conf = new Configuration();
String otherArgs = new GenericOptionsParser(conf,
args).getRemainingArgs();
if (otherArgs.length != 2)
{
System.err.println("Usage: wordcount <in> <out>");
```

```
System.exit(2);
}
Job job = new Job(conf, "word count");
job.setJarByClass(WordCount.class);
job.setMapperClass(TokenizerMapper.class);
job.setCombinerClass(IntSumReducer.class);
job.setReducerClass(IntSumReducer.class);
job.setOutputKeyClass(Text.class);
job.setOutputValueClass(IntWritable.class);
FileInputFormat.addInputPath(job, new Path(otherArgs));
FileOutputFormat.setOutputPath(job, new Path(otherArgs));
System.exit(job.waitForCompletion(true) ? 0: 1);
}
}
```

2. **Code Analysis**

Firstly, the source key-value pair < key, value > is formed, where the key is the line number of the document, and the value is the content (Fig. 3.17). The source key-value pair is used as the mapping input to get the middle key-value pair, as illustrated in Fig. 3.18.

The obfuscation operation is performed on the intermediate key-value pairs to obtain the intermediate key-value pair sets < “hello”, > , < “world”, > and execute

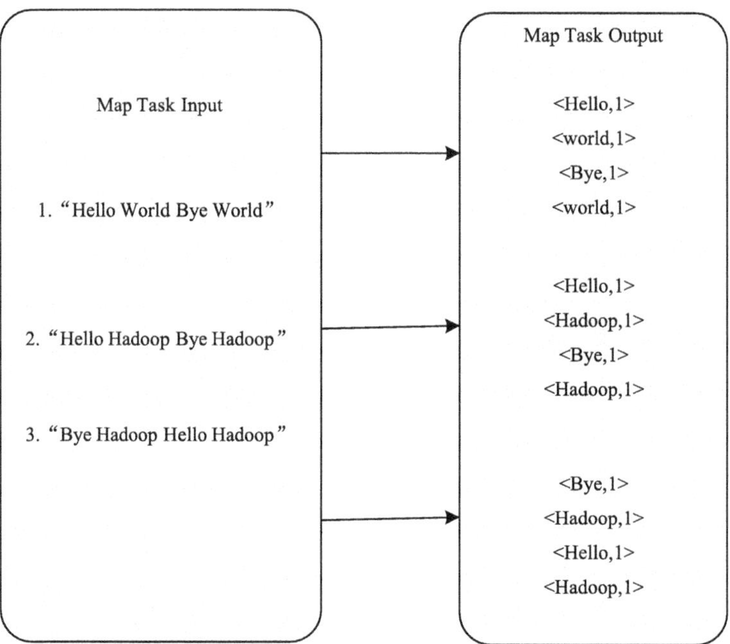

Fig. 3.17 Mapping process diagram

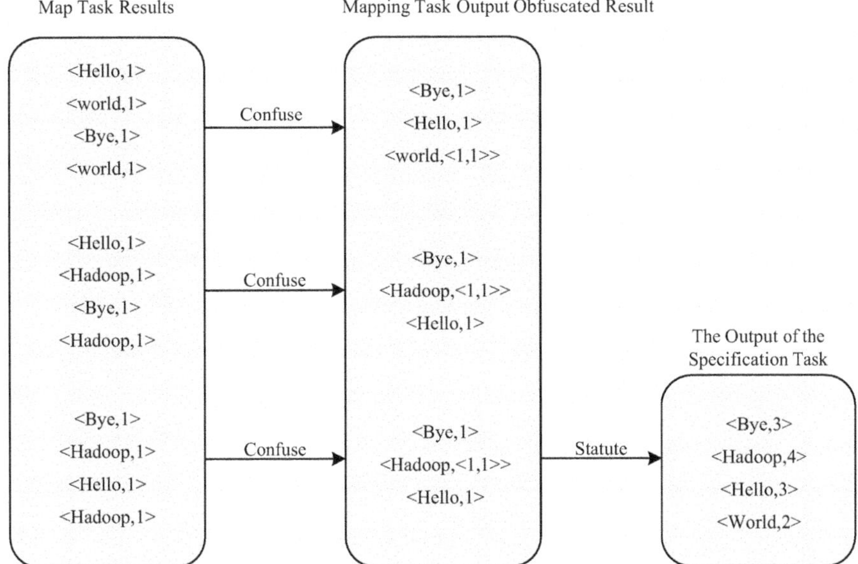

Fig. 3.18 Schematic diagram of the Reduce process

the reduce operation to get the final result < “by”, 3 > , < “hello”, 3 > and so on, as shown in Fig. 3.18.

Finally, the above process can be summarized as follows:

(1) MapReduce divides the input into *M* jobs.
(2) The master transfers each job to a mapping node in the idle state.
(3) The mapping operation is performed for each source key-value pair, and the intermediate key-value pair is written to the local disk. The file information is transmitted to the master.
(4) *R* protocol nodes get data from different mapping nodes’ partitions, sort them with keys, and perform the protocol operation for each unique key.
(5) Following the execution of the process, the result is returned.

3.3.7 Typical Case

Static batch processing, with its excellent data processing capabilities, has become a core component of a variety of data analysis platforms. This section takes telecom operators as an example to illustrate the application of the static batch computing mode in the big data analysis system.

Telecom operators have accumulated a large amount of data in operational services as a data pipeline, including operational data, telecommunication infrastructure data, derived budget, financial, and other data types. By mining these data resources,

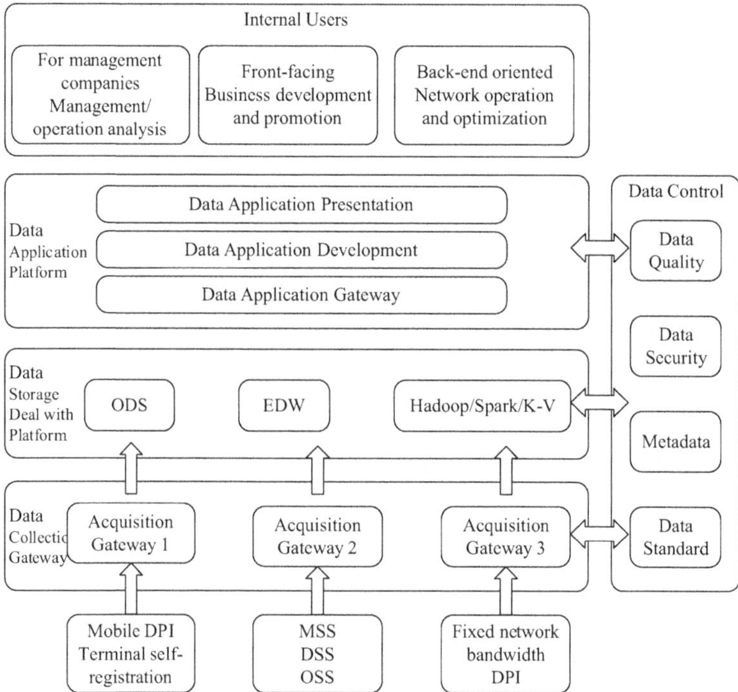

Fig. 3.19 A big data analysis framework of a telecom company in Shanghai

operators can improve operational efficiency, reduce operating costs, and find more business opportunities. Among them, the core structure of big data analysis of a telecom company in Shanghai is shown in Fig. 3.19 [14]:

The architecture is mainly composed of four modules: a data acquisition gateway, a data storage and processing platform, a data application platform, and a data management and control platform. The data collection gateway is mainly responsible for data collection, transmission, etc., using a distributed pre-deployment approach. The data storage and processing platform is responsible for integrating, correlating, and transmitting data, including mobile DPI and fixed-line broadband DPI. The data application platform is responsible for the management of encapsulating data and providing a unified data interface. The data management and control platform are responsible for the management and control of various subsystems, including data quality review and data security verification, to ensure the stability and efficiency of data operations.

In the aforementioned system, the static batch processing calculation mode is mainly adopted, wherein the data acquisition gateway initially performs data acquisition and integration. Subsequently, the data storage processing platform performs unified analysis and processing. The data storage processing platform is built on a Hadoop platform. The hardware configuration of the Hadoop platform includes 1 log collection server, 2 AAARadius collection servers, 4 fixed-line HTTPGet pre-collection servers, 4 fixed-line bidirectional DPI pre-collection servers, 11 fixed-line two-way DPI cleaning servers, 11 fixed-line HTTPGet cleaning servers, 93 Hadoop data node servers, and 2 Hadoop control node servers. Based on the above system, the data collection frequency can be maintained at 5 min/time, and the collection speed is about 500,000 records/sec. During the data aggregation period, each aggregation takes 4 h, the aggregation speed is 2.22 million items/sec, and the daily processing data volume is about 2 TB.

In the absence of a real-time service in the above application scenarios, the static batch computing mode can meet the above cases' data processing requirements. However, with the increasing amount of data and the expansion of business scope, the emergence of real-time requirements has become a general trend. MapReduce, however, has proven unable to meet new business requirements. Storm, a real-time stream computing, has gradually attracted the world's attention due to its excellent real-time stream computing capabilities. It has become the mainstream computing mode in real-time services.

3.4 Real-Time Stream Processing with Storm

3.4.1 Basic Knowledge

1. **Introduction to Flow Calculation**

In real-time applications, the generation of new data by businesses occurs continuously, resulting in a decline in the value of data decreases over time. It is necessary to process the data in real time to maximize its value. The offline batch processing mode described in the preceding section is no longer sufficient to meet the requirements of real-time processing. Stream computing enables the acquisition of valuable information in real time by analyzing a large number of data streams.

The characteristic of stream computing is that the data to be processed is not pre-stored in memory, but arrives in the form of a data stream, where some data is processed, rather than cached and processed in batches. In comparison to the offline batch model, the stream computing model exhibits the following characteristics: [15]

(1) The potential size of the data stream is infinite.
(2) The data stream elements arrive and are processed online instead of caching and batch processing.
(3) Once an element in the data stream is processed, it is either stored or discarded.

Based on the above characteristics, the flow calculation system should meet the following requirements:

(1) High performance: The system should be able to process large quantities of data.

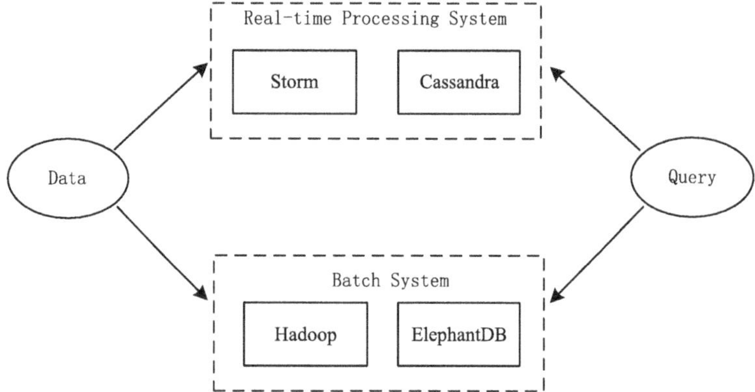

Fig. 3.20 Storm data processing architecture

(2) Real time: Fast processing speed, at least at second level.
(3) Distributed: The system should support the basic architecture of big data and have certain expansibility.
(4) Reliability: The system should be able to handle data streams reliably.
(5) Ease of use: The system should be designed in a way that allows developers to quickly use it for development work.

The flow calculation process is typically divided into three stages: real-time data collection, real-time data calculation, and real-time data query.

2. **Basic Concepts of Storm**

Storm is an open-source stream computing framework developed by Twitter in 2011. It is developed using Clojure and Java, which facilitate the handling of data flow in a straightforward and efficient manner. Besides, Twitter uses hierarchical data processing architecture to construct a batch processing system and a real-time system, respectively, as presented in Fig. 3.20.

In the hierarchical processing architecture, Twitter uses Storm and Cassandra (a hybrid non-relational database) to construct a real-time processing system, while Hadoop and Elephant are used to construct a batch processing system. The real-time system's processing results are delivered to the batch processing system for correction, thereby improving the reliability of the real-time processing system.

Storm framework exhibits the following features:

(1) Simple programming model: Storm provides multiple APIs to reduce real-time processing complexity.
(2) Support multiple languages: Clojure, Java, Python, and Ruby are supported by default, and other languages can also be supported by implementing a Storm communication protocol.
(3) High fault tolerance: Storm can automatically manage work processes and node failures.

(4) Strong reliability: Storm guarantees the complete processing of each message.
(5) Strong scalability: Storm computes in parallel across multiple threads and servers, so that it can run in a distributed cluster.
(6) Quick deployment: Storm is easy to install, configure, deploy, and use.

Based on the above characteristics, there are three main types of Storm application scenarios:

(1) Information flow processing: Storm can be used to analyze and process data in real time.
(2) Continuous calculation: Storm can perform a continuous query and feedback the customers' results in real time.
(3) Distributed remote procedure call: Storm can be used to process intensive queries in parallel.

3.4.2 Programming Model

Storm abstracts certain concepts. The following description of the Storm programming model provides a general overview of the tuple, message flow stream, message source spray, processor bolt, stream grouping, task, and strategy topology [16].

1. **Tuple**

Tuple is the basic data format utilized within the Storm framework. As the basic messaging unit within Storm tuples are used for packaging the actual data. Tuple can be regarded as an ordered list of named elements. Element types are arbitrary and compatible with an integer, byte, string, floating-point number, and other types. Tuple can also be understood as a key-value pair in MapReduce. In this context, the key refers to the tuple's name, while the value refers to the element of the tuple.

2. **Message Flow**

As shown in Fig. 3.21, in a Storm stream, the message flow represents an abstract stream of data. The tuples are generated and processed in parallel in a distributed fashion.

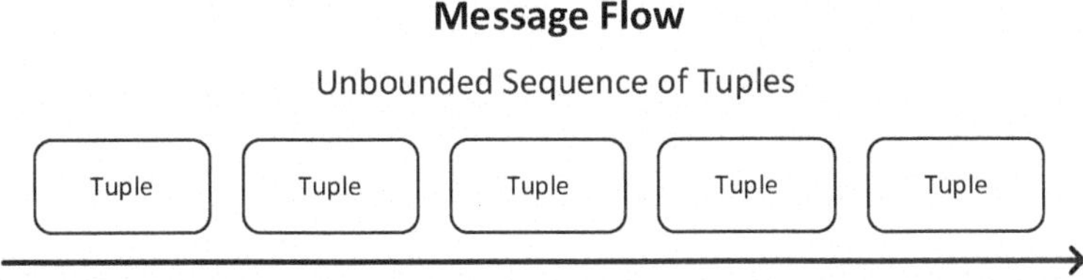

Fig. 3.21 Message flow

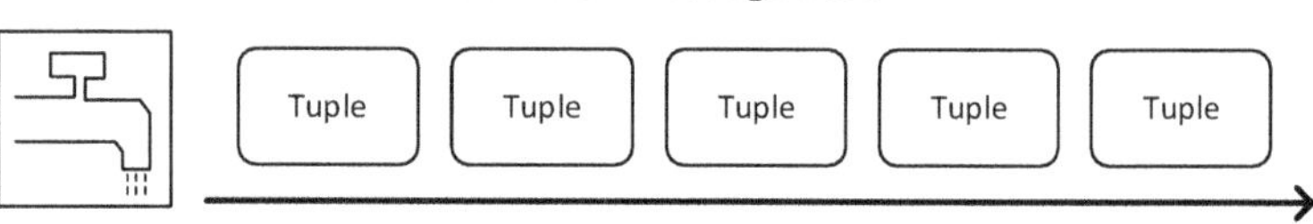

Fig. 3.22 Message source

3. **Message Source**

In Storm, the message source represents the source of data flow. Typically, the message source reads stream data from external data sources (such as message queues and databases) and encapsulates it into a meta-group form, subsequently sending it to the message flow. The message source provides an interface next tuple, in which the Storm framework will continuously call to read the stream data. The relationship between the message source and the message flow is shown in Fig. 3.22. Furthermore, the transmission mode of the message source is divided into two categories: reliable transmission and unreliable transmission. Reliable transmission mode calls ACK or fails function to inform users of the effectiveness of the transmission. In contrast, in the case of unreliable transmission mode, no error control mechanism exists.

4. **Processor**

As shown in Fig. 3.23, Bolt is the processor responsible for message flow, including function processing, filtering, flow merging, and local storage. A processor can receive and process an unlimited number of message flows, as well as outputting an unlimited number of message flows. Generally, a data processing task is divided into multiple processors, with each processor performing a portion of the data processing task. Concurrently, the parallel execution of multiple processors can improve the processing capacity and efficiency of the Storm framework, as well as the concurrency of the pipeline. When declaring a processor, it is necessary to specify which specific output streams the processor is to receive and process. The execute method represents the processor's most important interface, as it determines the processing logic and is written by the user. Upon receipt of the message, the processor will call this function, which implements user-defined processing logic.

5. **Stream Grouping**

Stream grouping is the process of transferring tuples between two components, such as a message source and processor, or between different processors, as shown in Fig. 3.24. It is similar to the routing function in a computer network, which determines the processing route of tasks. Stream grouping in Storm mainly includes the following methods:

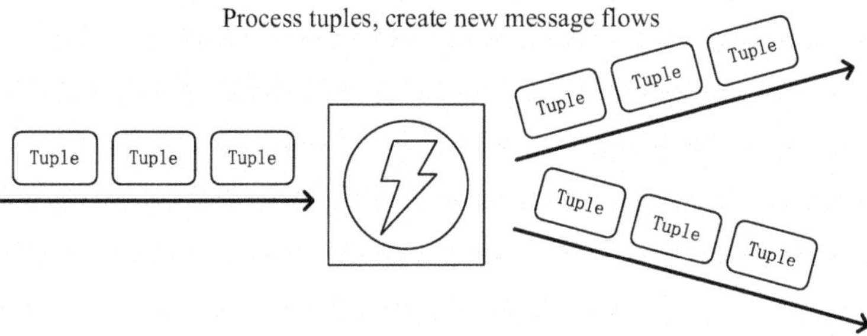

Fig. 3.23 Processor

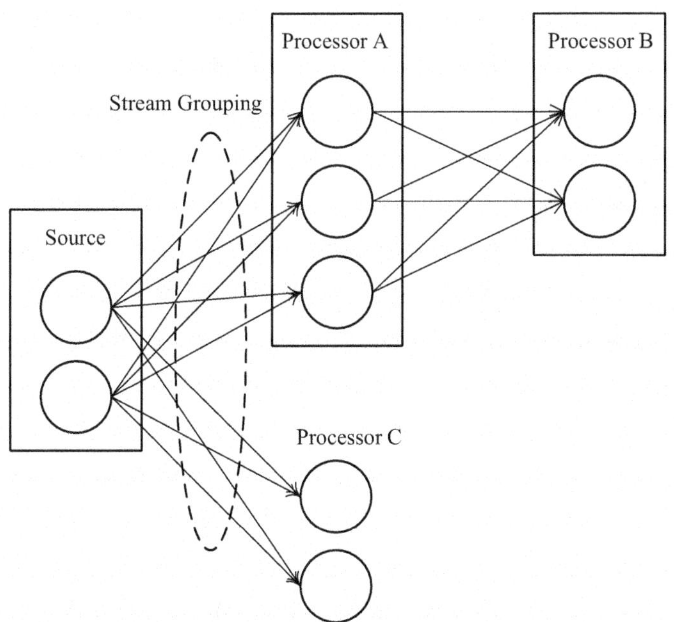

Fig. 3.24 Flow grouping

Shuffle grouping: Randomly grouping and randomly distributing tuples in the message flow to ensure that the number of tasks receiving each processor's tuples is approximately the same.

Fields grouping: Grouping by field to ensure that the same field's tuples are assigned to the same task.

All grouping: Broadcast packets, each tuple will be sent to all tasks.

Global grouping: All tuples are sent to the same task.

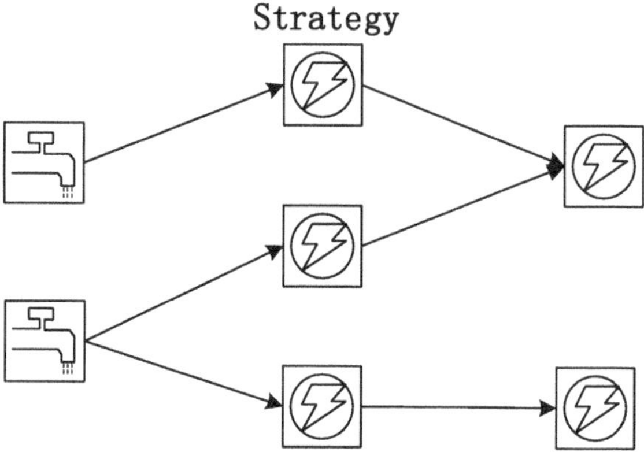

Fig. 3.25 Strategy

Non-grouping: Similar to shuffle grouping. The current task will be executed in the same thread as its subscribers.

Direct grouping: Directly specifies a task that performs tuple processing.

6. **Task**

Task is an execution unit in the Storm framework. Each message source and processor decompose the task into multiple tasks and runs them in parallel on the cluster. Each task corresponds to an execution thread. To manage the concurrency of each message source and processor, the Storm framework provides the setSpout and setBolt functions through TopologyBuilder API.

7. **Strategy**

Storm abstracts the network structure from the message source to the processor as strategy topology. The strategy is the processing logic of a Storm task. The Storm task is similar to the Job in MapReduce, except that the Job will end eventually, while the Storm task will continue to run until it is killed. Generally, an undirected loop graph is used to describe the strategy. As shown in Fig. 3.25, the undirected loop graph describes the topology formed by a stream grouping that connects the Spout and the processor. In an undirected loop graph, the lines between components indicate the direction of data flow, and each component runs in parallel.

Users are required to define the overall logic of the flow calculation task in the strategy, after which it is submitted to Storm for execution.

3.4.3 Architecture

Storm cluster adopts a master–slave operating mode. The Nimbus node is the master node, the supervisor node is the work node, and the worker is the work process. Each

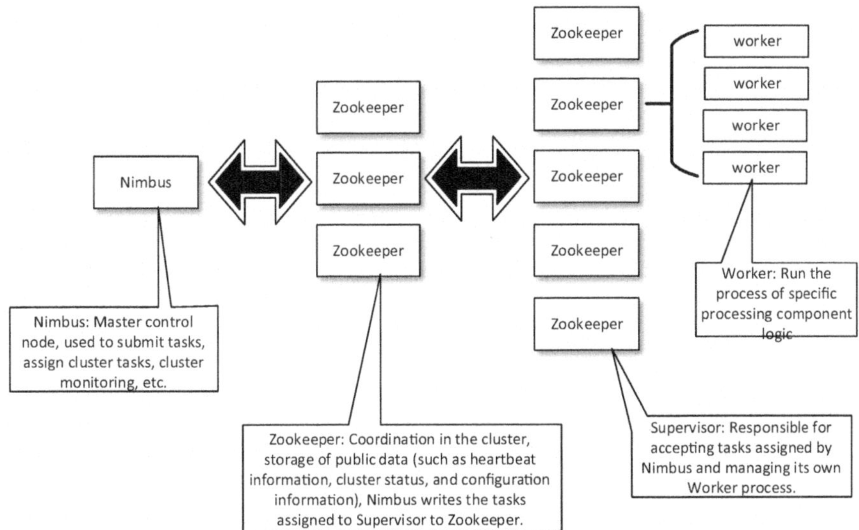

Fig. 3.26 Storm cluster architecture

Sput/Bolt thread in the worker process is called a task. The cluster architecture is presented in Fig. 3.26.

The Nimbus node is similar to the job node in Hadoop. It is responsible for distributing code within the cluster, assigning tasks to worker nodes, and monitoring for failures.

The supervisor node is responsible for monitoring the tasks assigned to its machine to start and stop the worker process according to the Nimbus's tasks. A supervisor node can run multiple worker processes simultaneously. The worker process executes the specific data processing logic, and a strategy can be executed in one or more workers. For example, if the strategy parallelism is set to 400 and executed with 80 worker processes, each worker processes five tasks.

Furthermore, Storm uses Zookeeper as a distributed coordination component, responsible for all coordination between Nimbus and multiple supervisors. Zookeeper does not engage in message transmission, rather, it provides a coordination service, and stores topology status and statistics. With the assistance of Zookeeper, the communication between the Nimbus node and supervisor node is separated, in which there is no direct communication between the two nodes. Moreover, the status information of the node is stored in the Zookeeper cluster. In the event of an unexpected termination of the Nimbus or supervisor process, the state information stored in the cluster can be used to restore the previous state and continue to work, which improves the stability of Storm (Fig. 3.27).

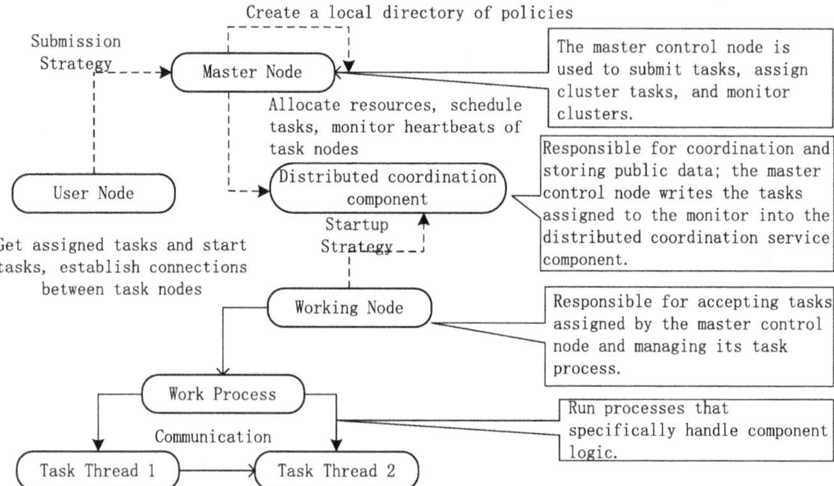

Fig. 3.27 Storm workflow

3.4.4 *Workflow*

Based on the above architecture design, storm workflow is shown in Fig. 3.28.

Firstly, the Storm client node submits a user-defined execution logic strategy to the master node for execution.

Secondly, the master node creates the local strategy directory and receives the strategy, divides it into multiple task nodes, and writes the task node's information and corresponding work node to the distributed cluster component.

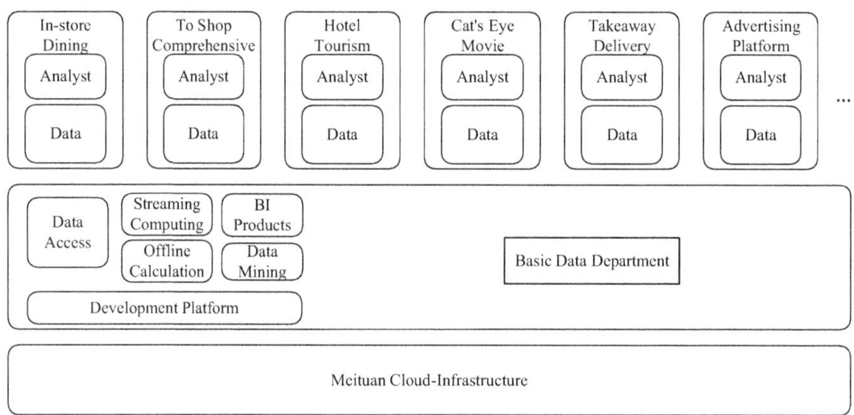

Fig. 3.28 Meituan big data analysis architecture

Finally, the work node claims its task node from the distributed cluster component cluster, and notifies its work process to execute the task thread. Moreover, the master node is engaged in a continuous monitoring process of the cluster's running status and the status of each task.

3.4.5 *Fault Tolerance*

The Storm fault tolerance mechanism includes architecture fault tolerance and data fault tolerance [17].

1. **Architecture Fault Tolerance**

The Nimbus and Supervisor processes are designed to terminate rapidly in the event of an anomalous occurrence. As the status information is stored on Zookeeper or the disk, the node's abnormality will not damage the status information, and the node can quickly restore to its previous operating state. In addition, in contrast to Hadoop, the Worker process in Storm will not be affected by the termination of Nimbus or Supervisor, while the termination of the job node in Hadoop will result in the forced termination of all tasks.

In the event that the Nimbus process terminates abnormally, if Nimbus is in process supervision mode, the process will be restarted without any adverse effects. In the event that the running Worker is unable to work, yet the Supervisor will always restart the current Worker because it cannot communicate with Nimbus. In other words, when Nimbus is in an abnormal state, the current topology can continue to operate as normal, but a new topology cannot be submitted.

Similarly, when the Supervisor process ends abnormally, and if it is in the process supervision mode, the process restarts without impact. If the aforementioned conditions are not met, the tasks assigned to this node will expire, and Nimbus will reassign these tasks to other nodes.

In the event that the Worker process terminates abnormally, the corresponding Supervisor will restart the Worker process. If multiple restarts fail, the Worker will not be able to maintain communication with Nimbus. Subsequently, following the expiration of the specified period, Nimbus will reassign the Worker to other nodes for execution.

2. **Data Fault Tolerance**

Each topology in Storm contains an Acker component. The task of this component is to track the processing of all tuples bound by each message ID that is transmitted from the Spout of a task. Suppose the maximum timeout time set by the user is timeout (we can use the Config.TOPOLOGY_ MESSAGE_ TIMEOUT_ SECS). If these tuples are not completely processed, Acker will call the Spout's fail method to indicate that the message processing has failed. Conversely, it will call the ACK method in Spout to inform Spout that the message has been successfully processed.

3.4.6 *Programming Example*

Realize a Topology, which includes a Spout and two Bolts. The Spout is responsible for the transmission of words. Bolt's function is to append a string "!!!" to the end of the input data. The data within the system passes through three nodes in turn. In other word, Spout sends data to the first Bolt. Subsequently, the first Bolt transmits the data to the second Bolt.

1. **Implementation Code**

```
(1) Topology part of the code:
TopologyBuilder builder = new TopologyBuilder();
builder.setSpout("words", new TestWordSpout(), 10);
builder.setBolt("exclaim1",new ExclamationBolt().
shuffleGrouping("words");
builder.setBolt("exclaim2", new ExclamationBolt().
shuffleGrouping("exclaim1");
(2) Spout part of the code:
public void nextTuple() {
Utils.sleep(100);
final Stringwords=new String{"nathan","mike","jackson","golda",
"bertels"
};
(3) Bolt part of the code:
public static class ExclamationBolt implements IRichBolt {
OutputCollector _collector;
public void prepare(Map conf, TopologyContext context, OutputCol-
lector collector) {
_collector = collector;
}
public void execute(Tuple tuple) {
_collector.emit(tuple, new Values(tuple.getString(0) + "!!!"));
_collector.ack(tuple);
}
public void cleanup() {}
public void declareOutputFields(OutputFieldsDeclarer declarer) {
declarer.declare(new Fields("word"));
}
}
The final result is: {"Nathan!!!","mike!!!","jackson!!!","golda!!!",
"bertels!!!"}.
```

2. **Code Analysis**

(1) Topology section:

First, create a Topology. Then set Spout and output as words. Finally, set up two Bolts, specify the data stream and output result of Bolt subscription. The data stream of the first Bolt subscription is words, and the output is exclaim1. The data stream of the second Bolt subscription is exclaim1, and the output result is exclaim2.

(2) Spout part:

Define the content of the data stream. Here, specify the data stream as an array of strings: {"nathan", "mike", "jackson", "golda", "bertels"}.

(3) Bolt part:

An ExclamationBolt is defined in Bolt, which implements the IRichBolt interface in the Storm framework. In the ExclamationBolt class, users need to implement four methods: prepare, execute, cleanup, and declareOutputFields. Among them, the execute method specifies Bolt's specific processing logic for the data. In this example, "!!!" is added to each data stream's tuple, and the ack method is used to ensure reliable data transmission.

3.4.7 Typical Cases

As a famous consumption platform, Meituan has many large flow-order businesses, including takeout, tourism, film, and advertising platform business. The comprehensive processing of such a diverse range of businesses requires a set of big data platforms that can handle high concurrency and large traffic volumes. This section takes Meituan big data analysis architecture as a case study to illustrate the application of real-time stream computing storm in big data analysis business [18].

Taking Meituan delivery and ordering business as an example, the overall data flow is considerable during the peak period of dining, which requires high data processing capacity and hardware load of the platform. Concurrently, it is necessary to build a set of multi-interface log collection platforms for the comprehensive management of its various platforms. The business characteristics determine that Meituan's big data platform must exhibit strong compatibility and the capability for high-level parallel processing and comprehensive data collection. Accordingly, based on application considerations, the functions of the Meituan big data platform can be broadly classified into three principal categories:

1. Streaming calculation of data.
2. Multi-business log collection for subsequent data mining and analysis.
3. Further data analysis and mining on the data collected by online businesses with the objective of providing feedback and recommendations for new businesses.

In terms of system architecture, Meituan's big data analysis architecture diagram is illustrated in Fig. 3.29.

The overall architecture is constructed based on the Meituan cloud, making an overall analysis of multiple services. Once the traffic data has been collected, it is then imported into the development platform. First, the flow calculation is carried out, and then the offline calculation is carried out. In the data collection phase, Meituan supports various data formats, and can receive millions of data per second during peak business hours.

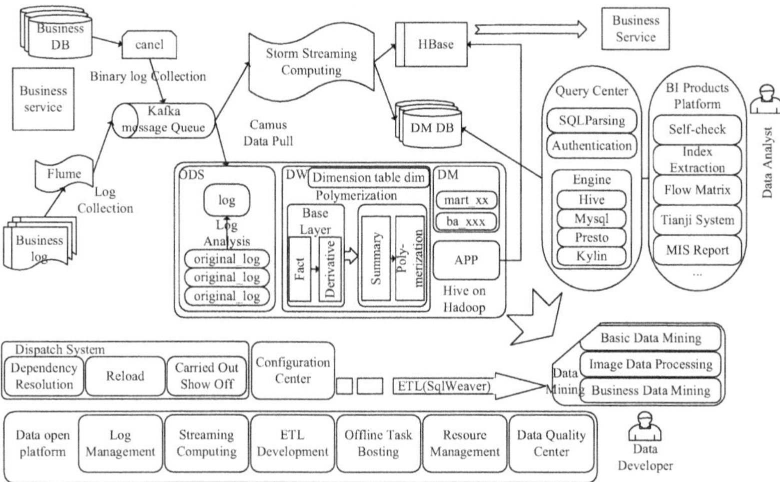

Fig. 3.29 Meituan data processing architecture

In regard to the calculation mode, Meituan adopts a combination of flow computing and offline computing. The data processing architecture is shown in Fig. 3.28.

The data is received in the form of the data stream, which is first calculated in real time and then calculated offline by Hadoop. Meituan divides the data into two categories: additional log data and relational data, which comes from business collection. Firstly, online business data is stored in the business database. Subsequently, the streaming data is transferred to the Kafka queue for processing. The combination of flow processing and batch processing enables Meituan to effectively support its extensive business scope and various application scenarios.

The comprehensive analysis conducted by the Meituan big data analysis platform supports the online business operations and independent business development of Meituan. Furthermore, the depth of data mining can provide a complete data mining platform in the era of successful data mining. The Storm computing framework provides Meituan with powerful real-time processing capacity, thereby ensuring the stable and efficient operation of the Meituan big data analysis platform.

3.5 Graph Computing with Pregel

3.5.1 *Basic Knowledge*

1. **Introduction to Graph Computing**

Graph computing is a computing mode for processing graph-structured data. The graph structure can effectively express the relationship between data, thus, many big data are presented in large-scale graphs or networks, such as social networks and traffic networks. These graphs are large in scale, often reaching billions of vertices and trillions of edges. Due to the large scale of data and the complexity of data relationships, traditional graph computing algorithms expose some problems:

(1) The poor locality of memory access.
(2) Less work is needed for a single vertex.
(3) The parallel degree of computation is often changed in the process of calculation.

In view of the above three problems, researchers have proposed several solutions, yet there are still some shortcomings:

(1) The disadvantage is that the generality is not good.
(2) The disadvantages of graph computing on the existing distributed computing platform are low performance and poor usability. For example, using MapReduce, a distributed batch processing system will significantly increase the calculation time and complexity.
(3) The disadvantage of using a single machine graph algorithm library is that it is constrained by significant limitations, which restrict the scale of the graph.
(4) Using existing parallel graph computing systems, such as Boost Graph Library (BGL), LEAD, and Stanford Graphbase, has the disadvantage of low fault tolerance.

Since the traditional graph computing schemes cannot adapt to large-scale data and complex data relationships, researchers have designed a graph computing software suitable for big data. These include Infinite Graph, DEX, and OrientDB, which use traversal algorithms, and Pregel, Hama, Giraph, which are based on message passing batch processing [19].

2. **Basic Concepts of Pregel**

Pregel is a BSP-based parallel graph processing system developed by Google. It is mainly used for graph traversal, shortest path, PageRank calculation, etc. Bulk Synchronous Parallel (BSP) is a whole synchronous parallel computing model, which is composed of a large number of interconnected processors. A BSP calculation is composed of a series of global supersteps, which is the iteration of calculation. A BSP calculation can be understood as a process of multiple iterations. Generally,

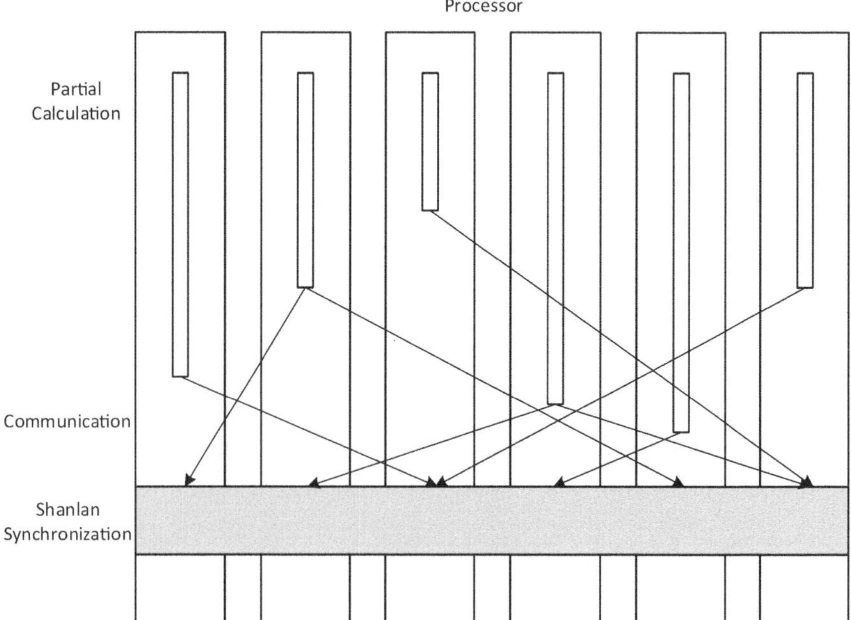

Fig. 3.30 Vertical structure

a superstep consists of three stages: local computing, message passing, and fence synchronization, as presented in Fig. 3.30.

(1) Local computing: Each processor reads data from local memory and performs its tasks. The computing tasks between different processors are asynchronous and independent.
(2) Message passing: Following the completion of the calculation by each processor, data is exchanged with the relevant processor.
(3) Fence synchronization: As a result of the differing processing tasks and speeds, the duration of each processor to complete the task varies. Hence, each processor requires synchronous completion at the end of each step. By setting the synchronization fence, each processor will be suspended until all have completed their calculation tasks. This ensures that all parallel computing is completed before the next superstep is executed to achieve processor synchronization. Each synchronization marks the end of one superstep and the beginning of the subsequent one.

3.5.2 Programming Model

The following describes the Pregel programming model through three aspects: vertices, message passing, and the calculation process [20].

Fig. 3.31 Vertex model

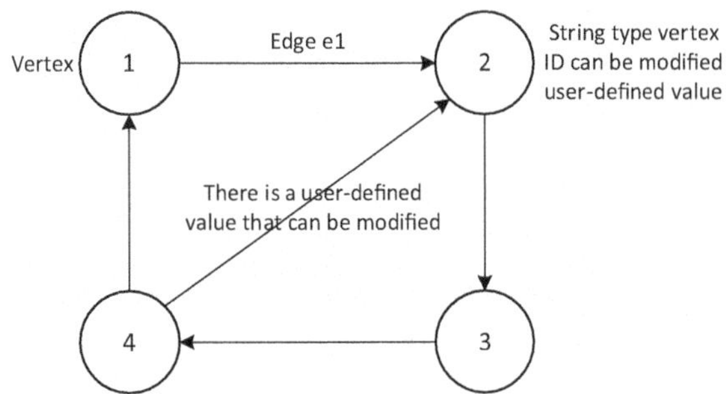

1. **Vertex**

A significant problem of the traditional graph computing model is that it lacks the ability to process vertices. The Pregel computing model has changed at this point. The model places a greater emphasis on the processing of vertices, with a corresponding increase in the processing of nodes. Pregel takes a directed graph as input. Each vertex of the directed graph contains two components, an ID of the String class and an arbitrary type of data. The ID is the unique identification of a vertex, while the data is a user-defined value that can be modified to participate in node operations. Each edge contains three components: the source vertex ID, the target vertex ID, and the user-defined data, as shown in Fig. 3.31. Each directed edge is associated with the source vertex, and the ID of the target vertex is recorded concurrently. Additionally, a modifiable user-defined value is presented on the edge.

In each superstep, all vertices in the graph will receive the information from the previous one, and execute the same user-defined function compute() in parallel. This function should specify the way the previous superstep's message is to be processed, including any modifications to its value and the state of the exit edge, as well as the transmission of messages to other vertices. Finally, the function indicates the state modification of this vertex, which will be explained in detail in the following calculation process.

2. **Message Passing**

Information exchange between Pregel vertices adopts a pure message passing model instead of remote data reading or shared memory. The message passing paradigm possesses expressive power, enabling the use of remote reading or shared memory. Furthermore, the asynchronous and batch message delivery of messages can mitigate latency and enhance overall system performance, as shown in Fig. 3.32.

3. **Calculation Process**

Pregel is a graph processing system based on BSP, so the calculation process of Pregel is also composed of multiple supersteps. In each superstep, the user-defined functions are executed for each vertex, and each vertex executes in parallel without interference.

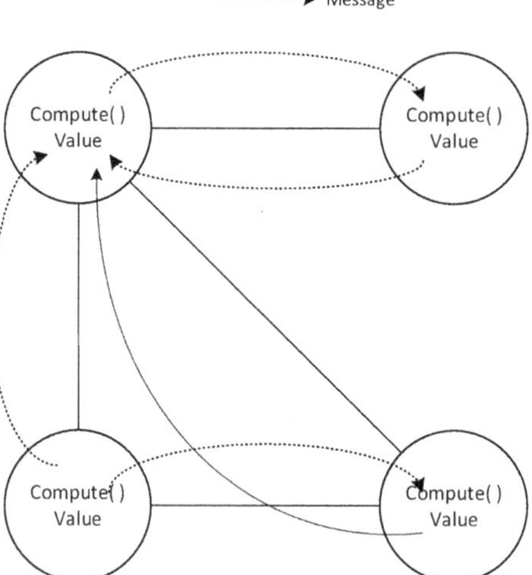

Fig. 3.32 Messaging model

The termination of the Pregel calculation process is related to the state of the vertex. Each vertex has two states, active and inactive. Active means that the vertex is in an active state, in which there are still calculations to be performed in the next superstep. In contrast, halt means that the fixed-point subject is inactive, indicating that a vertex no longer participates in the calculation process of the next superstep. The initial state of the vertex is active. Once all vertices have reached the halt state, the algorithm will terminate. A vertex in the halt state may still receive messages and be awakened, despite the fact that Pregel will no longer execute the vertex's function. In the custom function, the user must specify the circumstances under which the vertex performs a state transition. The state transition model is presented in Fig. 3.33.

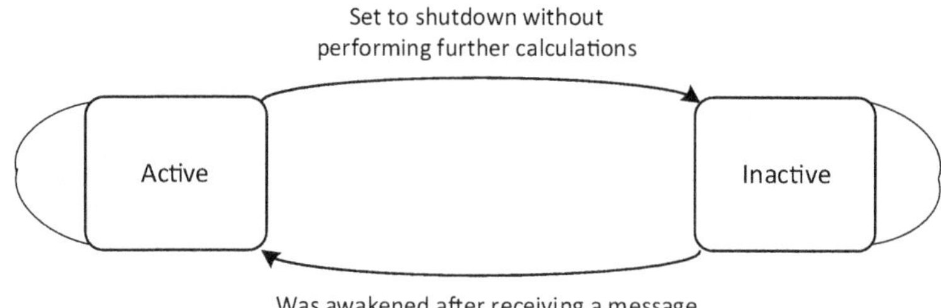

Fig. 3.33 State transition model

3.5.3 Architecture

Pregel also uses a master–slave structure. A graph computing task is divided into several subtasks, which are executed on multiple machines in parallel. One machine is the Master node, and the other machines are the sub-node Workers. Simultaneously, to facilitate task management, Pregel provides a name service system and identifies each task. [21]

1. **Name Service System**

Each task is given a logical name that is not related to its physical location. This will facilitate the identification of each task and the establishment of a connection between the Worker and the central node.

2. **Central Node Master**

Due to the considerable volume of graph data processed by Pregel and high complexity, the graph must be partitioned and divided into multiple subgraphs, as shown in Fig. 3.34. The central node is responsible for this task. The Master determines the subgraph to which a vertex is assigned based on its ID. The default rule is hash(ID) mod N. In this context, N is the total number of subgraphs, while ID is the vertex identifier. Furthermore, users are permitted to define the partitioning rules according to their specific requirements.

In addition, the Master is also responsible for user input division, task assignment, and the monitoring of Worker working status. To facilitate the Master's management, each Worker is required to register with the Master and obtain an ID. The Master is responsible for maintaining a Worker list, which contains the ID, address, and sub-map allocation of each Worker.

Like many other Masters with a master–slave structure, Pregel's Master does not directly process subgraphs. Instead, it monitors and coordinates Workers through a heartbeat mechanism. Upon receipt of an instruction from the Master, each Worker is required to respond within the specified timeframe. In the event that the Worker is overtime, the Worker will be considered invalid, and the task will be reassigned to other Workers. The specific steps will be explained in detail in the subsequent section on fault tolerance mechanisms.

3. **Child Node Worker**

A Worker is mainly responsible for the execution of computing tasks. The subgraph status information is recorded, including the current vertex value, the list of the exiting edge, the message queue, and the flag bit. In this context, since the current superstep is engaged in message processing, the vertex is also receiving messages from other vertices. Each vertex maintains two message queues, one to store information received in the previous superstep and the other to store information received in the current superstep. The former is used to message the current superstep, while the latter is delivered to the subsequent superstep. If the current superstep receives messages from other vertices, it can be inferred that there is a message to be processed

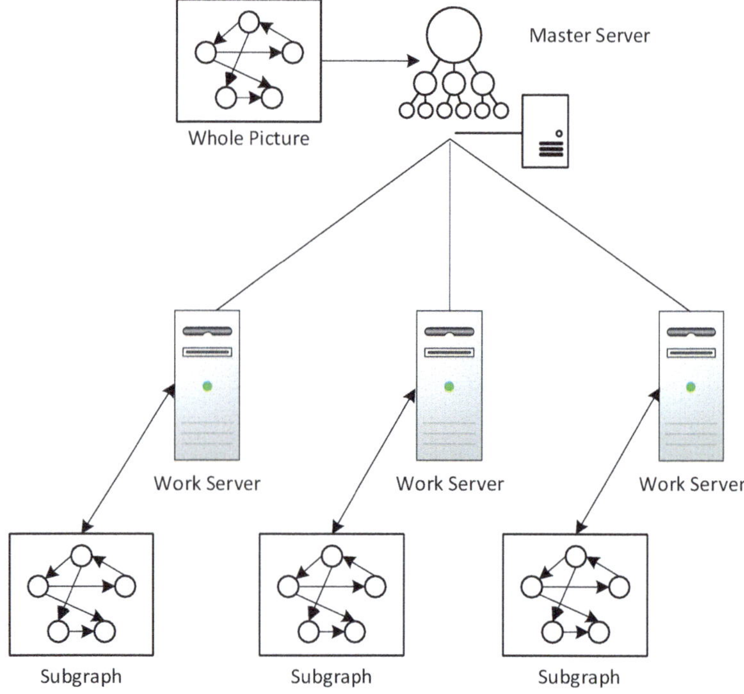

Fig. 3.34 Graph partition model

in the next superstep, in which the vertex is in the active state in the next superstep. Otherwise, it is set to the halt state (Fig. 3.35).

Worker uses the name service system to locate the location of the Master. The Master will assign an ID and a subgraph to each Worker. In each superstep, the Worker traverses each vertex in the subgraph under its jurisdiction and calls each vertex's Compute() function. When the Compute() function requests to send a message to other vertices, the Worker first confirms the location of the target vertex. If it belongs to the current subgraph under its jurisdiction, the Worker will write the message directly to the input message queue of the target vertex. In the event that the target vertex is in a remote area of the Worker, the Worker will first write the message to the cache. Once the cache has reached a pre-defined threshold, the message will be transmitted to the target vertex (Fig. 3.36).

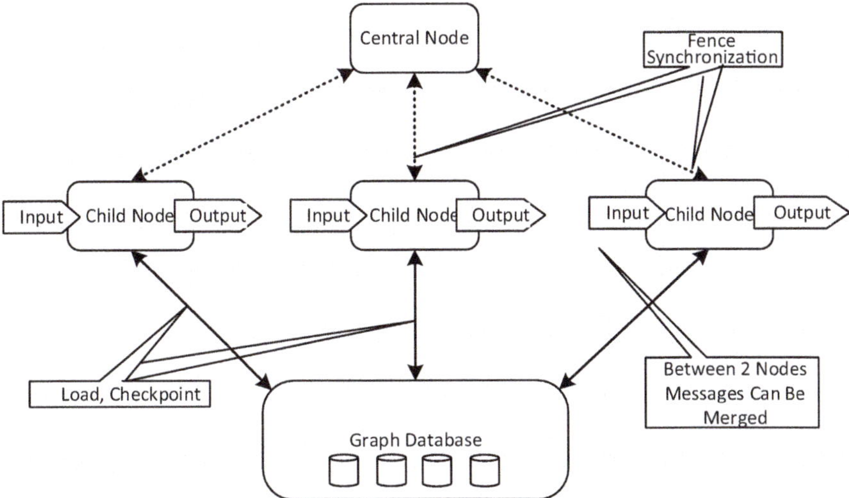

Fig. 3.35 Pregel workflow chart

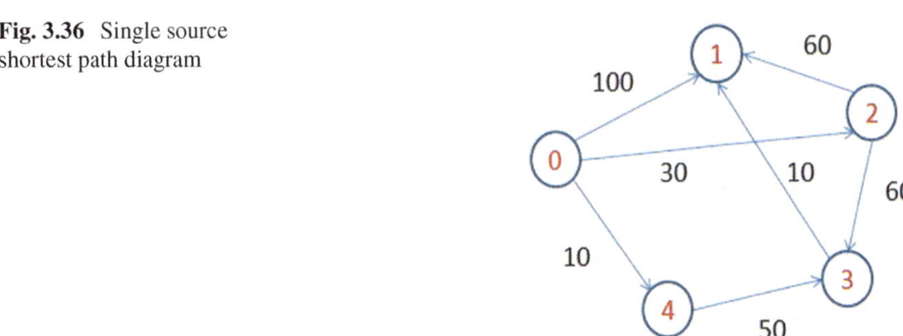

Fig. 3.36 Single source shortest path diagram

3.5.4 *Workflow*

The Pregel workflow can be divided into five stages, as displayed in Fig. 3.37.

1. **The First Stage**

Specify multiple machines in the server cluster to perform graph computing tasks, run a copy of the user program on each machine, and select one machine as the Master and the rest as Workers. The Worker locates the location of the Master through the name service system and sends registration information.

2. **The Second Stage**

The Master divides the graph into multiple subgraphs and assigns them to the registered Workers. Since the partitioning rules are identical, each Worker knows the distribution of the subgraph. Each Worker traverses its subgraph vertices in each

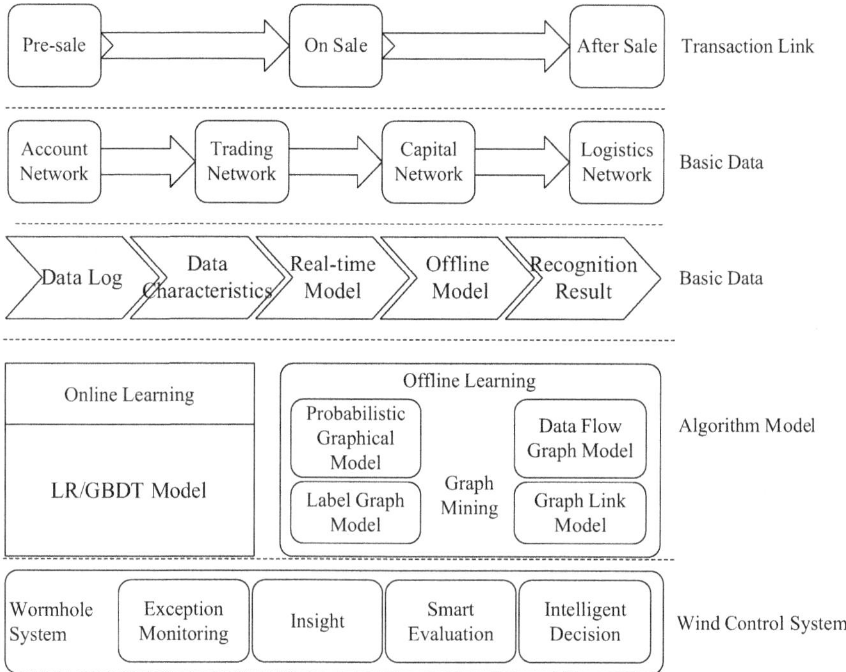

Fig. 3.37 Taobao anti-cheat system architecture

superstep, executes the Compute() function of the vertex, and sends and receives messages simultaneously.

3. **The Third Stage**

The Master divides user input into multiple parts, typically based on file boundary division. Each part contains multiple records, and each record contains a certain number of points and edges. Subsequently, the Master will assign the user's input information to the Worker. If the vertices that a Worker loads from the input content are the vertices in the subgraph it is assigned to, then the Worker will immediately update the corresponding data structure and mark the vertices as active. Otherwise, the Worker will send the information to the Worker responsible for the relevant subgraph based on the ID of the vertex loaded. Once all inputs have been successfully loaded, all the vertices in the graph are marked as active.

4. **The Fourth Stage**

The Master sends instructions to each Worker, after which the Worker starts to run an superstep. The Worker will then assign a thread to each subgraph under its jurisdiction. For each vertex in the subgraph, the Worker will send the message sent to that vertex in the previous superstep to the vertex, and call on the vertex in the active state. The Compute() function performs data operations. Once the above work has been completed, the Worker will notify the Master of the task's completion and report the number of vertices in the active state in the next superstep to the Master. Repeat the above steps until all vertices are in the halt state, no messages are being transmitted within the system, and the calculation process has been completed.

5. **The Fifth Stage**

Once the calculation process has been completed, the Master will send instructions to all Workers, requesting that they store their calculation results persistently.

3.5.5 Fault Tolerance Mechanism

Pregel uses the checkpoint mechanism of backward recovery to achieve fault tolerance [22]. At the beginning of each superstep, the Master will notify each Worker to checkpoint the subgraph related information (for example, the vertex value, edge value, and receipt of messages within the subgraph), and subsequently write this to a reliable storage device for fault-tolerant recovery. The Master will periodically send messages to the Worker, and the Worker is required to provide feedback within a specified timeframe. In the event that the specified time limit is exceeded, the Worker is considered invalid. Concurrently, if the Worker fails to receive instructions from the Master within a specified timeframe, the Worker will also stop working.

Should a worker fail, the subgraph's status information under its jurisdiction may be damaged or lost. The Master will reassign the failed Worker's tasks to other Workers and simultaneously instruct all Workers to restore their state from the latest checkpoint to recover. Pregel achieves fault tolerance through the mechanism of state regression.

Although the checkpoint mechanism is reliable, it has a significant execution overhead and delay. In order to improve the fault tolerance mechanism, researchers are developing Confine Recovery. The principle of Confine Recovery is to first recover through checkpoints, and then only to recalculate the missing subgraphs. The system will check the status information of the workers that are working normally and then compare this status information with that contained in the checkpoint. If the information is identical, the status will not be regressed. Confine Recovery is designed with the intention of reducing the computational burden associated with state recovery. Moreover, the ability of each worker to restore to more recent state information reduces the delay caused by failure significantly.

3.5.6 Programming Example

This section first introduces Pregel's API, and then explains how Pregel can be applied to the shortest path problem to help readers better understand the Pregel programming model.

1. **Pregel API**

A base class Vertex is pre-defined in Pregel:

```
template <typename VertexValue, typename EdgeValue, typename
MessageValue>
class Vertex {
public:
virtual void Compute(MessageIterator* msgs) = 0;
const string& vertex_id() const;
int64 superstep() const;
const VertexValue& GetValue();
VertexValue* MutableValue();
OutEdgeIterator GetOutEdgeIterator();
void SendMessageTo(const string& dest_vertex,const MessageValue&
message);
void VoteToHalt();
};
```

In the Vertex class, three value type parameters are defined: the vertex value (VertexValue), the edge value (EdgeValue), and the message value (MessageValue). In order to write Pregel programs, it is necessary for users to inherit the Vertex class and overload the virtual function Compute(). Additionally, they must write custom processing logic. Besides, the Vertex class also provides numerous interfaces. The GetValue() method is used to get the value of the current vertex, the MutableValue() method is used to modify the value of the current vertex, the VoteToHalt() method is used to change the state of the vertex, and the SendMessageTo() method is used to send a message to the specified vertex. Additionally, the Vertex class provides the OutEdgeIterator of the outgoing edge, through which users can modify the value of the outgoing edge through this iterator.

2. **Single Source Shortest Path Problem**

In the case of the single-source shortest path problem, for example, the Pregel model can effectively simplify the problem Table 3.1. The directed graph shown in Fig. 3.36 and the following uses the API provided by Pregel to address the single-source shortest path problem Table 3.2.

The initial value of the vertex is INF, and the initial state is active. In superstep 0, vertex 0 sends 100 to vertex 1, 30 to vertex 2, and 10 to vertex 4, the value of vertex 0 becomes 0, and superstep 0 ends.

Table 3.1 Vertex value and vertex state before the start of superstep 0

Vertex	Value	Status	Vertex	Value	Status
0	INF	Active	3	INF	Active Active
1	INF	Active	4	INF	Active Active
2	INF	Active			Active Active

Table 3.2 Vertex value and state at the end of superstep 0

Vertex	Value	Status	Vertex	Value	Status
0	0	Active	3	INF	Active
1	INF	Active	4	INF	Active
2	INF	Active			

In superstep 1, each vertex processes the message received in superstep 0 and updates the vertex value. Among them, vertices 1, 2, and 4 received messages in superstep 0, so thus remaining active in superstep 1. In contrast, vertices 0 and 3 had not received messages and were placed in the halt state. Concurrently, vertex 2 sends 60 to vertex 1 and 3, while vertex 4 sends 50 to vertex 3. At the end of superstep 1, the values and states of each vertex are shown in Table 3.3.

In superstep 2, each vertex processes the message received in superstep 1 and updates the vertex value. Among them, vertices 1 and 3 received messages in superstep 1, so the state is active in superstep 2. The remaining vertices have not received messages and are therefore set to the halt state. Concurrently, vertex 3 sends 10 to vertex 1. At the end of superstep 2, the values and states of each vertex are shown in Table 3.4.

In superstep 3, each vertex processes the message received in superstep 2 and updates the vertex value. Among them, vertex 1 received the message in superstep 2, so the state is active in superstep 3. The remaining vertices have not received the message and are therefore set to the halt state. No messages are transmitted to Superstep 3. At the end of superstep 2, the value and state of each vertex are presented in Table 3.5.

In superstep 4, each vertex processes the message received in superstep 3 and updates the vertex value. Since there is no message transmission in Superstep 3, all vertices in Superstep 3 are in the halt state. At this time, the calculation task ends. At the end of superstep 4, the values and states of each vertex are shown in Table 3.6.

Table 3.3 Vertex value and state at the end of Superstep 1

Vertex	Value	Status	Vertex	Value	Status
0	0	Halt	3	INF	Halt
1	100	Active	4	10	Active
2	30	Active			

Table 3.4 Vertex value and state at the end of Superstep 2

Vertex	Value	Status	Vertex	Value	Status
0	0	Halt	3	60	Active
1	90	Active	4	10	Halt
2	30	Halt			

Table. 3.5 Vertex value and state at the end of Superstep 3

Vertex	Value	Status	Vertex	Value	Status
0	0	Halt	3	60	Halt
1	70	Active	4	10	Halt
2	30	Halt			

Table. 3.6 Vertex value and state at the end of Superstep 4

Vertex	Value	Status	Vertex	Value	Status
0	0	Halt	3	60	Halt
1	70	Halt	4	10	Halt
2	30	Halt			

At this point, the single-source shortest path problem is solved, and the distance between each vertex and vertex 0 is shown in Table 3.6. The implementation code in Pregel is as follows:

```
class ShortestPathVertex:
public Vertex<int, int, int> {
void Compute(MessageIterator* msgs) {
int mindist = IsSource(vertex_id()) ? 0: INF;
for (; !msgs->Done(); msgs->Next())
mindist = min(mindist, msgs->Value());
if (mindist < GetValue()) {
*MutableValue() = mindist;
OutEdgeIterator iter = GetOutEdgeIterator();
for (; !iter.Done(); iter.Next())
SendMessageTo(iter.Target(),mindist + iter.GetValue());
}
VoteToHalt();
}
};
```

3.5.7 *Typical Case*

The Taobao anti-cheating system is a comprehensive set of monitoring and early warning, online analysis, and risk operation systems based on the data layer. It is designed to rapidly and efficiently observe the whereabouts of orders, block their profit points in a timely manner, and maintain the trading platform's fairness. The system includes four networks, including the account network, the transaction network, the capital network, and the logistics network[23]. Taobao combined multiple computing frameworks and computing models to form a complete and efficient anti-cheating system based on the four networks. This section takes Taobao's anti-cheat system as a case study to illustrate the typical application of graph computing in big data analysis. The overall architecture of Taobao's anti-cheat system is displayed in Fig. 3.37.

The anti-cheating algorithm framework integrates the big data from four networks: the account network, the transaction network, the capital network, and logistics network. The system covers multiple business links in the e-commerce: before shopping, during shopping, and after shopping. The system is developed based on a streaming computing framework, combined with a graph computing model. The data log is processed into some transaction attribute features based on a recognition algorithm, which is applied to two major computing modules: real time and offline. Real-time computing is designed to analyze and convert abnormal online data, such as abnormal product sales or abnormal seller reputation growth, which is quickly analyzed and converted into corresponding features. In contrast, offline computing is the feature processing and processing of full-link data. The combination of online and offline computing allows for the integration of long-term and short-term behavioral changes. The influence of short-term factors is comprehensively considered in the model calculation to further improve the timeliness and accuracy of recognition.

The system mainly applies two algorithms: large-scale graph mining and online learning. Large-scale graph mining enables the identification of cheating behavior from a global perspective, transcending the limitations of locality. The application of large-scale graph mining technology enables the extraction of common cheating behaviors in practical applications, as presented in Fig. 3.38.

Due to the complexity of the graph structure, typical batch processing and stream computing techniques are unable to efficiently process graph structure data. Consequently, in the Taobao anti-cheat system, the implementation of large-scale graph mining algorithms is primarily achieved through graph computing, which represents one of the most prevalent application scenarios for graph computing. Moreover, graph computing has many applications in other industries, including the financial and the Internet industries. In the financial industry, the complexity of the financial entity model often includes the equity relationship between enterprises and the family relationship between individual customers. Traditional graph-based cognitive analysis struggles with large-scale graphs, whereas graph computing compensates for the deficiencies of conventional analytical methods. It can mine the hidden information

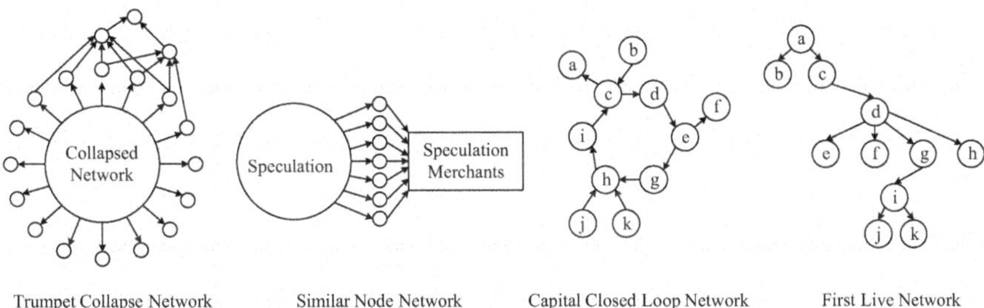

Fig. 3.38 Common cheating network diagram

of financial entity relationships, including financial risk control, customer marketing expansion, internal audit supervision, investment, and financial management. With the diversification of data in the Internet industry, there is a substantial increase in data volume and the breakthroughs in computing capabilities. Hyper-scale graph computing is playing an increasingly important role in big data companies, especially in the domains of deep learning and graph computing. The large-scale graphs combined with calculation are represented as a series of algorithms. Major companies are continuing to improve their graph computing capabilities, developing their graph computing platforms, such as Google Pregel, Facebook Giraph, Tencent Star Atlas, and Huawei's graph engine service GES.

3.6 Data Visualization

3.6.1 Introduction to Visualization

1. **The Concept of Visualization**

Visualization refers to representing the data in a large dataset in the form of graphics and images, as well as the use of data analysis and development tools to discover unknown information. Data visualization involves many fields, including graphics, image processing, computer vision, and computer-aided design. It has emerged as a comprehensive technology for studying various problems, including data representation, data processing, and decision analysis. The fundamental concept underlying data visualization technology is the representation of each data item in the database as a single graph element. A large number of datasets form a data image. Concurrently, each attribute value of the data is represented in the form of multidimensional data, which enables in-depth observation and analysis of the data from different dimensions [24].

2. **The Development of Visualization**

The concept of visualization can be traced back to the sixteenth century, when cartographers began creating maps to explore unknown areas. In the first half of the nineteenth century, there was considerable growth in the development of statistical and thematic graphics. Almost all known statistical graphics with properties were invented during this period. Until the advent of computers and the subsequent development of computer graphics in the 1950s, the use of computer technology to draw various graphs and charts on the computer screen was limited. However, the advent of visualization technology opened a new stage of development in this field. The advent of big data has led to the generation of vast quantities of data on a continuous basis. It is essential that data be analyzed in a timely, comprehensive, rapid, and accurate manner in order to present the data's value. The utilization of visualization technology is essential for the effective analysis of data. Visualization serves as the final link in the process of big data analysis, and it is of paramount importance to users.

3. **Visual Classification**

The field of visualization is often divided into four categories: data visualization, scientific computing visualization, information visualization, and knowledge visualization.

(1) Data visualization:

Data visualization technology is used to describe the process of converting data into graphics or images that can be displayed on the screen. This is achieved through the use of computer graphics and image processing technology, in conjunction with interactive processing theories, methods, and technologies.

(2) Scientific computing visualization:

The field of visualization of scientific computing refers to the theory, method, and technology of using computer graphics and image processing technology to convert data generated through engineering measurements and scientific computing, as well as calculation outcomes, into graphic images that can be displayed on a computer screen. This process is often accompanied by interactive processing.

(3) Information visualization:

Information visualization is the use of a computer-supported, interactive, and visual representation of abstract data to enhance the cognitive processing of abstract information. Information visualization is a process of extracting, transforming, mapping, highly abstracting, integrating the feature values of non-spatial data information objects, and expressing the content features and semantics of the information objects through graphics and images, and animation. Information objects include text, images, video, voice, and other types. The visualization of these objects is achieved through the use of diverse model methods Table 3.7.

(4) Knowledge visualization:

Knowledge visualization is a relatively new research area that has developed based on scientific computing, data, and information visualizations. The field uses visual representation methods to promote the dissemination and innovation of group knowledge.

Knowledge visualization studies visual representations in improving knowledge dissemination and innovation between two or more individuals. In this sense, knowledge visualization refers to all graphical methods that can be used to construct and convey complex knowledge. In addition to the transmission of factual information, the objective of knowledge visualization is the transfer of insights, experiences, attitudes, values, expectations, opinions, and predictions, facilitating the accurate reconstruction, recollection, and application of this knowledge by others.

Furthermore, visualization and data mining are mutually reinforcing. Both are aimed at deriving information from data, yet employ different methodologies. Data

Table. 3.7 Comparison of visualization types

	Data visualization	Scientific computing visualization	Information visualization	Knowledge visualization
Visualization object	Spatial numeric data	Spatial data	Non-spatial non-numeric	Human knowledge
Visualization purpose	Express abstract data in an intuitive way	Express abstract data in a three-dimensional and intuitive way	Discover some new information from a large amount of abstract data	Promote the dissemination and innovation of group knowledge
Visualization technology	Parallel coordinate graph, scatter graph, etc.	Volume and surface rendering	Outline drawing, cone tree	Knowledge graph
Interaction type	Human–computer interaction	Human–computer interaction	Human–computer interaction	Human–computer interaction

mining discovers hidden information in data through algorithms, and data visualization allows users to discover valuable information through intuitive graphics and images. It can be understood that data mining and data visualization are two distinct approaches to data processing and analysis. Data mining is a purposeful exploration of the correlation between data and other information, while data visualization does not make any pre-assumptions about the data model. It is, therefore, an exploratory analysis of data.

3.6.2 *Visualization Method*

The five most common methods of visualization are concept maps, mind maps, cognitive maps, semantic networks, and thinking maps [25].

1. **Concept Map**

The concept map is a graphical method that uses nodes to represent concepts and links to represent relationships between concepts. It was initially proposed by Dr. Novak from Cornell University and others, based on David Ausubel's meaningful learning theory. It serves as a means to organize and represent knowledge. The tools are depicted in Fig. 3.39.

The process of constructing the concept map is as follows: first, the relevant concept of a certain topic is placed in the circle and the box, and then the connection is used to link the related concept and the proposition. The connection shows the meaning relationship between the two concepts.

The most significant advantage of this kind of knowledge visualization method is that it allows for the clear expression of the knowledge system structure (concepts and

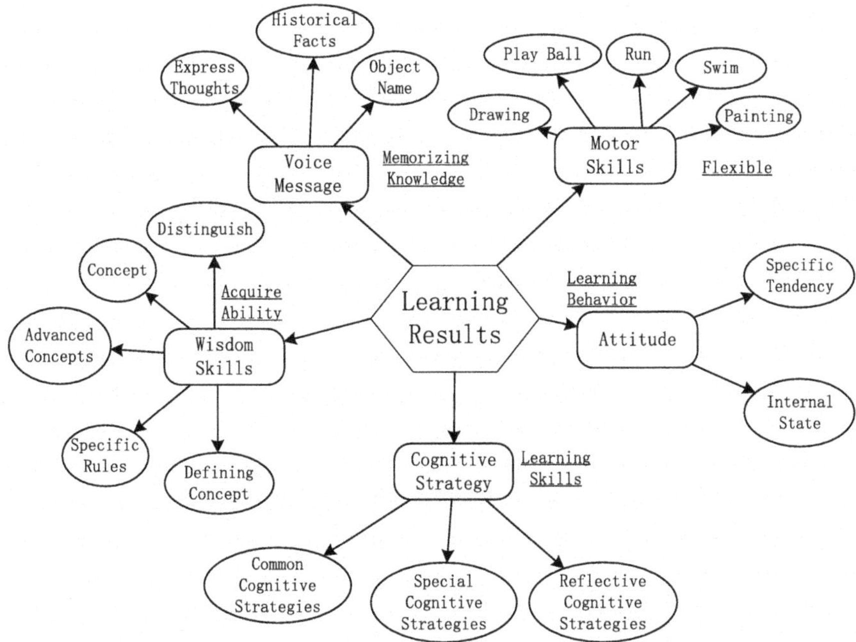

Fig. 3.39 Concept map example

the relationship between the concepts) while highlighting the knowledge system's hierarchical structure.

2. **Mind Map**

The mind mapping is a method of visualizing thinking that employs a graphical representation of interconnected words, ideas, tasks, or other related items in a radial line through a central keyword or idea. The technique was developed by British researchers in the 1960s. Tony Buzan was inspired by the concept of radioactive thinking as it relates to the human brain. Radioactive thinking represents the natural method of thought in the human brain. Any data that enters the brain can become a thinking center, and numerous additional centers can be generated from the original center. The theme diverges outward once more, exhibiting a three-dimensional radioactive structure that can be conceptualized as a mind map, as illustrated in Fig. 3.40.

Tony Buzan believes that traditional methods of drafting and note-taking have four major disadvantages: the obfuscation of keywords, difficult to remember, time-consuming, and fail to effectively stimulate the brain. The ability to take concise,

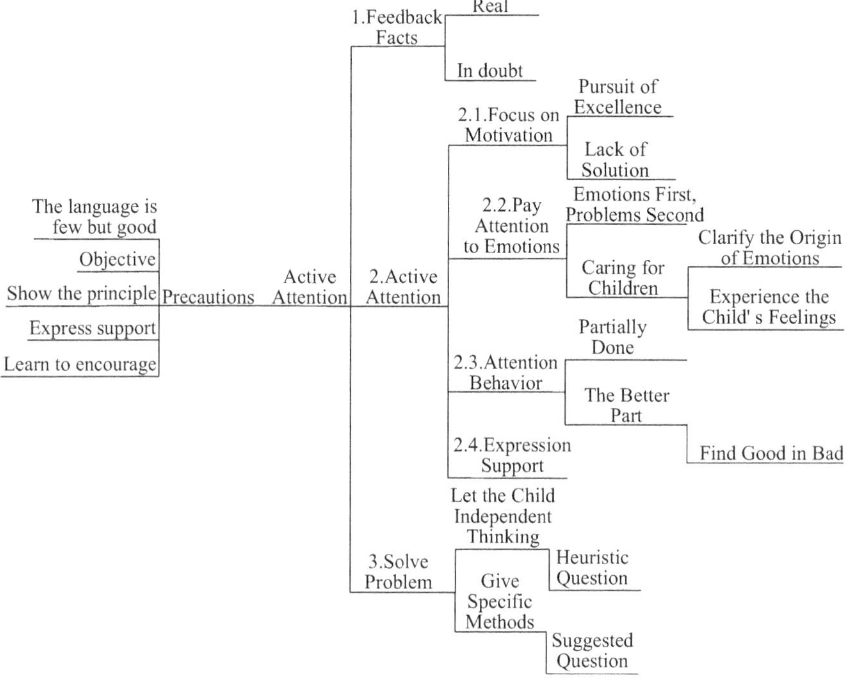

Fig. 3.40 Mind map example

efficient, and active notes is crucial for effective note-taking. Mind maps have gradually replaced draft notes due to their characteristics of increasing returns, and have become a new way of recording and thinking.

3. **Cognitive Map**

The cognitive map, also known as cause-effect map, is a model based on experience that represents "ideas" as nodes, and connects them to form a map-like model. It is mainly used to help people plan their work and promote group decision-making.

Ackerman and Eden, the authors of cognitive maps, believe that "ideas" are distinct from concepts, which are usually expressed as sentences or paragraphs. The cognitive map is based on the theory of individual construction. The ideas in it are connected by connecting lines with arrows, but there are no connectives on the connections. The implicit meaning of the connecting lines is "causality" or "causation." Furthermore, there is no level restriction.

4. **Semantic Map**

The semantic map, also known as semantic network, is a directed graph that represents the structure of human knowledge structure in a network format. It is used to express the semantic similarity or degree of relevance of words or concepts. The

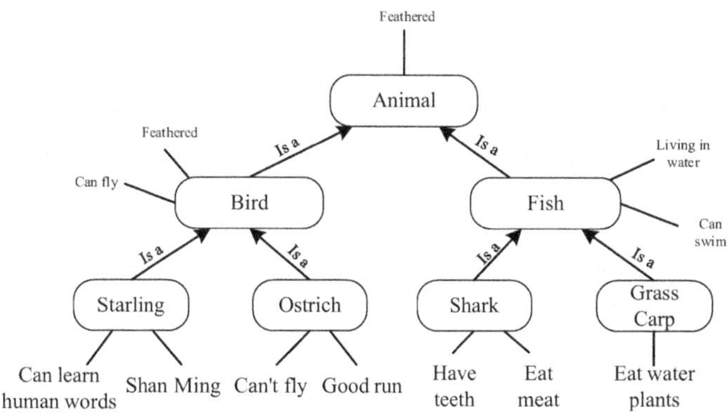

Fig. 3.41 Semantic network example

semantic network is composed of nodes and arcs between nodes. Nodes represent concepts (events, things), and arcs represent the relationship between them, as shown in Fig. 3.41.

A distinguishing feature of the semantic network is that it can express knowledge at a deep level, including entity structure, hierarchy, and causal relationships between entities. Therefore, the semantic network can directly and clearly express the semantic relationships between concepts, thereby simulating human semantic memory, and association methods. In addition, the semantic network can be used for the efficient retrieval and reasoning of information. However, the semantic network is not suitable for quantitative or dynamic knowledge, nor can it express procedural or controlled knowledge.

5. **Thinking Map**

The thinking map is a language to help learning developed by Dr. David N. Hyerle. It is designed to assist students in improving their reading comprehension, problem-solving in the writing process, and thinking skills. The language employs a total of eight pictures, which are based on basic cognitive processes, including comparison and contrast, parallelism, classification, and causal reasoning. As depicted in Fig. 3.42, the eight types of diagrams are as follows: bracket diagram, bridge diagram, bubble diagram, circle diagram, double bubble diagram, flow chart, complex flow chart, and tree diagram.

The thinking map is a brand-new thinking mode that combines the whole brain's concept, including the logic, order, regulations, text, numbers of the left brain, the image, imagination, color, and space of the right brain. It can greatly increase reading speed, memory, organization, and creativity. In addition, the thinking map is a simple and effective instrument that is widely used in both academic and professional settings, improving learning and work efficiency, and expeditiously resolving occupational challenges.

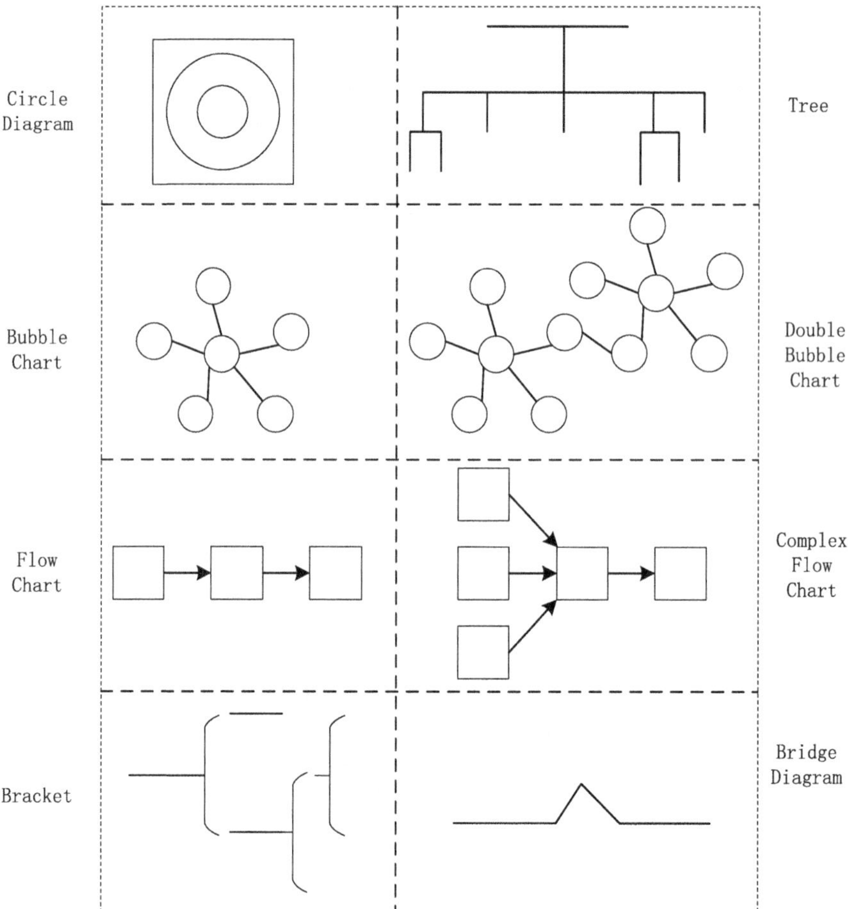

Fig. 3.42 Thinking map example

3.6.3 Visualization Technology

Visualization technology is typically classified into four categories: geometry-based technology, pixel-oriented technology, icon-based technology, and layer-based visualization technology [26].

1. **Geometry-Based Technology**

Geometry-based technology refers to the use of geometric drawing or geometric projection to represent data within the database. It is the most commonly used technology, such as the parallel coordinate method.

The parallel coordinate method is one of the earliest visualization techniques to represent multi-dimensional data in a two-dimensional form. The basic idea is to map the n-dimensional data attribute space to a two-dimensional plane through n equidistant parallel axes. Each axis represents one attribute dimension, with the value range of the axis uniformly distributed from the minimum to the maximum of the corresponding attribute. Consequently, each data item can be represented by a polyline segment on n parallel axes according to its attribute value.

2. **Pixel-Oriented Technology**

The pixel-oriented technology was proposed by Daniel K of the University of Munich, Germany, who developed the VisDB visualization system. The basic concept of pixel-oriented technology is to associate the data value of each data item with a colored screen pixel, and to represent distinct data attributes in distinct windows. One of the key advantages of pixel-oriented technology is its ability to display a comprehensive range of relevant data items on the screen.

3. **Icon-Based Technology**

The fundamental concept of icon-based technology is to represent n-dimensional data attributes with various simple icons. This technique is applicable to certain datasets with n-dimensional values with good expansion properties on a two-dimensional plane. The dendrogram method represents a typical example of icon-based technology. As shown in Fig. 3.43, the dendrogram method first selects two attributes of the multi-dimensional data set as the basic X–Y plane axes. Subsequently, it uses the difference in the length or angle of the twigs to show the changes of other attribute values on this plane.

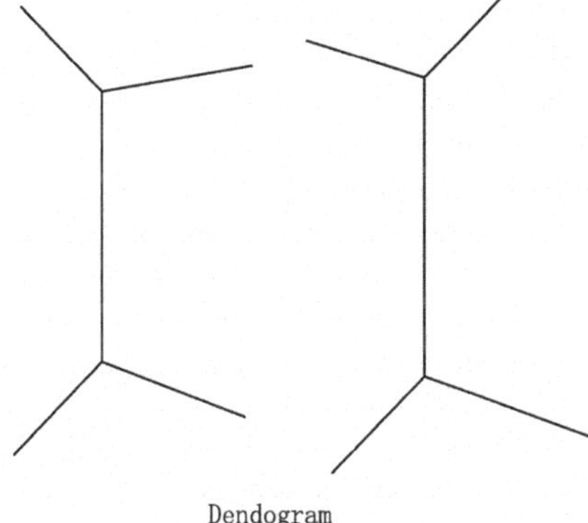

Fig. 3.43 Icon-based visualization technology

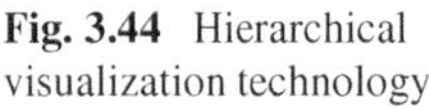

Fig. 3.44 Hierarchical visualization technology

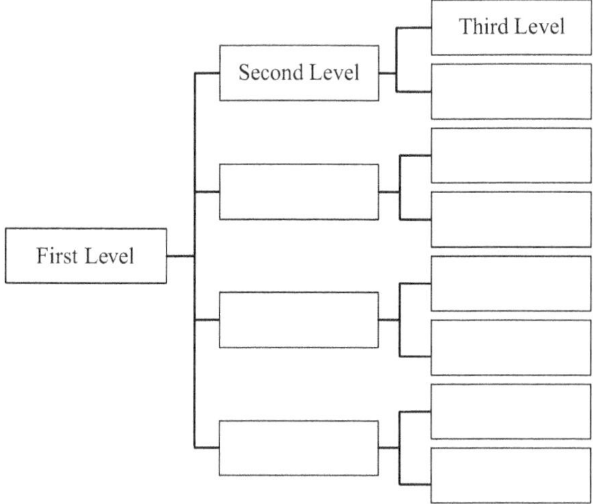

4. **Layer-Based Technology**

The objective of the level-based visualization technology is to facilitate the analysis of data with a hierarchical structure in the database system, including personnel organization, file catalog, and population survey data. Its fundamental concept is to divide the n-dimensional data space into multiple subspaces, which retain their hierarchical structure and are represented graphically, as shown in Fig. 3.44.

The tree diagram adopts a typical hierarchical visualization technology. It divides the screen into multiple rectangular subspaces according to the data's hierarchical structure, with the size of the subspace determined by the node size. The tree map hierarchy is based on the order from the root node to the leaf node, both horizontally and vertically. Space is divided horizontally at the next level, and the next level is divided vertically. For each divided rectangle, the corresponding color matching or necessary instructions can be carried out.

3.6.4 Visualization Tools

According to the functions and applicable scenarios of visualization tools, visualization tools are divided into entry-level tools, infographic tools, map tools, timeline tools, and advanced analysis tools [27].

1. **Entry-Level Tools**

Microsoft Excel is a typical entry-level tool that can perform various data processing, statistical analysis, and auxiliary decision-making operations. It has been widely used in management, statistics, and finance.

2. **Infographic Tool**

Infographics are visual representations of information, data, and knowledge. The use of the human brain's characteristics allows for the more efficient, intuitive, and clear conveyance of information through the medium of graphics, as opposed to the more complex process of interpreting textual information. It is used in the fields of computer science, mathematics, and statistics. The most common tools utilized for the creation of information icons are the Google Chart API, D3, Visually, Tableau, and Big Data Magic Mirror.

The Google Chart API is a mapping service interface of Google designed for the generation of statistical data and the generation of visual representations thereof, including the creation of charts. It is a user-friendly tool that enables the online viewing of statistical charts through a browser. D3 is one of the most popular visualization libraries. It is a JavaScript function library for the creation of web pages and the generation of interactive graphics. D3 can provide a large number of complex chart styles other than line graphs and bar graphs. These include Voronoi diagrams, tree diagrams, circular clusters, and word clouds. Visually is a very popular infographic creation tool that is relatively easy to use. The software allows users to create custom, beautiful, and visually impactful infographics without any design-related knowledge. Tableau is the simplest business intelligence tool software available for desktop systems. It is suitable for enterprises and departments to perform daily data reporting and data visualization analysis. Tableau represents the optimal integration of data operations and beautiful charts. The user needs to only drag and drop a large amount of data onto the digital "canvas" to create various charts in a blink of an eye. Big Data Magic Mirror is an excellent domestic data analysis software. Its rich data formulas and algorithms facilitate a comprehensive understanding, exploration, and analysis of data. The software enables users to create interactive charts and data mining models through an intuitive drag-and-drop interface.

3. **Map Tools**

It is more common to utilize map tools in data visualization. These tools have strong expressive power in the display of data based on spatial or geographic distribution. Moreover, they are capable of intuitively depicting various analysis indicators' distribution and regional characteristics. When the topic to be expressed by the indicator data is related to the region, it is possible to choose to use the map as the general background, with the intention of helping users to understand the overall data situation in a more intuitive manner. Additionally, the geographic location of a region can be identified with relative ease, allowing users to access detailed data.

Typical map tools include Google Fusion Tables, Modest Maps, and Leaflet. Google Fusion Tables presents data in the form of charts, graphs, and maps, thereby facilitating the discovery of patterns and trends that may be obscured by the data itself. Modest Maps is a small, extensible, and interactive free library that provides a set of APIs for viewing satellite maps, with a size of only 10 KB. Additionally, Modest Maps is an open-source project with a robust community support, making it an optimal choice for integrating map applications on a website. The Leaflet framework

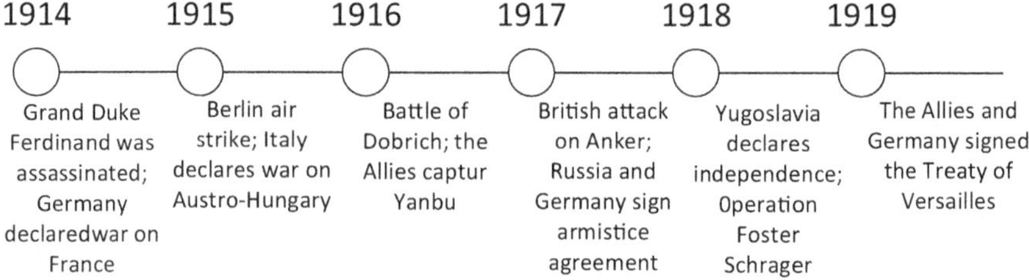

Fig. 3.45 Timeline of World War I

is a miniature map system that addresses the needs of mobile web pages through its miniaturization and lightweight design.

4. **Timeline Tools**

The timeline is an effective way to express the evolution of data in the time dimension. It connects one or more events through Internet technology according to the time sequence to form a relatively complete record system, and then presents it to users in the form of graphics and text. The timeline can be used in different fields, and its biggest function is to systematize, complete, and refine the past. Since Facebook released the function of organizing content in a timeline format at the F8 conference in 2012, timeline tools have become popular on social networking sites at home and abroad. Figure 3.45 is a timeline of the main events of World War I, drawn with the timeline tool.

Typical online editing timeline sites include Timetoast and Xtimeline. Timetoast offers personalized timeline services. The flash platform allows users to add events to the flash timeline, define their times, names, images, and descriptions of each event, and finally display the event's development in the time series on the timeline. Xtimeline supports the construction of timetables by adding event logs.

5. **Advanced Analysis Tools**

Typical advanced analysis tools include R language, Weka, and Gephi. The R language is a free and open-source software belonging to the GNU system. It is an excellent tool for statistical calculation and statistical mapping, although it is difficult to use. The R language is both simple and powerful. Its functions include data storage and processing systems, array operation tools, complete and coherent statistical analysis tools, and excellent statistical charting tools, which can be used for big dataset statistics and analysis. Weka is a free, Java-based open-source machine learning and data mining software that supports data analysis and chart generation. Gephi is mainly used for visualizing social graph data and generating beautiful visual graphics.

3.6.5 Visualization Case

There are innumerable examples where visualization technology facilitates the statistical analysis process. The Kaspersky Cyber Threat Map offers a comprehensive global view of the frequency of hacker attacks globally, and it is one of the most comprehensive maps currently available. Furthermore, it provides a user-friendly graphical interface. Each line on the map represents an attack activity, thereby enabling the user to comprehend the number of such infiltrations that have occurred globally on a daily, hourly, or even a second-by-second basis. Upon opening the map, it is capable of detecting the user's current location and subsequently displaying statistics for their country. Additionally, the application includes information on the country that was most recently attacked and infected country from the previous week. It is evident that the utilization of visualization technology enables a more comprehensive comprehension of the "invisible dispute."

Furthermore, Ramio uses the programming language from Freebase to maintain the table's data and to draw the influence relationship between programming languages. As depicted in Fig. 3.46, each node in the figure represents a distinct programming language, and the connections between them represent the influence of the programming language on other languages. Influential languages will establish connections with multiple languages, and the corresponding nodes will also become increasingly larger. The figure contains a total of 1,184 programming language nodes and 972 kinds of relationship connections, which provide a clear understanding of

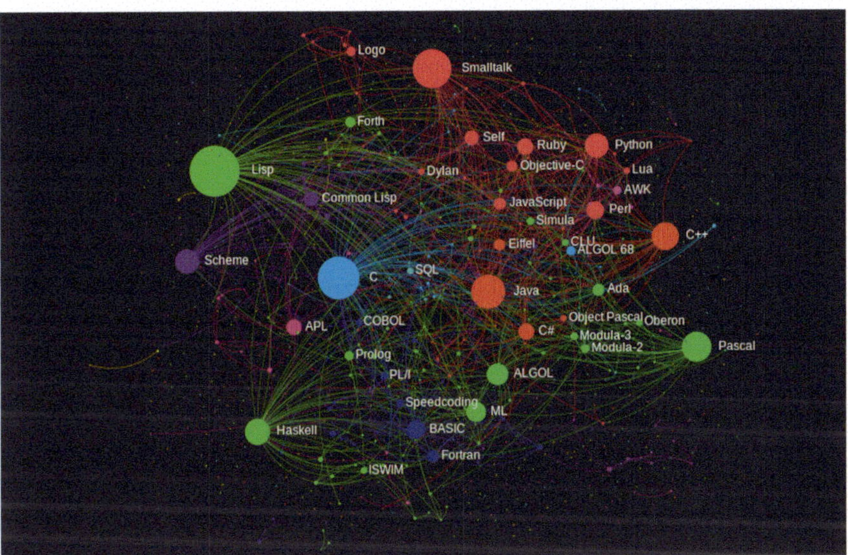

Fig. 3.46 Programming language influence diagram

the relevance and influence of each programming language. The languages with relatively larger nodes are Lisp, C, Java, Smalltalk, and Pascal. Ramio uses the robust expression capability and outstanding visual effects of visualization technology to visualize the relationships and influences between various programming languages.

3.6.6 Development Trend of Visualization

The primary limitations of visualization can be attributed to three key factors: the scalability of computing power, the limitation of cognitive ability, and the limitation of display capability [28].

1. **Scalability of Computing Power**

In the era of big data, data is frequently subjected to operations such as data cleansing and conversion, which significantly increases the computational complexity. The ability of machines to process vast quantities of data in a timely manner has become an important factor in the limitations of data visualization.

2. **Limitation of Cognitive Ability**

Human cognitive ability and attention are limited, and the ability to maintain a high concentration of attention for an extended period is limited. As a result, the visual search can only be performed for a short time. The maintenance of cognitive ability is also a limiting factor in the development of data visualization when confronted with a substantial quantity of complex data.

3. **Limitation of Display Capability**

The objective of data visualization is to present useful information to users efficiently. However, hardware devices have limited display capabilities. The display of too much information at the same time will be a source of visual confusion, and less information will reduce efficiency. Only by improving the display capability can information be effectively transmitted based on high data volume.

To overcome the limitations of the above three points, the development direction of visualization technology is mainly in the following three points:

1. Improving the display capabilities of hardware devices

In order to satisfy the demands of virtual and physical reality visualization, it is necessary for hardware devices to improve the spatial and temporal resolution of visual display, increase the capacity, and improve the capability of real-time processing and the efficient display of big data.

2. Combining related disciplines to develop visual analysis systems for various fields

The integration of visualization technology with related disciplines enables the transcending of cognitive and display limitations. Related disciplines not only refer to data mining and computer graphics, but also psychology. These disciplines must analyze and process a substantial quantity of multidimensional data. The use of visualization technology facilitates the presentation of data processing results, thereby enhancing the ability of relevant personnel to identify key problems within the data.

Nevertheless, in practice, it remains challenging to apply visualization methods to earthquake relief, financial security, and social networks. These application fields involve a variety of disciplines, necessitating the effective combination of diverse types of data and their real-time presentation. A significant development direction of visualization technology will be the creation of a comprehensive visual analysis system for multiple fields.

3. Realizing the visualization of the dynamic evolution process

In the current era, there is a growing emphasis on the dynamic evolution of data, rather than on static data characteristics. The majority of existing visualization algorithms and tools are used to display static structure, yet these cannot accurately describe the continuous evolution that occurs in reality. Furthermore, they fail to meet the needs of users. Consequently, the visualization of dynamic evolution process is also worthy of significant attention.

If the issue of dynamic visualization technology can be resolved, it will be of great significance to show the evolution process. This can assist individuals in analyzing the characteristics and knowledge mining of a time-varying situation more effectively, especially in the fields of network attack analysis and epidemic spread analysis.

3.7 Summary

This chapter introduces the most common approaches to big data calculations and corresponding tools, as well as the related concepts of data visualization.

Static batch processing MapReduce orchestrates the processing by marshalling the distributed servers, running the various tasks in parallel, managing all communications and data transfers between the various parts of the system, and providing for redundancy and fault tolerance. MapReduce abstracts the complex parallel computing process that runs on large-scale clusters into two functions, Map and Reduce. This significantly reduces the difficulty of distributed parallel programming and makes it easy for programmers to complete batch processing of massive dataset calculations.

Storm, a real-time stream computing method of big data, has demonstrated the potential for more powerful real-time stream computing capabilities by abstracting concepts such as data streams and execution strategies. Concurrently, it integrates the master–slave operating mode. The master node is responsible for task assignment and cluster status monitoring. The distributed coordination component is responsible for task scheduling, while the work process in the working nodes completes various tasks, thereby effectively improving the computing efficiency and fault tolerance. It is suitable for all kinds of real-time analysis services.

Pregel, a graph computational model, is well suited to the execution of large-scale graph calculation tasks, including graph traversal, shortest path problems, and PageRank calculation. A distinguishing feature of Pregel is that the vertex serves as the central unit of computation, which increases the computational load on the node. The Bulk Synchronous Parallel Computing Model (BSP) is used to achieve local computing, message transfer, and task synchronization. This approach effectively addresses the limitations of traditional graph computing methods, which are difficult to adapt to large-scale graph data computing, lack versatility, and exhibit low efficiency. Furthermore, Pregel incorporates a checkpoint mechanism to improve fault tolerance and facilitate rapid recovery.

Data visualization is an effective means for improving the efficiency of user data analysis. There is a wide range of tools available to realize the potential of visualization technology. Section 3.6 provides an overview of the most commonly used methods, technologies, tools, and typical applications. Currently, the limitations of visualization technology can be classified into three categories: the scalability of computing power, the limitation of cognitive ability, and the limitation of display ability. In the future, the development trend of visualization technology can be summarized in the following three aspects: first, improving the display capabilities of hardware devices; second, combining related disciplines to develop visual analysis systems for various fields, thereby overcoming the limitations of cognitive and display capabilities; and third, realizing the visualization of the dynamic evolution process to accurately describe the true situation of continuous evolution in reality.

From the perspective of the development process of big data analysis and calculation models, the problems of big data processing are complex and diverse, and a single calculation model cannot cover all computing needs. The existing calculation models are only applicable to the current development stage of big data. As technology evolves, the big data computing model will continue to evolve as well, introducing new models to better adapt to future business needs.

3.8 Exercise

1. Briefly describe the general process of pattern recognition.
2. Briefly describe the common methods of data mining.
3. List typical big data analysis and calculation modes. Briefly explain application scenarios of various calculation modes, and point out commonly used programming models.
4. Briefly describe the MapReduce architecture and explain the functions of each component.
5. Briefly describe the MapReduce shuffle phase's main tasks from the mapping task and the reduction task, respectively.
6. Briefly describe the fault tolerance mechanism of MapReduce.
7. Briefly explain the basic unit of Storm messaging.

8. Briefly describe the functions of Stream Grouping in the Storm programming framework, and the common grouping forms of parallel sentences.
9. Explain the concept of "superstep" in Pregel, and briefly describe the process of superstep.
10. Briefly describe Pregel's fault tolerance mechanism.
11. List common methods of visualization technology.
12. Briefly explain the constraints on the development of visualization technology, and explain how to break through these constraints.

References

1. Bishop CM. Pattern recognition and machine learning. Berlin: Springer; 2006.
2. Wikipedia. Pattern Recognition. https://en.wikipedia.org/wiki/Pattern_recognition. Accessed 04 Jun 2019.
3. Chao X. Overview of pattern recognition theory and its applications. China Sci Techonl Inform. 2006;06:171–2.
4. Han J, Data mining: concept and technology. China Machinery Industry Press, Campbell;2012
5. Wikipedia. Data mining. https://en.wikipedia.org/wiki/Data_mining. Accessed 04 Jun 2019
6. Wang G, Jiang P. Summary of data mining. J Tongji Univ. (Nat Sci Edn) 2004, 32(2): 246–52.
7. Big data application. What is big data analysis. https://zhuanlan.zhihu.com/p/39833121. Accessed 04 Jun 2019.
8. Li J, Cui J, Wang P, et al. Overview of MapReduce parallel programming model research. 2011.
9. Dean J, Ghemawat S. MapReduce: simplified data processing on large clusters. Commun ACM. 2008;51:107–13.
10. Julong Y. The full solution of big data technology: basic, design, development and practice. China Inform Technol. 2014;6:40.
11. Wikipedia. MapReduce. https://en.wikipedia.org/wiki/MapReduce. Accessed 04 Jun 2019
12. Guangzhong S, Feng X, Xi X. Research on the scheduling and fault tolerance mechanism of MapReduce model. Microelectron Comp. 2007;24(9):178–80.
13. Xie J, Yin S, Ruan X, et al. Improving Mapreduce performance through data placement in heterogeneous hadoop clusters. In: 2010 IEEE international symposium on parallel & distributed processing, workshops and Phd forum (IPDPSW). Piscataway: IEEE;2010. p. 1–9.
14. Yu F. Analysis of typical cases of big data application in telecom operators. Inform Commun Technol. 2014;6:63–9
15. Sun D, Zhang G, Zheng W. Big data streaming computing: key technologies and system examples. J Softw. 2014;4.
16. Quan C, Qianni D. Cloud computing and its key technologies. Comp Appl. 2009;29(9):2562–7.
17. Duan Z. Research on load balancing and fault tolerance mechanism of big data streaming system. North China Electric Power University;2017.
18. Meituan Technical Team. Meituan Comments on Storm-Based Real-Time Data Processing Practice. https://tech.meituan.com/2018/01/26/realtime-data-measure.html. Accessed 04 Jun 2019.
19. Liu L, Guo Q, Chen Y. Large-scale graph data processing technology in cloud computing environment. 2011.
20. Zhang J. Big Data Daily Knowledge Record: Architecture and Algorithm. Publishing House of Electronics Industry;2014.

21. Malewicz G, Austern MH, Bik AJ, et al. Pregel: a system for large-scale graph processing. In: Proceedings of the 2010 ACM SIGMOD international conference on management of data. New York: ACM;2010. p. 135–46.
22. Kaiyue L. Overview of big data overview on big data. Comp Sci Appl. 2018;8(10):1503.
23. Deep learning and data mining combat. Large-scale graph search and real-time computing in the application of Ali anti-cheat system. https://cloud.tencent.com/developer/article/1365696. Accessed 04 Jun 2019.
24. Ren L, Du Y, Ma S, Zhang X, Dai G. Overview of big data visual analysis. J Softw. 2014;(09):1909–36
25. Lin Z. Principle and application of big data technology. People's Posts and Telecommunications Press;2015.
26. Liu Ka, Zhou X, Zhou D. Research and development of data visualization. 2002.
27. Usama F, Georges GG, Andreas W. Information visualization in data mining and knowledge discovery. San Francisco: Morgan Kaufmann; 2002.
28. Cardno AJ, Ingham PS, Lewin BA, et al. Methods, apparatus and systems for data visualization and related applications. U.S. Patent 9, 870,629, 16 Jan 2018.

Chapter 4
Big Data and Cybersecurity

4.1 Introduction

Cyberspace is becoming increasingly relevant to economic development, social stability, and national security. It is now regarded as the "fifth space" after land, sea, air, and sky. Its position in the international economy and politics has become increasingly important, and it is a significant concern among major countries. It is the battlefield of objects and interactive games. The quantity of data in cyberspace is immense and growing exponentially due to advances in information technology [1]. The mining and application of big data have generated significant resources for society. Big data technology is driving transformative change in various fields, including government, enterprises, public service departments such as healthcare and education, and business, science, and technology. The entire society is also actively promoting the advancement and utilization of big data technology. In this context, big data analysis technology is expected to provide novel solutions to traditional challenges in cyberspace security. Furthermore, the security and privacy concerns surrounding big data itself have emerged as a key focus area for the industry. Based on the realization of the security of big data itself, the use of big data analysis technology to conduct security analysis and address the challenges posed by network security issues has emerged as a prominent topic in the field of information technology in the contemporary era.

This chapter aims to present the cognitive basis of big data cyberspace security by including information network knowledge base, basic concepts of information network security confrontation, and basic theoretical knowledge of information security confrontation. Furthermore, this chapter discusses the various types of big data resources in cyberspace, including user data, industry data, traffic log data, network public opinion data, and some public datasets used for security analysis in network security scenarios. Moreover, the chapter introduces the application and case analysis of big data analysis techniques in network security, including security incident correlation analysis, network anomaly detection analysis, data content security analysis,

S. Luo and L. Pan, *Big Data Analysis*, Big Data Management,
https://doi.org/10.1007/978-981-95-5725-7_4

and security situation awareness analysis. Finally, this chapter addresses the significant security threats and attacks that big data faces today, as well as the mainstream protection technologies related to protecting big data security. Additionally, it will present case studies of how large Internet companies themselves have constructed their big data security systems.

4.2 Foundations of Cybersecurity

This section outlines the fundamental concepts and theoretical knowledge involved in big data and cyberspace security. First, the fundamental concepts and definitions of information networks are explained from four aspects: complex networks, information networks, cyberspace, and cyberspace security. Subsequently, the fundamental concepts of information security confrontation are introduced. Finally, based on the above two dimensions, the basic theory of information security confrontation is discussed from three perspectives, namely the basic-level principles, the system-level principles, and the system-level methods.

4.2.1 Basic Concepts of Information Network

The fundamental knowledge of information networks includes the fundamental concepts of complex networks, information networks, network spaces, and network space security. This section presents the definitions and interconnections of these concepts.

A. **Complex Networks**

Qian Xuesen proposed a more restrictive definition of a complex network. According to this definition, a complex network refers to a network with self-organization, self-similarity, attractor, small world, scale-free, partial, or total properties.

- Complexity performance

  In essence, the complexity of a complex network is mainly manifested in the following aspects:

  - Complex structure: reflected in the vast number of nodes and diverse characteristics of the network structure.
  - Network evolution: manifested in the emergence and disappearance of nodes or connections. For example, in a worldwide network, web pages or links may appear or disconnect at any time, resulting in continuous changes in the network structure.
  - Connection diversity: the connection weights between nodes may differ, and there may be directionality.

- Dynamic complexity: the node-set may belong to a nonlinear dynamic system. For example, the node changes in a complex manner over time.
- Node diversity: nodes in a complex network can represent a wide range of entities. For example, a complex network node formed by interpersonal relationships can represent a single individual, and a complex network node created by the World Wide Web can represent different web pages.
- Multi-complexity integration: the above-mentioned multiple complexities influence each other, leading to more unpredictable outcomes. For example, the design of a power supply network necessitates consideration of the network's evolution process, which determines the network's topology. In the event of frequent energy transmission between two nodes, the connection weight between them will increase accordingly, and the network performance will be gradually improved through the process of continuous learning and memory.

The findings of research indicate that the Earth is becoming increasingly interconnected, with the potential for it to be perceived as a global village or a small world.

Moreover, the cluster is the concept of the clustering coefficient. For example, there are always circles of acquaintances or circles of friends in social networks, where each member knows other members. The significance of the degree of agglomeration is the degree of network grouping; this is a kind of cohesion tendency of the network. The concept of a connected group reflects the distribution and interconnection of small networks in an extensive network. For example, it can reflect the mutual relationship between this circle of friends and another circle of friends.

In addition, the concept of the degree distribution of a power law can be employed. Degree is defined as the number of relations between a vertex (node) point (equivalent to an individual) and other vertices in the network (expressed by edges in the network). Degree correlation refers to the closeness of the relationship between vertices. At the same time, betweenness is an essential global geometric quantity. The betweenness of vertex u is defined as the number of shortest paths in the network passing through u. It reflects the influence of vertex u (that is, connected individuals in the network). The characteristics of a scale-free network are primarily determined by the concentration of agglomeration.

B. **Information Network**

The network is composed of nodes and connections, representing many objects and their interconnections. In mathematics, a network is a graph that is generally regarded as a weighted graph. In addition to the mathematical definition, the network also has a specific physical meaning, which can be understood as a model abstracted from some practical problems of the same type. In the field of computer science, the network is a virtual platform for the transmission, reception, and sharing of information. Information pertaining to various points, areas, and bodies is linked together to facilitate the sharing of these resources. The Internet is the most significant invention in human history, facilitating the advancement of technology and human society.

Prior to 1999, the prevailing view was that the structure of the network was random. Nevertheless, following the discovery of the scale-free and small-world characteristics of the network by Barabasi and Watts in 1999, which was subsequently published in the world-renowned "Science" and "Nature" magazines, the complexity of the network became apparent. A network is defined as a collection of multiple nodes and links that are physically and/or logically connected according to a specific topological structure. It is a system comprising unstructured nodes and interaction relationships.

In contrast, a computer network is defined as a system comprising communication lines and communication equipment that connects multiple computer systems, each with independent functionalities, distributed in different locations. It enables data communication and resource sharing between the connected systems via the support of network software. From the perspective of logical function, a computer network is a collection of computer systems that connect multiple computers with communication lines for the fundamental purpose of transmitting information. It consists of transmission media and communication equipment. From the user's perspective, a computer network is a network operating system that can be automatically managed for them. It responds to user requests by locating and utilizing the requisite resources, and the entire network, in essence, becomes transparent to users.

The Internet originated in the United States in 1969. It is a vast network composed of a multitude of interconnected networks. These networks are interconnected by a set of common protocols, thereby forming a logically single, substantial international network. This method of interconnecting computer networks can be called "network interconnection." Consequently, a global interconnection network covering the globe has been developed, which is commonly known as the Internet. This network structure is connected to other networks. The Internet is distinct from the World Wide Web. The World Wide Web is a global system based on hypertext interconnections and is one of the services that the Internet can provide.

As previously stated, information represents the state of movement of objective things, and the network is a system composed of unstructured nodes and interaction relationships. The information network refers to the physical or logical network that carries information, including essential functions such as information collection, transmission, storage, processing, management, control, and application. In addition, it is necessary to consider the characteristics of the network, information, and network information. The Internet is a kind of information network. Similarly, radio, television, and mobile communications are also examples of information networks. Virtual networks, such as VPNs built on the Internet, are also considered to be an information network.

C. **Cyberspace**

Cyberspace is defined as follows:

- The Online Oxford English Dictionary defines cyberspace as: "Cyberspace: An imaginary environment for communication based on computer networks."

- Li Naihe "Cyber Space and Cyber Confrontation": "Its basic meaning refers to the Cyberspace or virtual space created by computers and modern communication technology, which is different from the real space. Cyberspace or virtual space is created by A huge 'artificial world' composed of images, sounds, text, codes, etc., which is created and supported by computers and communication networks all over the world."

The media have become an essential and sufficient condition for Cyberspace. It is characterized by three key attributes: real-time interaction, holography, and hyperspace. (1) Real-time interaction is an essential feature of cyberspace interactivity. The speed of interaction is mainly determined by the propagation speed of information across space and the calculation speed of massive complex information. (2) Holographic: The integration of various media in the past and the powerful information processing capabilities of computers and the Internet are two defining characteristics of cyberspace. For the first time in human history, it uses many different forms of information to "holographically" construct the image of things, and subsequently create various other "realities" comparable to the real world. These "realities" may be considered holographic representations of the original real world, exhibiting characteristics. (3) Time and space: The media of cyberspace surpass the limitations of temporal and spatial limitations of natural media. In the context of natural media, it is a fundamental requirement that both parties involved in real-time interactivity and the dissemination of a large amount of information are in close proximity in both close space and time.

D. **Cyberspace Security**

Cyberspace security is dependent upon the virtual cyberspace. Computer networks are based on their storage and interaction characteristics. All information and the external hardware within this space are protected from damage, modification, and threats. The objective of cybersecurity is to guarantee the security, credibility, reliability, and control of information and information systems. The primary factor is the communication system utilized for the transmission and conversion of data. Information communication systems integrate Internet communication systems, various broadcast communication systems, numerous computer systems, and diverse industrial control systems. The data transmitted by the information communication system plays a pivotal role in the context of cyberspace security. Data are a kind of acousto-optic electromagnetic signal that is used for expression, storage, processing, and transmission. These signals are transformed into data and information through the processes of generation, storage, processing, transmission, and display within an information communication technology system. The issue of cyberspace security in data transmission represents a pivotal concern for the overall security of communication systems.

Cyberspace sovereignty is an integral aspect of the broader concept of national sovereignty. To be precise, the country has jurisdiction over information and communication activities. Cyberspace sovereignty includes four fundamental rights. The first right is that of equality, which implies that all countries are equal in the context of international network interconnection. The second right is the right to self-defense, which means that a country is not constrained by other countries when operating information and communication infrastructure within its territory. Furthermore, there is the concept of independence, which refers to the capacity of each country to safeguard its cyberspace from external interference. Finally, there is the right of jurisdiction, which allows each country to regulate data and activities within its own cyberspace.

4.2.2 Basic Concepts of Information Security Confrontation

The section outlines the fundamental principles of information security countermeasures across five key dimensions: problems, features, security, attack, and defense.

A. **Information Security Issues**

Safety is defined as the antonym of injury and damage. Information is the representation and description of the state of things. Information security is defined as the state in which the information has not undergone destructive changes. This implies that the representation of the movement state of things has not undergone destructive changes, such as tampering, deleting, or substituting false information for authenticity. Information security problems can be caused by a multitude of factors, and the potential for such problems to arise is influenced by a number of variables. In general, information security problems are a highly complex matter. With regard to the manipulation of data, including tampering, deleting, and substituting accurate information with false information, the form of information representation plays an important role. For example, if the important digital part of the information content is expressed in the Arabic decimal system, the alteration of the decimal point position will have a significant impact on the value. Tampering with the decimal point position may cause a profound impact. Nevertheless, the aforementioned issue will not arise if a number is expressed in the Chinese capital, but this approach is highly inconvenient. As another example, the addition of a digital watermark to the information work or the use of a hash function to form a content summary allows for the auditing of the information content.

The discussion of the security of information or information systems is related to many content areas and many disciplines. It is an open and complex issue. This chapter focuses on fundamental issues, such as the basic concepts of information security and the key points.

B. **Features of Information Security**

The maintenance of the characteristics of information security, that is, the prevention of their destruction, is closely related to the information system's performance quality and security level. The primary characteristics of information security can be summarized as follows:

- The confidentiality of information is designed to prevent the unauthorized acquisition of data by individuals who do not have authorization to access it.
- The integrity of information ensures that the information is not subject to tampering, deletion, or unfair forgery.
- The controllability of information refers to the ability to monitor and control the flow of information within an information system.
- The authenticity of information refers to the assurance and confirmation of the information source, and non-repudiation of the information publisher during the interactive operation.
- The availability of information is defined as the operation and use of information in an orderly manner to ensure that legitimate users obtain legitimate information within the scope of authorization in a timely manner.

The availability of information also includes features such as authentication, notarization, verification, non-repudiation, and trust. In specific instances, it may be necessary to implement multiple features concurrently to guarantee the security of information and information systems. There are numerous such examples. For example, a key management center is essential for confidential communication, yet it cannot be set up at will. A notarized authority must authorize its establishment. During communication, users apply for keys to the key management center, and the center verifies the user's identity. Communication can only proceed upon successful verification. These are all issues of availability. A further example is the availability of digital signatures, mainly reflected in aspects such as non-repudiation.

C. **Information System Security**

The information system is a vital instrument for human beings, comprising information as the core element of the system. Information cannot be transformed into a service function if it is isolated from the information system. In the absence of information, the system is unable to operate or serve humanity. Information and information systems are inextricably linked and inseparable. This feature is also closely related to information security issues. Information security issues can be classified into the following three primary categories in the context of information systems.

The first category includes instances of information security issues that originate within the information itself. The content of information and information works has been tampered with, deleted, and replaced by fakes. Although security issues are directly reflected in the information or information works, the process of occurrence is reflected in the information system's operation, which is inseparable from the information system as the operating platform. This demonstrates the inextricable link between information and information systems in terms of information security.

In the second category, information security issues manifest within the operation of the information system. The occurrence of information security issues in the information system results in the destruction of the relevant operating order of the system, which is primarily caused by conscious human behavior in the case of confrontation. This destruction of normal functions and severe impact on the application is reflected in the occurrence of certain information at a specific moment, which causes damage. Furthermore, other factors, such as the failure of the information to reach its intended destination and excessive transmission delays, will also impact the application. Similarly, the leakage of information will have a detrimental impact on the application. It should also be noted that the security issue with information pertains to the specific threat of an attack on or destruction of a specific piece of information or an information-based process. The occurrence of security issues within an information system (in the absence of timely corrective action) indicates the emergence of a problematic situation. For example, a considerable delay in the transmission of information implies that all transmitted data is subject to a similar delay. There are numerous factors that contribute to the security issues observed in information systems. It is generally accepted that the development of information systems and their applications must involve contradictory movements. The security confrontation problem is a manifestation of numerous contradictions. Such contradictions are numerous; for example, the support of science and technology for information systems' functions is not yet complete. A certain technical measure may have positive effects but may also produce negative effects. Although information systems have self-organizing functions, they are still inseparable from necessary management, so they need to be set up with necessary management measures. It is also possible for attackers to use the administrator's entrance to manage the system, which represents an entrance for attacking information systems. Due to the inherent complexity of software, as well as the NP problems, errors are inevitable. Consequently, the interface that is used to correct these errors (patching) must remain accessible. It is therefore possible for an attacker, once they have gained access to this interface, to launch attacks against information systems.

In the third category, information security issues occur in information systems. The challenge of information security is that the attacker can directly damage the information system in a subtle and indirect manner. The method may not directly pertain to the domain of information, but rather, it may originate from other fields. For instance, they use anti-radiation missiles to destroy radar, cables to disrupt communication systems, nuclear explosions to inflict a range of damage to information systems, and chemical energy to convert into electromagnetic solid energy to destroy various information systems, etc.

D. **Information Attack and Confrontation**

A significant proportion of the causes of information security issues can be attributed to human factors. Subjective intentions can be used to categorize these actions into two types: negligence, which is related to people's omissions and mistakes; and intentional and planned actions. Various actions can result in the destruction of information, information systems, and the operational order of information systems with the intention of achieving a specific objective. Such incidents are commonly referred to as information attacks. It is evident that the victim will not remain passive in the face of such attacks. In response to information attacks, it is inevitable that various measures will be taken to resist them. These may include preventive and emergency measures, with the aim of making the attack difficult to work, reducing its losses, and even punishing the attacking party and counterattacking the other party. This kind of confrontational action is called information confrontation.

Information confrontation is the development process of a group of opposing, contradictory movements with complicated causes. The process is a dynamic, multi-stage, multi-principle method that measures the intervention of opposites' unity. Although information confrontation is not a beneficial phenomenon for the information system application side, it should be understood as an inevitable event in a rational sense. It is a process of contradictions, and contradiction is an inherent aspect of the development of human society. From a dialectical perspective, a negative phenomenon plays a pivotal role in the advancement of its own opposite. The concept of "development is the last word" must be embraced in light of the inevitability of this phenomenon.

The information confrontation process is inherently complex and can be represented by a six-element relationship group in time and space, as follows:

$$\text{Confrontation process} \leftrightarrow R^n[G, P, O, E, M, T],$$

where n is the number of confrontation rounds, P is the parameter domain (an important parameter that prompts the two sides to confront each other), G is the target domain, O is the object domain, E is the constraint domain, M is the method domain, T is the time, and R^n is the six-element relationship group complex interrelationships. The relation is another intuitive name for the combination of operation and mapping. It also includes the mutual change rate of the elements: $\frac{\partial \mathrm{O}}{\partial \mathrm{P}}, \frac{\partial^2 \mathrm{O}}{\partial \mathrm{P}^2}, \frac{\partial^3 \mathrm{M}}{\partial t \partial \mathrm{O} \partial \mathrm{E}}$. These three rates of change represent continuous multiple changes, while discontinuous changes are commonly used, such as sequence and difference equation.

It is challenging to describe a complex confrontation process quantitatively and comprehensively in detail, although many important relationships have been found in natural science and mathematics. For example, in functional analysis, the generalized distance relationship between sets of elements constitutes a distance space, the size measurement relationship constitutes a Norm space, and some operational relations between sets (with certain constraints) constitute an inner product space; in algebra, there are isomorphic homomorphisms; and in physics, there are a series of important relationships between forces and motion. However, as far as the six-element complex relationship in the field of confrontation is concerned, it is difficult to directly use the relationship due to its extensiveness and complexity (including uncertain specific conditions and time-varying factors). The method of combining qualitative and quantitative approaches is used to determined $R^n[G, P, O, E, M, T]$.

4.2.3 Basic Theory of Information Security Confrontation

In this section, the basic theory of information security confrontation is described from two levels: basic-level principles and system-level principles, and the relation is introduced.

A. **Basic-Level Principle**

Principles of maintaining and exploiting the particularity of information systems and attacking countermeasures: in various information systems, their operating laws and principles can be broadly understood as the maintenance and conversion of certain "particularities" based on universalities (relativity), such as the storage, exchange, transmission, and processing of information. "Security" can be understood as the orderly maintenance and operation of "particularity," and various "attacks" can be understood as the purposeful destruction, change, and even infiltration of the original order and "particularity" to achieve the purpose of the attack "particularity." At the level of abstraction and generalization, information security and confrontation revolve around particularity, and information security is mainly about maintaining and using particularity.

The principle of relative authenticity of information security and countermeasures: along with the existence of the state of motion, there must be corresponding "information." Moreover, due to the complexity of the environment, specific "information" can represent the motion in various forms, and has a relative authenticity. As the representation of the state of motion, information exists objectively, but it cannot be absolutely hidden, imitated, or falsified, which is formed by the objective existence of motion and the indestructible nature of motion, and the existence of information is relativistic.

Generalized Spatio-temporal information interweaving representation and finite measurement principle: all kinds of specific information exist in time and a generalized space, which represent their existence in a certain form interwined with time. The specific form of information is limited by the size of the generalized space and the length of the time dimension. In the field of information security, information can be transformed and/or processed in the time and space domains to meet the needs of information confrontation. For example, the commonly used interception signal with a low probability in information hiding is the principle that the small volume of information and signals in the generalized space and time dimensions is difficult to detect and intercept by the other party.

On the basis of the common way, the opposite of this principle is an important transformation and embodiment of the law of the unity of contradictions and opposites in the field of information security. "To share the same path" is the foundation and premise, and is also part of the law of confrontation. In the field of information security confrontation, the "adversarial" stage, with "acting against the other path" as the core, is the main stage of confrontation. "Tao" is a summary of the mechanism, measures, and methods used to achieve the goal of one's confrontation. Applying this principle to study information security countermeasures can be transformed into using this law to study the interaction of complex dynamic relations in a set of relations. The opposite mechanism is manifested in the transformation of opposites into each other, such as using one's strength to counter another's strength, etc., which is the dialectical movement process of contradictions and struggles produced by things under the control of the law of unity of opposites.

On the basis of a common path, in information security confrontation, the two parties can be regarded as "positive" and "negative" to each other. In form, the opposite party is the main principle, and the principle of achieving the effect of reverse confrontation (adversity) is called the principle of acting in the same way. The opposite principle: "To weaken something, you must first strengthen it; to abolish something, you must first promote it; to take something away, you must first give it." In the field of information security confrontation, the principles of "success" and "abnormal" often have flexible and diverse connotations. For example, the attacker often organizes multi-level attacks, in which feints often attract the attention of the other party, and the main attack is easy to succeed in covering. The counter-attacker also often feints to attract the opponent's main attack to appear as soon as possible and then hit him hard.

Contention system against information rights, rapid establishment of countermeasure response principles: according to the definition of information and the principle of the relativity of the existence of information, every action taken by the two sides in the confrontation process must be accompanied by "information," which is called "confrontational information." It is very important for both parties. Only through it can we judge the "Tao" of the opponent's attack, and then provide a basis for opposing resistance to "do the opposite." Otherwise, it will not be "the opposite," let alone "the opposite is true." The struggle between the two sides over "confrontational information" is a complex struggle in the space and time domains. In addition to the confrontational struggle between "confrontational information" hiding and anti-concealment embodied in space, there are also "preemptive" and "reactive" in the time domain. The struggle in the sense of "early" is equally important. The time and space intertwined the two sides to form a complex "information confrontation" struggle, which became the frontline focus of the first round of the struggle between the two sides in the information security confrontation process, and played an important role in achieving victory.

B. **System-Level Principle**

Active and passive positions and the principle of winning in the process of partial struggle for the main power: This principle shows that the attacking party takes the initiative in the overall situation. In theory, it can attack any information system and any part at any time, with any attack method, and the attack preparation work can be carried out in secret. In this sense, the attacked party is in a passive state that cannot be changed. What the attacked party can do is to strive for the partial initiative under the overall passive state. The main measures to strive for the partial initiative are: ① hiding important information as much as possible; ② continuously analyzing the vulnerability of one's information system in the confrontation environment in advance, and making systematic remedial plans for possible attacks in advance; ③ dynamic monitoring system operation, fast capture and analyze attack information, make scientific decisions and quickly take effective measures against attacks; ④ conducting comprehensive operations to gain the initiative in the fight against the information; ⑤ using false information to set traps to induce attackers to launch attacks and kill them.

Putting information security issues at the top level of information system functions: information security is a very important function embedded in the functions of an information system, but it is not, after all, all functions, but only serves as a guaranteed service. Therefore, the security function should be dealt with scientifically and comprehensively according to the specific situation, and placed in the appropriate "degree" range. However, it must be emphasized that, especially for systems with high security function requirements, information security issues must be considered at the beginning of the system design.

The transfer of technological core measures constitutes a serial chain structure and forms the principle of vulnerability: the implementation of any technology is relatively conditional. It must rely on the establishment of its necessary and sufficient conditions, and the "condition" as a thing indispensably depends on the establishment of its required conditions. In the face of technical measures implemented to achieve the "objective" that is in the process of gradual implementation of the effect from the direct measures to achieve the objective, the security measures must follow the rule that the core of the technology is gradually transferred to the ordinary technology, thus forming a serial chain structure law.

Principles of information countermeasures based on symmetric and asymmetric transformation: "Transformation" can refer to the transformation of interaction, which can be regarded as the transformation of the "representation" of the attributes of things from one way to another. It can also be regarded as the transformation between relationships, which is the transformation relationship. In mathematics, transformation can be regarded as a kind of mapping, and it can be regarded as a kind of "reduction" in the method of thinking. This principle can also be used in the field of information security confrontation by using symmetrical transformation to maintain its function, while using the opponent's lack of symmetrical transformation conditions to weaken the opponent to achieve its goal of confrontation and victory.

Equivalence principle of indirect confrontation under the condition of a multi-level and multi-section dynamic combination: suppose the system configuration can be divided into a hierarchical structure $L_0, L_1, L_2, \cdots, L_n$, and $L_0 \subset L_1 \subset L_2 \subset \cdots \subset L_n$. When the L_i layer subsystem is attacked by information, the performance of the L_i layer can be reduced when certain measures are taken. However, effective measures are supported at the L_{i+j} layer to make the high level of confrontation win and thus win on a larger scale. Therefore, the opposing party bypasses a certain level of direct confrontation and chooses a higher and more core level to conduct a more effective indirect confrontation, which is called the principle of indirect confrontation equivalence.

C. System-Level Method

In the operational struggle of information security confrontation issues, the basic-level and system-level principles are applied. These principles are often intertwined and complement each other, rather than acting in isolation. It is crucial to use these principles to observe and analyze the root cause of the problem, and then implement a solution. The key paths and various methods used to achieve a certain goal are known as "methods." The generation of "methods" is to discover some specific realization paths and methods according to the mechanisms and laws of things. Consequently, corresponding to the "principles" set of production methods, it is based on "methods." In information security and confrontation, the key challenge is to develop the following range of flexible solutions by applying the relevant principles to actual situations.

(1) Opposition serves as the antithesis to the core strategy, a method that not only steers thought processes but also operates as the central mechanism within a strategic framework. The principle of "opposite yet complementary" adeptly harnesses a multitude of elements, including the adversary's "strengths," to devise potent countermeasures.
(2) An integrative approach that fuses the method of mutual opposition, the "relativity principle of information existence," and the "multi-tiered interwoven representation and finite measurement principle theorem of generalized time–space and time-dimensional information" offers a sophisticated framework for analysis and understanding.
(3) An innovative methodology that integrates the concepts of "reciprocal opposition," the method of contraries, with the "competitive framework for information rights" and the "swift establishment of systemic countermeasure and response protocols" provides a robust strategy for navigating complex informational landscapes.
(4) An advanced methodology that integrates the concept of "opposites being two sides of the same coin" with the "principle of contending for information rights and the expedited formulation of systemic countermeasures." Furthermore, it encapsulates the notion that 'the transfer of technological core measures forms a serial chain structure, thereby establishing the principle of vulnerability.'
(5) An integrative approach that harmonizes the "method of contraries," where opposites are both opposing and complementary, with the "principle of transformation application," encompassing both symmetric and asymmetric transformations.
(6) Adhering to shared principles while engaging in divergent actions represents a crucial and practical approach. The concept of "being contrary to each other" can be expanded to illustrate two dynamics: at one level, a process is interdependent with another; at another level, it stands in opposition. The two entities involved in this "contrary" relationship are not necessarily identical. In the course of

real-world interactions, both parties in a conflict may employ the strategy of 'unifying in commonality while diverging in approach.'

(7) Various countermeasures have been developed to address the issue of compound attacks. A compound attack refers to an attack mode in which the attacker organizes multi-level, multi-section time and space attacks. Its characteristic is that the information system security problem is generated in addition to the attack's effectiveness at each level and section. The implementation is also reflected in the attack on the other side. It is necessary to implement countermeasures to form a new additional attack, which is a serious attack that automatically forms a chain of attacks. The objective of combating compound attacks is to use the discrepancy between the attacker's attack sequence (time and space) to defeat or make the counter-attack measures ineffective by creating a chain reaction of attacks.

4.3 Big Data Resources in Cybersecurity

This section will introduce basic big data resources and some public datasets related to network security. The basic resources include personal data, industry data, security traffic data, security log data, and security public opinion data. The public application dataset includes the Knowledge Discovery and Data Mining Cup (KDD CUP), the Information Security and Object Technology (ISOT), the European Conference on Machine Learning and Principles and Practice of Knowledge Discovery in Databases (ECML-PKDD), the Spanish National Research Council (CSIC), the Czech Technical University in Prague-13 (CTU-13), the Drebin Dataset (DREBIN), and the Android Malware Dataset (AMD).

4.3.1 User Data

User data refers to a range of information data that can be used to identify a specific individual, either on its own or in combination with other information sources. It reflects a specific individual's activities, including data related to user account and user behavior. User account data includes user name, date of birth, ID number, address, communication contact information, communication records and content, account password, property information, credit information, accommodation information, health, physiological information, etc. In particular, this encompasses email accounts and passwords, online shopping accounts and passwords, etc. User behavior data refers to human network social behavior, including logs, pictures, audios, and videos generated by chat, shopping, search, etc. This data stored on various media is vast and diverse. It includes call records, online shopping records, personal browsing preferences, website browsing traces, IP addresses, software usage traces, and geographic location.

4.3.2 *Industry Data*

Industry data refers to data that is generated during economic development. This data includes security, environmental protection, urban management, traffic, pension, medical, community, education, energy, family planning, social security, etc. This section will introduce two types of data: Internet medical data and Internet financial data.

A. **Internet Medical Data**

The sources and scope of Internet medical data are diversified. Even if it is challenging to identify specific individuals after desensitization, the collection of such information is likely to be identified as important data closely related to national security, economic development, and public interests. The recently established national standard, "Information Security Technology-Guidelines for Data Cross-Border Transfer Security Assessment," contains important data related to the medical industry. The primary focus of the standard is on population health data and drug data.

Population health data includes the following:

(1) Personal privacy, patient data, and reporter information obtained during the reporting and monitoring of adverse drug and contraceptive reactions.
(2) Personal privacy, related disease, and epidemiological information of infectious disease patients, their families, and close contacts obtained during the monitoring of public health emergencies and infectious disease epidemics.
(3) A variety of diagnostic, therapeutic, and health data information, including personal electronic medical records, health files maintained by medical institutions, and health management service institutions.
(4) Personal information of human organ donors, recipients, and applicants for human organ transplantation in the medical service of human organ transplantation.
(5) Personal information about sperm and egg donors, users in human-assisted reproductive technology services, and applicants for human-assisted reproductive technology services.
(6) Personal privacy is involved in the process of family planning services.
(7) The genetic information of individuals and families.
(8) Life registration information.

Drug data includes the following:

(1) Drug experimental data submitted in the process of drug approval for drugs involving national strategic safety, such as pharmacological, toxicological, stability, pharmacokinetics, and other experimental data on animal models, clinical trial data in human body, and test data related to drug production process and production facilities.
(2) Clinical trial data/reports of Type II and Type III medical devices.
(3) Critical (emergency) information about drug safety, including the time and location of the incident, the current situation, the degree of harm, the preliminary

treatment, the development trend, the progress of the incident, the follow-up response measures, the investigation details, and the cause analysis.

B. **Internet Financial Data**

Internet finance is a new field that combines the traditional financial industry with the Internet. The spirit of "openness, equality, collaboration, and sharing" of the Internet has penetrated the traditional financial industry, making traditional financial services more transparent, involved, better collaborative, lower intermediate costs, and easier to operate. Internet financial data mainly includes transaction data, text data, and financial service or product supply data.

(1) Transaction data: Internet finance represents the evolution of traditional finance toward electronic information. The main activities of Internet finance are inextricably linked to user transactions. Internet financial companies provide an online platform for users to access a range of financial services. To ensure transaction security, improve the service quality of enterprises, and facilitate backtracking and obtaining evidence, the system will record the user's transaction process through the Internet platform, such as payment and transfer information. The long-term accumulated transaction data can be used to analyze the user's trading preferences and detect any unusual or irregular trading behavior. This can then be used to provide a basis for preventing trading risks.
(2) Text data: as a platform for information transmission, the Internet offers a wealth of evaluations, messages, communication, and exchange of information, which reflect the trends of public opinion. The foundation of financial operations is credit and expectations. This feature makes it more susceptible to social credit and expected public opinion. Financial public opinion can have a real impact on the operation of Internet finance through a specific mechanism of action. In the event that the attention and the response to minor financial public opinions are not paid in a timely manner, there is a possibility that a significant financial crisis may ensue.
(3) Financial service or product supply data: includes information on the country's macroeconomic operation, price levels, import and export, and industry development.

C. **Internet Basic Resource Data**

Internet basic resource big data includes a wide range of data types (such as domain name registration information, IP address, and autonomous system number) generated during the registration, analysis, and application support for Internet basic resources, related Internet physical facilities, and Internet application data.

Domain names, IP addresses, and the associated service systems represent fundamental resources of the Internet, providing key Internet core services. The related systems serve to address Internet resources and can be considered the "central nervous system" and the Internet's key infrastructure. The strengthening of the management of basic Internet resources, such as domain names and IP addresses, and the improvement of the service capabilities, technological innovation capabilities, and security assurance capabilities of critical Internet information infrastructure are an important part of my country's "Thirteenth Five-Year" National Informatization Plan.

In recent years, there has been a notable expansion of global domain names, IP addresses, and Autonomous System (AS) numbers, accompanied by a corresponding surge in the generation of related data. The Internet's basic resource, big data, has reached a considerable scale, which is of considerable and growing significance for the Internet and the entire economy and society.

4.3.3 Traffic Log Data

Traffic log data represents the quantity of information transmitted on the network over a given period of time. It is comprised of continuous data packets that contain the same communication quintuple information (source IP address, source port, destination IP address, destination port, and transport layer protocol) between two terminals data. Traffic logs a multitude of log types, including operating system logs, access logs, WAF logs, and HIDS logs. The operating system security log contains information such as the command being executed in the current host, the user logged in to the current host, and the login operation's IP address. The access log contains the web request access logs of both normal users and abnormal users.

The data characteristics of traffic logs are primarily classified into three categories: network traffic characteristics based on message headers, network flows, and connection graphs. Network traffic characteristics based on message headers generally include IP addresses, port addresses, etc. It is based on network flow that mainly uses statistical data related to network traffic as characteristics, such as packet length and packet arrival interval. These can be further divided into single-flow and multi-flow characteristics. The single-flow feature uses only the statistical characteristics of all the packet sets comprising a network flow. These characteristics are considered network flow features and include the packet arrival time, packet size, mean/variance of packet size, number of data packets in the network flow, etc. Multi-flow characteristics are statistical characteristics formed by multiple network traffic flows with certain similar characteristics. The network traffic characteristics are based on connection graphs that use graph characteristics combined with network traffic characteristics to form the aforementioned characteristics.

4.3.4 Internet Public Opinion Data

Security public opinion big data encompasses the vast digital landscape of opinions and attitudes expressed through social media platforms like Weibo, WeChat, online forums, news portals, and various other network channels. This includes a diverse array of user-generated content such as text, pictures, audio, video, alongside behavioral data reflecting interactions like follows, mentions (@), likes, shares, forwards, and searches, offering a comprehensive view of societal sentiment and engagement. Public opinion big data has the general characteristics of big data, such as massive, multi-source, heterogeneous, and high-speed. However, there are also particularities that distinguish it from other forms of big data.

The principal characteristics of public opinion big data are as follows:

(1) Social networking. It incorporates attributes such as time, space, social identity, social relations, social rationality, and social emotions. These attributes are combined and superimposed to make the public opinion big data that have multi-dimensional, multi-level, and complex related social network attributes.
(2) Real-time interaction. The continuous dynamic updates over time and the massive, high-frequency real-time interactions make public opinion big data present significant public opinion information value in the process of opinion expression, opinion analysis, and emotional collision.
(3) Multimedia. It contains a large amount of unstructured data, including text, images, audio, and video, which enables a more comprehensive description of event phenomena, characteristics, and laws.

Security public opinion data serves as the foundation for social security incident intelligence perception. The theoretical basis for social security incident intelligence perception based on public opinion big data is the social network attribute of public opinion big data. In the initial stage, intelligence perception can facilitate interactive decision-making and command through data fusion analysis and intelligence early warning, thereby preventing the explosive evolution and secondary derivative events of social security events. In the post-event stage, intelligence perception can provide early warning of the recurrence of similar events.

4.3.5 Some Datasets of Cyberspace Security Big Data

This section will introduce some public datasets of cyberspace security big data, which can be applied to computer network intrusion detection, botnet detection, malware detection, and other related fields.

A. **KDD**

In 1998, the Defense Advanced Research Projects Agency (DARPA) and the Air Force Research Laboratory (AFRL) jointly sponsored the first batch of standard data from MIT Lincoln Laboratory to evaluate computer network intrusion detection systems. The KDD Cup 1999 was a part of the data collected from the MIT Lincoln Laboratory, which included the tcpdump and BSM list files. This dataset is based on data captured in DARPA's 1998 IDS evaluation program, compiled by Stolfo et al. [2]. Furthermore, this dataset is regarded as the benchmark data for evaluating intrusion detection systems. The four main types of attacks are as follows: Denial of Service (DoS), User-to-Root (U2R), Remote-to-Local attacks (R2L), and Probing attacks. The dataset comprises three content features and 38 digital features. These characteristics include the basic characteristics of each TCP connection, the content characteristics within the connection suggested by domain knowledge, and the traffic characteristics calculated using a two-second time window. The KDD 99 dataset is one of the most commonly used datasets for the assessment of anomaly detection methods.

B. **Information Security and Object Technology**

The Information Security and Object Technology (ISOT) dataset was jointly developed by Ericsson Labs and Ericson Research, a traffic laboratory from Hungary. It is a combination of various publicly available botnets and a common dataset containing 1,675,424 total traffic. For malicious traffic in ISOT, the French chapter of the honeynet project, comprising the free Storm and Waledac botnets, is used to collect non-malicious traffic. In addition to web browsing via HTTP and traffic generated by the Azureus BitTorrent client, this dataset also contains regular traffic for many applications.

C. **ECML-PKDD**

The ECML-PKDD dataset was created for the 2007 European Conference on Machine Learning and Knowledge Discovery. The ECML/PKDD Discovery Challenge is a data mining competition held concurrently with the 17th European Conference on Machine Learning (ECML). The dataset consists of three primary components: context, class, and query. The context section encompasses information regarding the operating system running on the web server. The class section includes details about the HTTP server to which the request was directed. It also specifies whether the server supports XPATH technology, if there is an LDAP database on the web server, and any information about the web server's SQL database.

D. **CSIC**

The CSIC dataset is generated and developed automatically by the CSIC Information Security Institute (Spanish National Research Council). The dataset can be used to test web attack protection systems. The data includes 6,000 normal requests and more than 25,000 abnormal requests. The HTTP requests are classified as either normal or abnormal. Abnormal request refers to a comprehensive range of application-layer attacks. The dataset comprises three types of attacks: static, dynamic, and unintentional illegal requests. Dynamic attacks include SQL injection, CRLF injection, cross-site scripting, and buffer overflow. Static attacks include outdated files, session IDs in URL rewriting, configuration files, default files, etc. It is important to note that unintentional illegal requests are not malicious, but they do not adhere to the typical behavior of web applications. Such values lack the same structure as normal parameter values, for example, a phone number composed of letters.

E. **CTU-13**

The CTU-13 dataset was developed by the Czech Technical University and is a combination of 13 distinct malware samples captured in a non-fictional network environment. The objective of this dataset is to capture real mixed botnet traffic, infected hosts generate botnet traffic, and verify that normal hosts generate normal traffic. The CTU-13 dataset includes the capture of 13 distinct botnet samples, which are also referred to as scenarios. Each of the scenarios is executed with a specific malware that uses various protocols and performs multiple operations. This dataset is one of the largest datasets established by CTU University of Prague in the Czech Republic in 2011 and has been designated as an existing dataset.

F. **DREBIN**

The DREBIN dataset contains 5,560 applications from 179 distinct malware families, as provided by the MobileSandbox project. The samples were collected from August 2010 to October 2012. The malware in the dataset originated from the Google Play Store, the Chinese market, and the Russian market. Moreover, it is important to note that the sample in question is situated within a nebulous area between the classifications of malware and benign functions, rendering it a less straightforward case than other adware samples.

G. **AMD**

The AMD dataset contains 24,650 malware samples, collected between 2010 and 2016, and classified into 135 distinct malware types and 71 malware families. The dataset provides the most recent malware in the current state of Android and consists of three components: standalone, repackaged, and library. AMD provided a detailed description of malware behavior through manual analysis, produced a detailed manual analysis result document, and reported current status observations of Android malware.

4.4 Big Data Security Analysis

This section introduces the following technologies: security event correlation analysis technology, network anomaly detection and analysis technology, data content security analysis technology, network security situation awareness analysis technology, and big data security analysis application cases.

By examining the interconnections between disparate security incidents and events and effectively correlating them with the contextual nuances of the operational environment, security event correlation analysis transforms initially isolated data through processes of filtration and aggregation. This sophisticated approach reveals the underlying, often concealed, relationships embedded within the data. Consequently, it equips administrators with information that is more comprehensive, reliable, and of intrinsic value. Network anomaly detection analysis establishes a normal behavior model by using normal behavior data. It is assumed that the behavior of the intruder is distinct from that of a normal individual. A comparison of the detected data with the normal behavior model shows that the detected data differs from the normal behavior model. This discrepancy serves as the basis for the detection of network abnormalities. Data content security analysis ensures network content security by analyzing and identifying abnormal information within the network. Network security situational awareness technology encompasses a diverse range of security analysis technologies. These include situation understanding, situation assessment, situation prediction, situation visualization, and other techniques aimed to provide users with comprehensive intelligence support for decision-making. Examples of big data security analysis applications include the Prism Project, the Einstein Project, and Israel's Big Data Anti-Terrorism initiative [3].

4.4.1 Security Event Correlation Analysis

Security event correlation analysis uses effective correlations between security events and between security events and the context of the operational environment to filter, aggregate, and de-authenticate relatively isolated data, discover the true connections obscured by these data, and provide more complete, comprehensive, reliable, readable, and more valuable information. The subsequent section will present three security incident analysis methods based on correlation analysis: the prerequisite analysis method, the similarity analysis method, and the time series analysis method.

A. **Analysis of Prerequisites**

The prerequisite-based correlation method involves correlating two security events by comparing the result of an earlier security event with the prerequisites of a later security event. The requirements of an intrusion refer to the conditions that must be met for an intrusion to be successful, while the result refers to the outcome of an attack that has been successful. Such information may be obtained by the attacker or may result from the victim's experience. The super-alarm class is used to represent the prerequisites and results of an alarm type. The following presents a variety of super-alarm class definitions to illustrate the correlation method based on the prerequisite analysis.

Definition 1: a super-alarm class is defined as a triple (fact, prerequisite, result). The fact is a collection of a series of attributes, each of which is associated with a threshold. This includes all information that is known or obtained in the alarm (or when detecting an attack). The prerequisites are a series of logical connections of predicates, which describe the conditions necessary for the attack to succeed. The variables associated with these predicates are all included in the facts. The result is a series of logical connections of predicates that describe the information obtained after the attack is successful, and its variables also exist in the facts.

Definition 2: a super-alarm class T = (facts, prerequisites, results) is defined. The prerequisite set is denoted as $P(T)$, and the result set is denoted as $C(T)$. All predicates are represented in the prerequisite set or result set, and corresponding attribute values replace their variables in the h tuple. For example, for an instance h of a super-alarm class T, its prerequisite set is denoted as $P(h)$, the result set is denoted as $C(H)$, and each element is associated with the timestamp of the corresponding tuple in h.

Definition 3: for two super-alarm classes $h1$ and $h2$, if $p \in p\ (h2)$ and $c \in C\ (h1)$ exist, for all $c \in C$, c. end time < p. start time, and in C, the predicates in all contain p, then $h1$ is an alarm to prepare for $h2$. The determination of this preparation relationship is conducive to the discovery of the causal relationship between super-alarm classes. Obviously, if the result of attacking $h1$ provides great convenience for attacking $h2$, then $h1$ is prepared for $h2$.

B. **Similarity Analysis**

The correlation method based on similarity is a real-time security event correlation analysis method. This method calculates the probability of similarity between the alarm information by determining the probability of similarity of the attribute values and subsequently associates the alarm information. Associating similar alarms (alarms originating from the same source or destination address) is an effective method of enhancing the efficiency of the correlation process. It is unnecessary to be aware of the associated information of each report in advance. The alarm information is automatically associated with key attribute values, and the associated sequence is automatically generated. The alarm information of different alarm types can be prevented from being erroneously associated due to deception on unimportant attributes by setting a minimum probability matching standard.

The similarity-based correlation method is a process that associates events through the following steps: first, calculate the similarity between various event attributes. The value of the similarity is between 0 and 1. The attributes of the event include the attack source address, the attack source port, the attack destination address, the attack destination port, the attack type, the timestamp, etc. For each attribute of the event, a corresponding similarity calculation function is defined. This function returns a similarity between 0 and 1. A similarity of 1 means a particular match. Secondly, upon the arrival of a new event, it will be compared with the corresponding attribute values of all the existing event threads, and the similarity between them will be calculated. The similarity between events is defined as the weighted average of the similarities of different attributes. This can be expressed as follows:

$$\mathrm{SIM(X, Y)} = \frac{\sum S_j SIM\left(X_j, Y_j\right)}{\sum_j S_j}$$

where X represents an existing event thread, Y represents a new event, j represents the index value of the event attribute, S_j represents the similarity expectation of attribute j, and X_j, Y_j represent the value of attribute j in events X and Y.

Finally, the security event with the greatest similarity between the event and the event thread and exceeding the set value is merged into the security event thread. If it does not exceed the set value, a new security event thread is generated. A table will be generated for different attributes of security events to store the fused information during the fusion process.

C. **Time Series Analysis**

The correlation algorithm based on time series analysis uses the historical alarm information that has been received to establish a prediction model about the alarm event. Subsequently, the trained prediction model is used to calculate the attack sequence that is most closely aligned with the sequence currently under analysis. This process serves to complete the correlation of the entire event.

The algorithm first defines the time interval T, then divides the time interval into N parts, and uses the clustering method to aggregate the alarm events that occurred in the time period i into the event A_i, thereby generating the alarm event set $\{A_1, \ldots A_i, \ldots A_n)$. In the article, the event $A_i (1 \leq i \leq n)$ is defined as the time series variable in the time series analysis. Subsequently, the AR model is introduced:

$$y(k) = \sum_{r=1}^{p} \theta_t y(k-i) + e_0(k)$$

The observed value $y(k)$ at any time in this equation depends on the observed value at the previous p times plus an error $e_0(k)$, where $e_0(k)$ is the white noise sequence with mean 0 and variance σ^2. Subsequently, the formula $g = \frac{(R_0 - R_1)/p}{R_1/(T-2p-1)} - F(P, T-2P-1)$ is used, where $R_0 = \sum_{k=1}^{r} e_0^2(k), R_1 = \sum_{k=1}^{r} e_1^2(k)$. The value of g between the event sequence variable $y(k)$ is calculated corresponding to the new alarm event and the time sequence variable $x(k)$ of the latest alarm event is calculated. If the value is between $y(k)$ and $x(k)$, and if the value of g is the largest, the alarm event corresponding to $y(k)$ and the alarm event corresponding to $x(k)$ have the greatest possibility of correlation, thereby completing the entire work of event correlation.

4.4.2 Network Anomaly Detection and Analysis

The general assumption underlying anomaly detection is that the network behavior indicative of an attack is anomalous in comparison to the normal activities observed in the network. Consequently, the objective of network anomaly detection is to identify subjects in the network data that deviate from the expected normal network behavior. This section presents an overview of the techniques employed in network anomaly detection and analysis, with a focus on social network anomaly detection, industrial control network anomaly detection, and communication network anomaly detection.

A. **Social Network Anomaly Detection**

The objective of social network anomaly detection is to detect irregularities in user behavior, such as false advertising, network fraud, zombie fan attacks, and the repeated posting of sensitive information, with the intention of protecting ordinary users' rights and interests and ensuring the stability and security of social networks. The creation, maintenance, and functioning of abnormal user accounts on social networks are a dynamic process that includes the creation, development, and application stages. Abnormal users can be divided into four categories based on their different manifestations at different stages: zombie users, spammers, Sybil users, and hijacked users.

Zombie users are accounts created by attackers through automated tools with the intention of simulating normal users. These accounts are used to post messages, add friends, and perform other operations. In most cases, zombie users adhere to similar naming rules during the batch creation process. Spammer is a fake user created by an attacker to post URL information, such as advertisements, phishing, pornography, or a fake user who changes reputation in social networks and engages in other malicious behaviors through mutual followers, adding friends, and likes. Such users exhibit obvious behavior characteristics. They use online social networks to spread harmful information on a large scale, interfere with the normal use of platforms, and threaten Internet security. Sybil users are also called witch attackers. This category of user is responsible for the creation of a multitude of illicit and fictitious identities, which are collectively referred to as witch nodes. Witch nodes increase their influence within the network by frequently interacting with other nodes, thereby making it easier to initiate social networks' deception in social networks, interfering with social individuals' relay selection, intercepting and tampering with data, and other attacks. The hijacked user is typically a regular user of the social network, exhibiting typical behavioral patterns and a substantial number of connections. The attacker employs a variety of methods to steal the account and subsequently engage in malicious behavior.

The social network anomaly detection methods are mainly divided into two categories: those based on graph models and those based on node characteristics. The detection method based on the network connection graph model mainly identifies the anomalous structures or nodes within the graph according to the different topological structures of anomalous users and normal users. This enables the detection of anomalous users. The node feature-based analysis method extracts one or more features from social network nodes. These features may include registration attributes, posting content, activity behavior, connection relationships, etc. Subsequently, these features are used to construct multi-dimensional feature vectors, which are then subjected to analysis in order to identify any anomalous users. Finally, supervised learning or unsupervised learning is employed to detect anomalous users.

This section takes the GCN (Graph Convolutional Network)-based social network spammer detection method as an example (as depicted in Fig. 4.1) to introduce a

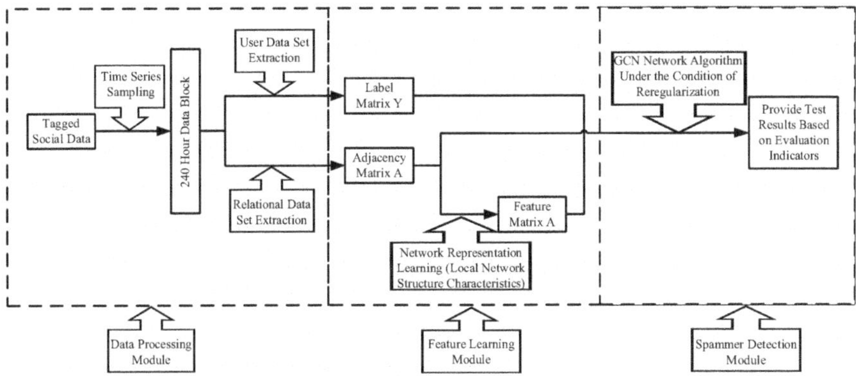

Fig. 4.1 Spammer detection method of social network based on GCN

social network anomaly detection model based on node features. This method first performs data processing on the personal information of social accounts, including demographic data such as gender, age, ID, and anonymity days. The processed social data is then input into a network representation learning algorithm, which extracts the local structural information of the network and generates a feature vector for each network node. The feature vector is input into the GCN algorithm with the re-regularization technology enabled, allowing for the detection of spammers within the network.

B. **Industrial Control Network Anomaly Detection**

The objective of industrial control network anomaly detection is to identify anomalous attack behaviors by monitoring and analyzing the network traffic of industrial production equipment, thereby ensuring the safety of industrial control systems and critical national infrastructure. The primary techniques employed in the field of industrial control network anomaly detection technology include rule matching, machine learning, and genetic algorithms. The following is a brief description of the general process of industrial control network anomaly detection with machine learning technology as an example.

The schematic diagram of the industrial control network anomaly detection based on machine learning technology is presented in Fig. 4.2. The training data includes network data traffic exhibiting known attack behaviors and security data traffic with no attack behaviors collected in the process of simulated attacks by using attack samples and real attack scenario data from research. The network attack event recognition model can be obtained through data preprocessing, feature extraction, and training of a Long Short-Term Memory (LSTM) model.

In practical applications, the first step is to collect real-time network data during industrial control equipment operation through a data acquisition device. This data is then processed with modules such as preprocessing and feature extraction to obtain numerical feature vectors. Subsequently, the vectors are input into the previously trained recognition model, which calculates the attack threat score according to the

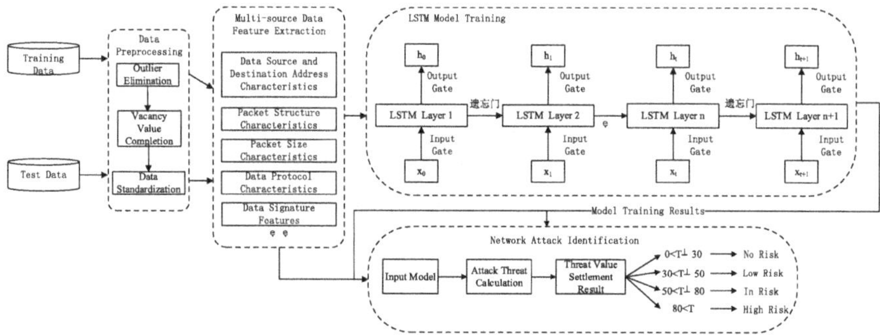

Fig. 4.2 Schematic diagram of industrial control network anomaly detection based on machine learning

model's output. Finally, the model's output score is used to identify whether there is a network attack in the current network environment in real time. The degree of threat faced by the network is then evaluated, with the aim of providing information support for industrial control network security protection decisions.

C. Communication Network Abnormality Detection

The objective of communication network anomaly detection is to monitor communication networks, detect communication network attacks, implement protection measures for communication networks, establish a security defense system for communication networks, and ensure the safety of communication systems and critical national infrastructure.

The methods of communication network anomaly detection mainly include rule-based, statistics-based, and machine learning-based methods. The rule-based method involves the establishment of a rule set describing normal access behavior through manual analysis, which is then utilized to detect anomalies. The statistical method is to perform numerical feature extraction, analysis, and modeling of normal communication network data, including the mean and standard deviation of the length of the URL parameter value, the number of parameters, the character distribution of the parameter, and the access frequency of the URL. A mathematical model based on statistical methods is to be established in order to achieve the detection of abnormalities in communication networks. However, the aforementioned two methods necessitate a considerable degree of manual involvement and a wealth of expert knowledge, rendering automatic detection a challenging endeavor. Machine learning involves the extraction of characteristics from communication network data, the application of a machine learning model to classify the communication network data, and the generation of an automatic detection result indicating the presence of an anomalous communication network. Accordingly, this section uses the communication network anomaly detection method based on machine learning as an example to illustrate the process of communication network anomaly detection.

The schematic diagram of the communication network anomaly detection method based on machine learning technology is shown in Fig. 4.3. This method first collects

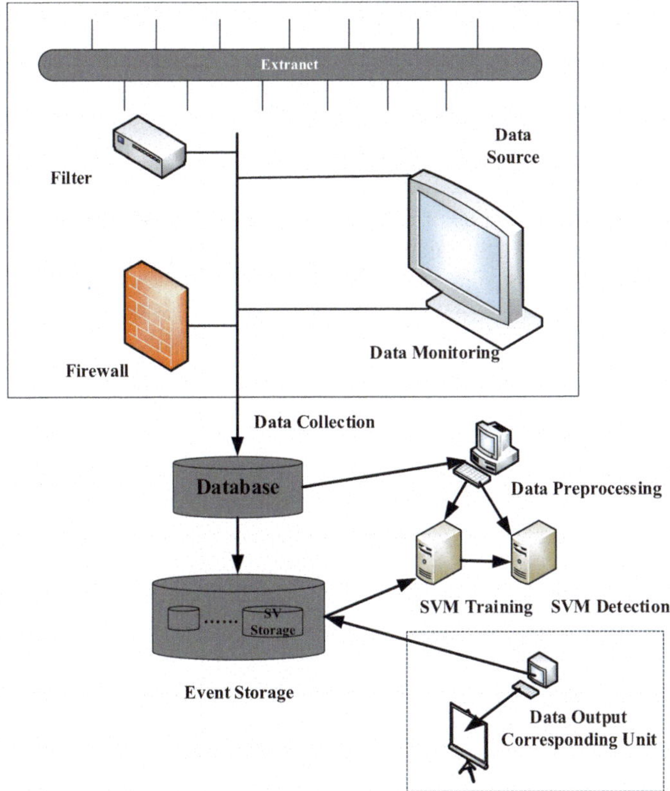

Fig. 4.3 Schematic diagram of the communication network anomaly detection method based on machine learning technology

communication network data (such as packet headers, header fields) through filters, firewalls, and traffic monitoring. Subsequently, the data is subjected to preprocessing, which involves the identification of anomalous values, dimensionality reduction, and standardization in order to facilitate the training of the classifier. Finally, the trained classifier is used to detect the communication network and determine whether the communication network is abnormal.

4.4.3 Data Content Security Analysis

Data content security analysis is the process of classifying and filtering online content to analyze legal and illegal content. This content includes incitement to subvert state power, incitement to hatred based on ethnicity or religion, terrorism, pornography, violence, and blood. The objective of data content security analysis is to maintain political ideology security and standards information dissemination order while ensuring a positive and healthy network ecological environment. Data content security analysis is mainly divided into two technical approaches: the traditional content analysis method based on manual review and the content analysis method based on artificial intelligence technology. The advent of the big data era has led to the emergence of data content that is characterized by immediacy, massiveness, and polymorphism. Artificial intelligence-based content security analysis technology has become the mainstream approach. This section will present an overview of the data content security analysis technology, divided into two categories: illegal text detection based on artificial intelligence technology and sensitive image detection.

A. **Violation Text Detection**

The illegal text refers to texts that exhibit one or more of the following characteristics: sensitive political tendencies (or anti-ruling party tendencies), violent tendencies, unhealthy colors, and uncivilized texts. The objective of illegal text detection is to accurately identify such texts within the massive text content transmitted on the network. This is achieved through technical means that allow for the supervision of illegal texts, thereby purifying the network environment and ensuring social stability and unity.

The existing methods for detecting illegal texts can be broadly classified into four categories: keyword-based filtering methods, keyword-based grammar filtering methods, machine learning methods, and hybrid methods combining keywords and machine learning. The keyword-based filtering method is highly efficient, but due to the ambiguity of word segmentation, misdetection is caused, the generalization ability is weak, and the lexicon's maintenance cost is high. In contrast, the filtering mode based on keyword grammar takes into account the context of keywords. In contrast to keyword filtering, it has the capacity to disambiguate. However, the manual summarization of keyword grammar represents a significant drawback. Furthermore, the continuous emergence of variants makes the control of the labor cost associated with mining interception grammar an unattainable objective. Additionally, the practical application of machine learning methods in variant-related scenarios presents significant challenges. The combination of keyword detection with machine learning offers an effective solution to the aforementioned issues. Consequently, this section will provide an example of an illegal text detection method based on Tencent Cloud Tianyu's hybrid strategy.

The schematic diagram of the illegal text detection method based on the hybrid strategy is shown in Fig. 4.4. The input data is UGC comment text data, which contains illegal vocabulary or content that is prohibited by law for online transactions. This method first identifies the input data as white (normal content), gray (suspected abnormal content), and black (violating text) through keyword filtering and machine learning models. Subsequently, in the event that the input data is identified as black, the business rejection strategy is used. In the event that the data is identified as white, a combination of statistical analysis, machine learning, and manual assistance will be employed to identify anomalies, discover variants, and submit the data to the audit

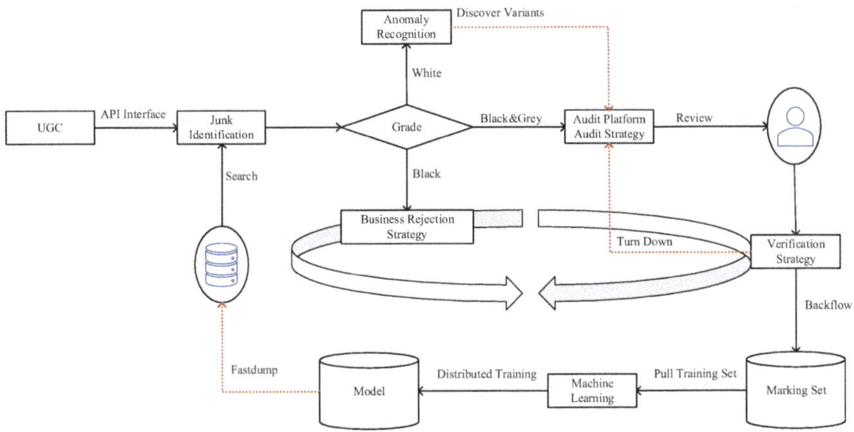

Fig. 4.4 Schematic diagram of the illegal text detection method based on a hybrid strategy

platform for review. In the event that the input data is identified as black and gray, it will be submitted directly to the audit platform for review. Subsequently, the audited irregularities identified in the white and black and gray data are incorporated into the marking set. Finally, the machine learning model is distributed using the marking set, and the machine learning model in the garbage identification link is updated, waiting for the next detection.

The FastText text classification algorithm is a machine learning algorithm used in a hybrid strategy-based illegal text detection method. The algorithm uses the hierarchical Softmax method to encode labels, which reduces the number of model prediction targets. Besides, FastText establishes a tree structure for characterizing categories. The depth of the tree structure of frequently occurring categories is smaller than that of infrequently occurring categories, which improves computational efficiency.

B. **Sensitive Image Detection**

Sensitive images include not only images of explosion fire, riots, blood, military weapons, murder, corpses, terrorists, police forces, privacy, but also images containing slogans and rumors that attack a political party or government. Additionally, text images of illegal and bad information, such as the slogan of the split organization and illegal advertising information, are also considered sensitive. The objective of sensitive image detection is to accurately identify sensitive images from the massive image content transmitted over the network through technical means, and to supervise the sensitive images, thereby purifying the network environment and ensuring social stability and unity.

Sensitive image detection is mainly divided into two categories: those based on traditional skin detection algorithms and those based on deep learning. The detection method based on the skin color algorithm calculates the area ratio of the skin

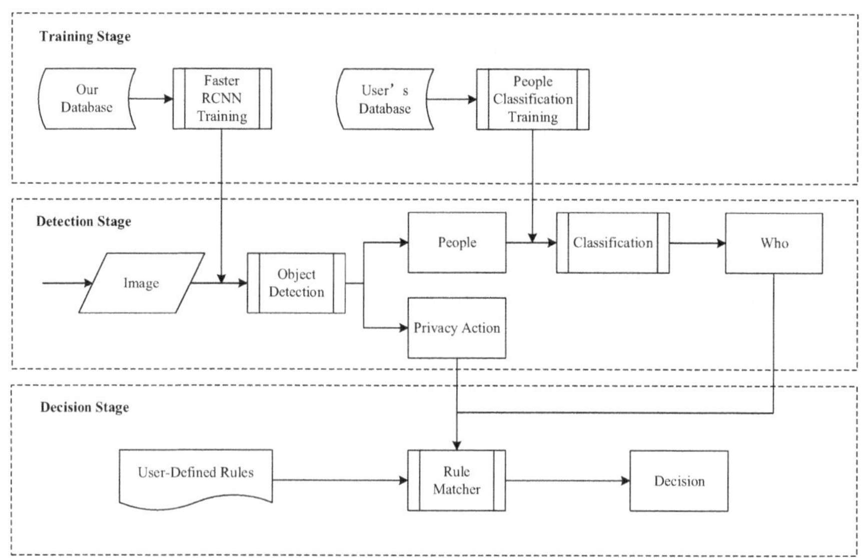

Fig. 4.5 Schematic diagram of a privacy person image detection and filtering system

color area in the image. It sets a threshold to determine whether the image is sensitive. However, due to the influence of background color, lighting conditions, and image quality, such methods are prone to misjudgments and missed judgments. Deep learning can simulate the method of human visual perception through the network, and automatically learn more hidden features, thereby constructing the network to obtain the overall perception of the target. Therefore, this section will take the privacy person image detection and filtering system as an example to introduce the sensitive image detection method based on deep learning.

The private person image detection and filtering system is depicted in Fig. 4.5. The whole system includes three stages: the training stage, the detection stage, and the decision-making stage.

In the initial training phase, the image data belonging to private individuals who engage in private behaviors such as kissing, hugging, and holding hands is divided into a training set and a test set according to a ratio of 1:1. Subsequently, the training set is used to train the Faster RCNN model to detect private behaviors and private person images. The test set is then utilized to assess the accuracy of the model.

In the detection phase, the model obtained in the training phase is employed to detect the privacy behaviors (e.g., kissing, hugging, holding hands, etc.) within the image and the corresponding probability score. The probability score is a decimal between 0 and 1. The score is divided into three levels: "high," "medium" ++ is used to record the number of such security events that occurred at a certain moment.

$$\Gamma(\text{what, when, where}) = \begin{cases} \overline{POP1} \\ Num[P0]++, 0 < \rho(when) \end{cases}, when = 0 \tag{4.1}$$

4.4.4 Safety Analysis Application Cases

This section will present a number of case studies illustrating the application of big data security analysis. These include the Prism Project, upstream monitoring projects, and the Israel big data anti-terrorism cases.

A. **Prism Plan**

The Prism Project (PRISM) is a top-secret electronic surveillance program implemented by the National Security Agency (NSA) in 2007. The official name is "US-984XN." The plan is designed to facilitate the collection of big data by controlling various network key infrastructures. The program is capable of correlating large quantities of data through the use of various algorithms, thus enabling the monitoring of social group behavior and the positioning of specific target groups. As reported, the disclosed documents indicate that the PRISM project involves in-depth monitoring of instant messaging and existing data. The permitted subjects of monitoring include any customers who use the services of participating companies outside the United States or any U.S. citizens who communicate with foreigners. Data such as emails, video and voice conversations, videos, photos, VoIP conversations, file transfers, login notifications, and social network details are available to the NSA in the PRISM program. The comprehensive intelligence document "President's Daily Briefing" used data from the PRISM program in 1,477 plans in 2012. The report on PRISM was released shortly after the revelations that the US government had continued to secretly require Verizon to provide the NSA with daily phone records of all its customers. The top-secret documents were leaked by Edward Snowden, an NSA contractor, and subsequently made public by the Guardian and the Washington Post on June 6, 2013 (Fig. 4.6).

The Prism project is built on a hardware and software foundation. The hardware foundation is supported by the client operating systems, email functions, instant messaging, network access services, and other functions provided by the nine major Internet companies, including Microsoft, Apple, Google, Facebook, YouTube, and AOL. It offers services to users simultaneously, while also providing convenience for the authorities to obtain information. All data from various servers must be transmitted via routers. Cisco routers have the functionality of monitoring and eavesdropping on these data transmissions. Furthermore, the authorities may unintentionally obtain data from server centers such as Microsoft through Cisco routers. The software foundation uses the database, Hadoop infrastructure, and machine learning as analytical tools to assist the authorities in the rapid analysis and processing of data. It offers significant advantages in large-scale data analysis, enabling rapid retrieval of data structures.

The Prism Project involves the network construction of key networks for various government departments, including customs, postal services, finance, railways, civil aviation, medical care, military police, and telecom operators. In the event of a war, the United States can control all computers connected to the Internet in the target country, paralyzing the target country's key systems such as finance, communications, and transportation.

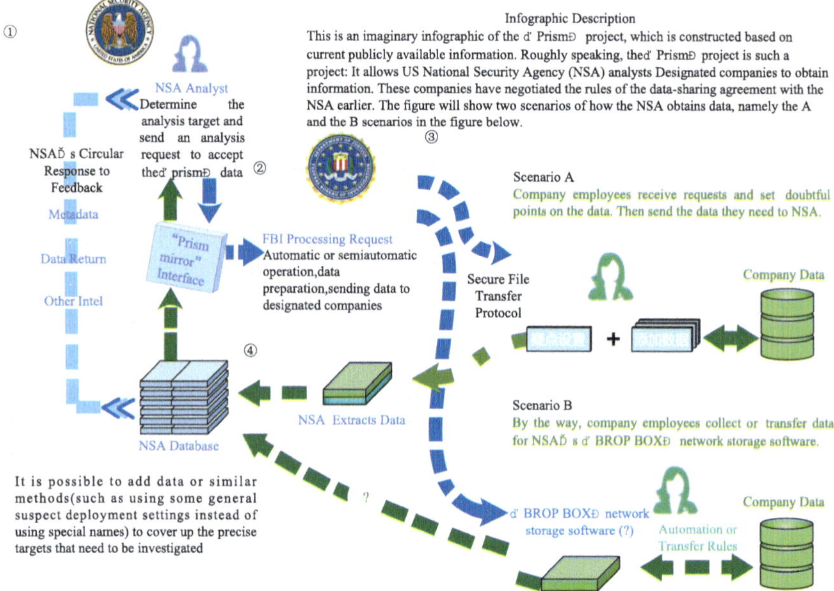

Fig. 4.6 Prism plan operation infographic

Although the disclosure of the U.S. "Polyprism" project system was unintentional, it was also expected by all countries. The Storm has prompted our country to accelerate the construction of cyberwarfare forces as soon as possible. The current defense-specific forces are unable to adapt to the changing nature of warfare. The incorporation of cyberwarfare into the composition of the offensive force of "soldiers without blood" is a strategy that has been adopted by many countries. Furthermore, our military must also proceed from actual combat, with a focus on our cyberwarfare capabilities.

B. **Einstein Plan**

The Einstein Plan is an automated network security monitoring project led by the US federal government, developed by the United States Computer Emergency Response Team (US-CERT) under the Department of Homeland Security (DHS). The plan's objective is to use DFI, DPI, and DCI technologies as a starting point, relying on big data technology. The core of the plan is threat intelligence, which is used to achieve continuous monitoring, early warning, and response to the US federal government's Internet export cyber threats. This improves the federal government's national security. The network's situational awareness and survivability establish a national security protection system architecture (as presented in Fig. 4.7).

In 2009, the US government launched the Comprehensive National Cyberspace Security Plan (CNCI). The Einstein plan was intended to be merged into CNCI and renamed National Cybersecurity Protection System (NCPS), but it is still referred to as the Einstein plan. Currently, NCPS has been deployed and operated in 23 other US

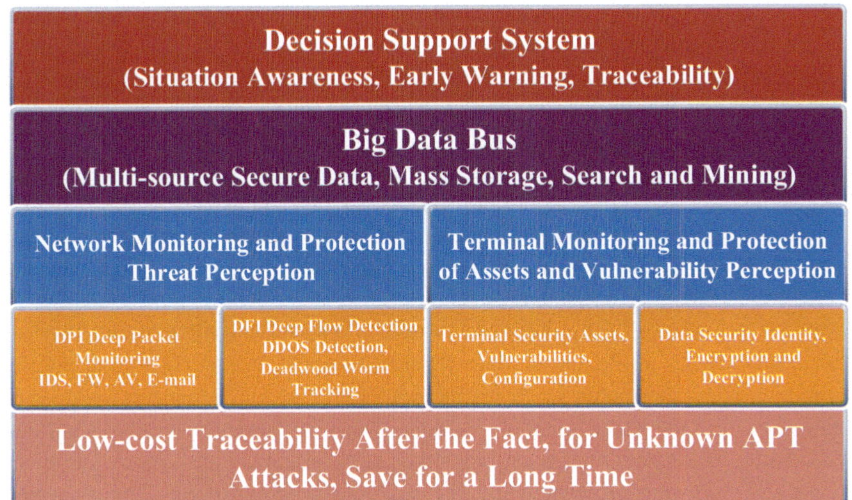

Fig. 4.7 National security protection system architecture

government agencies, with the exception of the Department of Defense and related departments.

The Einstein Plan is divided into three stages, as depicted in Table 4.1. The plan has four capabilities, including intrusion detection, intrusion prevention, data analysis, and information sharing.

Table 4.1 Einstein plan phase timetable

Codename	Deployment time	Objectives	Description
Einstein 1	2003	Intrusion detection	By deploying sensors at the internet outlets of government agencies, a set of processes for automated collection, correlation, and analysis of network traffic information captured by sensors is formed
Einstein 2	2009	Intrusion detection	By monitoring the internet connection of federal government agencies, matching the preset signatures of known malicious behaviors with them, identifying malicious behaviors, and issuing alerts to US-CERT
Einstein 3	2013	Intrusion detection Intrusion prevention	Through the deployment of intrusion prevention and threat-based decision-making mechanisms by the ISP, the malicious network behavior indicator developed by DHS is used to identify malicious behaviors and automatically block malicious traffic entering and leaving federal government agencies

The Einstein 1 Plan started in 2003 with the objective of using Deep Flow Inspection (DFI) technology to identify irregularities in network traffic data and to analyze overall trends. By installing sensors at the Internet outlets of government agencies, a series of automated collection processes, correlation, and analysis of network traffic information captured by sensors is established.

The Einstein 2 Plan started in 2008. The objective of the Einstein 2 Plan is to use IDS technology to perform Deep Packet Inspection (DPI) on TCP/IP communication packets to detect malicious behaviors such as attacks and intrusions. The Einstein 2 plan is an enhancement of the Einstein 1 plan. Based on the original analysis of abnormal behaviors, it has added the ability to analyze malicious behaviors. It is still an intrusion detection system in essence. The Einstein 2 Plan is designed to scan all Internet traffic and copies of government computers (including private communications) to identify malicious computer code features that may be used to obtain or harm government computer systems. When the Einstein 2 system identifies the communication data containing malicious code characteristics, it automatically obtains and stores the entire message, including the email's content. The identified and stored information is then reviewed by government officials from the computer network defense department. All of this work is conducted without the supervision and guarantee of the court or any external administrative department.

The Einstein 3 Plan started in 2013 with the objective of implementing intrusion prevention and threat-based decision-making mechanisms through Intrusion Prevention System (IPS), and using malicious network behavior indicators developed by the Data Handling System (DHS) to identify malicious behaviors and realize malicious traffic in and out of federal government agencies, thereby enabling the automatic blocking of such traffic. The Einstein 3 Plan represents a supplement to the Einstein 2 Plan, which is designed to enhance the detection capabilities of intrusion detection systems. Former Secretary of Homeland Security Chertoff stated that if the Einstein 2 system could be conceptualized as "a policeman holding a speed radar on the side of the road, with the capacity to use the phone to issue warnings to individuals who are driving under the influence or exceeding the speed limit." The Einstein 3 System, therefore, could be conceptualized as a policeman who can "arrest suspects" and "stop attacks." The Department of Homeland Security has played a crucial role in the Einstein 3 project. It maintains close cooperative relations with private networks and communication companies, and frequently shares information with them. The hardware and software of these companies constitute the backbone of the Internet, acting as the connection nodes of the Internet.

C. **Israel's Big Data Counter-Terrorism**

The objective of Israel's big data counter-terrorism is to use big data analysis to track and prevent terrorist actions. By collecting network information, monitoring telephone calls, and intercepting emails from governments, organizations, and even individuals, using big data technology for mining, analysis, and comparison, it can be refined into very useful action intelligence clues to provide support for intelligence analysts to make judgments.

In the Israeli big data anti-terrorism process, real-time network monitoring software and location-based search operations are two typical applications. Real-time network monitoring software can use data maps and link analysis to monitor social media dynamics, such as those observed on Facebook and Twitter, in real time. This software can process 300,000 posts per minute, allowing for the identification of the individuals who have posted and those who have liked a particular post. Additionally, it can be used to find out the most active members of a social media platform, track their geographic location, and observe their participation in specific actions. It is not necessary for all operations to involve hacking; instead, information is extracted from email and credit card payments using big data to obtain and analyze information on the open network. Location-based search operations can be used to analyze digital data. This kind of calculation functions as a computerized Sherlock Holmes, which reduces the information from telephone records into a set of variable numbers. The result of the calculation is a possibility map, which is used to predict the future activities of a potential terrorist and track the possible location of the terrorist in real time.

During the Gaza War, the Israeli army and *Xin Bet* used big data technology to track and locate several Hamas leaders. The massive, disorganized, and seemingly irrelevant video files, pictures, texts, and speeches collected by intelligence agencies are refined into highly useful action intelligence clues through the mining, analysis, and comparison of big data and are then submitted to high-level intelligence analysis. The personnel then make a judgment, and the Israeli military finally uses the intelligence to track and kill the Hamas leaders. Israel has conducted a series of "targeted removals" in the Gaza Strip against senior leaders of the Hamas military organization, with the majority of these operations relying on big data technology.

In early 2014, Israel successfully captured three terrorists who had attempted to commit suicide bombs at the U.S. Embassy in Tel Aviv and the Jerusalem Conference Center by means of cyber-surveillance. One of the three arrested was a member of the "Al Qaeda" organization. He used an online forum to promote the "Global Jihad" and to recruit the other two individuals into the attack plan. This forum has been closely monitored by the "Internet Brigade" for an extended period, and as a result, Israel was able to successfully capture the three individuals based on the clues left by them on the Internet forum.

4.5 Big Data Security Protection

Big data plays a vital role in the current academic and industrial circles. It is considered to be a significant change in the manner in which we live, work, and think. However, big data has many security risks in terms of data security and personal privacy. The privacy leakage caused by this brings troubles to individuals and enterprises, and the accompanying false information will also cause problems based on big data security analysis technology. The analysis results bring serious deviations. Consequently, big data security and its protection technology play a key role in cyberspace security research. This section commences with an overview of the principal risks associated with big data, followed by an examination of the technological solutions designed to safeguard this data. Finally, the construction of interconnected companies is used as a case study to illustrate the practical application of data security protection technology.

4.5.1 Threats and Attacks of Big Data

The integration of big data and machine learning technologies has the potential to provide novel solutions and insights into the challenges posed by cybersecurity. In addition, the severe security threats faced by big data itself represent a crucial aspect of cyberspace security. Big data threats and attacks mainly include big data platform security threats, big data infrastructure security, machine learning model attack threats, and advanced persistent attack threats [4].

A. **Data Platform Security Threats**

Big data platform security is guaranteed by the protection of the transmission, storage, computing, and other resources and functions of the platform. This includes the security of data transmission and exchange, storage security, computing security, platform management security, and platform infrastructure security. Transmission and exchange security refers to the assurance of the security and control of data exchange with external systems. In order to verify the legitimacy of the external system, it is necessary to implement interface authentication and other mechanisms. Furthermore, it is essential to employ channel encryption and other means to ensure the confidentiality and integrity of the transmission process. Storage security refers to setting up a backup and recovery mechanism for data within the platform, as well as the establishment of a data access control mechanism to prevent unauthorized access to data. It is imperative that computing components implement corresponding identity authentication and access control mechanisms to ensure that only legitimate users or applications are able to initiate data processing requests. Platform management security includes the security configuration of platform components, resource security scheduling, patch management, security audits, etc. Furthermore, the physical security, network security, and virtualization security of the platform's hardware and software infrastructure serve as the foundation for the safe operation of the big data platform.

The diagram of the big data platform system architecture is depicted in Fig. 4.1. The security threats it faces can be grouped into three categories. One such risk is the potential for unknown security vulnerabilities to emerge due to the integration of various new software, hardware, and protocols. As big data platforms continue to expand, new security vulnerabilities have emerged as a consequence of incorporating new software and hardware systems that are characterized by cloud computing. These vulnerabilities present a significant threat to the security of big data platforms. First, the existing security protection technology is unable to monitor the new technology's unknown vulnerabilities in real time. Second, the security guarantee mechanism of the big data platform is inadequate. One of the significant challenges associated with the implementation of a big data platform based on the Hadoop framework is the potential for security threats to arise. This is due to the fact that identity authentication, authority control, and security audits are not perfect, which in turn reduces network security. Third, the computing model, which is characterized by distributed computing and storage, blurs the security boundary. The relatively complex underlying protocols of the big data platform, coupled with the open storage and computing framework, render it challenging to define network security boundaries and pose threats to big data security.

B. **Infrastructure Security Threats**

The security threat posed by big data infrastructure represents a crucial component of big data security threats. The collection, storage, and analysis of big data all rely on related basic software and hardware facilities. Big data infrastructure includes storage devices, computing devices, all-in-one computers, and other basic software (such as virtualization software). The security threats of these infrastructures can be broadly classified into the following five categories:

1. Unauthorized access
2. Information leakage and loss
3. The integrity of damaged data during network infrastructure transmission
4. Denial of service attack
5. The propagation of a network virus.

C. **Model Attack Security Threats**

Machine learning models are learned through big data training, and the potential for attacks on machine learning models represents a significant challenge to data security. Most existing machine learning models are designed and implemented with a very weak threat model, failing to consider the potential actions of an attacker. Although these models can perform perfectly in the face of natural input, these machine learning models will encounter a large number of malicious users and even attackers in the real environment.

(1) The potential for a machine learning model to be compromised by an attacker during the training phase (learning phase) or the prediction phase (inference phase) is a significant concern. During this process, the attacker may attempt to alter the input and output of the model or gain unauthorized access to the internal components of the model in order to steal the model's parameters. This could result in the model's confidentiality, integrity, and availability being compromised. Attacks against machine learning models can be broadly classified into two categories: tentative attacks and adversarial sample attacks.

In the contemporary era, machine learning has emerged as a distinct business model. Notable IT companies, such as Google, Amazon, and Microsoft, have launched machine learning platforms. This machine learning-as-a-service model provides data holders who lack the requisite training capabilities. The data facilitates the training of prediction models for specific application scenarios. The computing resources available on the platform, along with the models or algorithms provided by machine learning as a service, are used to train their prediction models based on the data provided. Subsequently, they may develop their prediction interfaces externally and charge on demand. Although machine learning as a service offers users significant convenience, it also presents a potential vulnerability in that it may expose the private data of data holders to potential attacks by third parties.

A tentative attack may involve the theft of the model through a specific method or the recovery of a portion of the data used to train the machine learning model through some means, with the intention of inferring sensitive information belonging to a particular user. The threat of tentative attacks to machine learning models can be generally divided into two categories: data theft and model extraction.

Data stealing refers to obtaining the general content of the entire training dataset, as well as the statistical distribution of that data. Additionally, it includes the process of testing whether a given piece of data is present in the training dataset. It can be reasonably inferred that the attack is a method of data theft. In this attack, the attacker can use the model's API and the data records to infer whether the data records are part of the model training set. The utilization of the aforementioned method with patient data would inevitably result in the leakage of the training data. The method initially uses the training data, the predicted probability vector, and the label returned by the target model to train a shadow model with a similar architecture to that of the target model. This enables the determination of whether a piece of data belongs to the ground truth of the shadow model training set. The input data are input to the target model, which is then trained to classify using the predicted probability vector and the label returned by the API, along with an assessment of whether the binary label is included in the training set. The classification model outputs a determination as to whether the data is included in the training dataset. In the event that a specific item is to be evaluated in the future, the data must be input into the predicted probability vector and label returned by the target model API (Application Programming Interface) and then sent to the trained classification model (Fig. 4.8).

Model extraction involves obtaining the model's internal parameters or constructing a substitute model that is analogous to the target model in the black-box condition. In the absence of any prior knowledge about the model (training data, model parameters, model type, etc.), the attacker only uses the public access interface to gain access to the black box of the model, thereby constructing a model that is highly similar to the target model. At present, model extraction is primarily conducted through an equation-solving method, as shown in the figure below. The equation-solving method is designed to address logistic regression, support vector machines, and neural network algorithms. As these algorithms' models are function mappings, the input of the model is the input of the function. The output of the function is the direct output of the confidence of the output, which is a function of a series of parameters. Consequently, this method first uses the confidence level and input data to construct an equation, after which the function's parameters are solved in order to generate a model that is as close as possible to the target (Fig. 4.9).

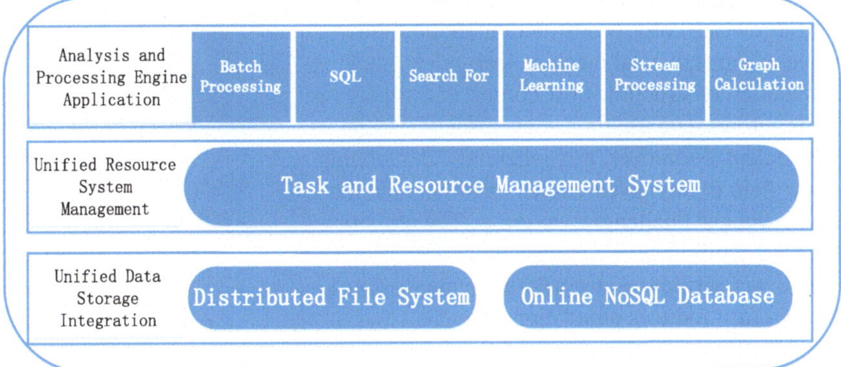

Fig. 4.8 The diagram of the open-source big data platform system architecture

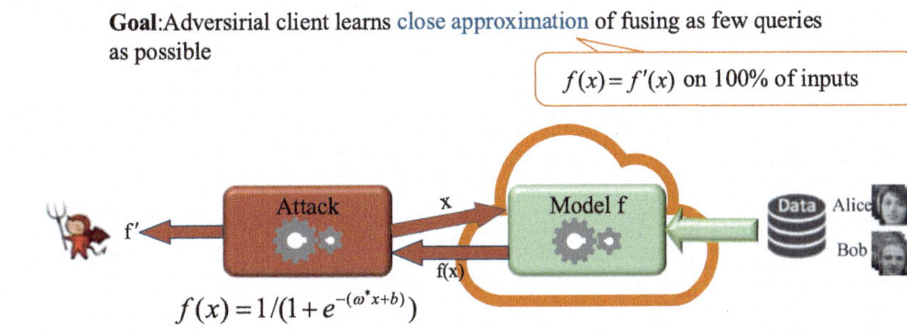

Fig. 4.9 Schematic diagram of model extraction based on the equation-solving method

(2) Adversarial sample attack

At present, machine learning technology is widely used in various fields, such as human–computer interaction, recommendation systems, and security protection. Specific applications include voice and image recognition, credit evaluation, fraud prevention, malicious mail filtering, resistance to malicious code attacks, network attacks, etc. Additionally, attackers attempt to circumvent these safeguards through

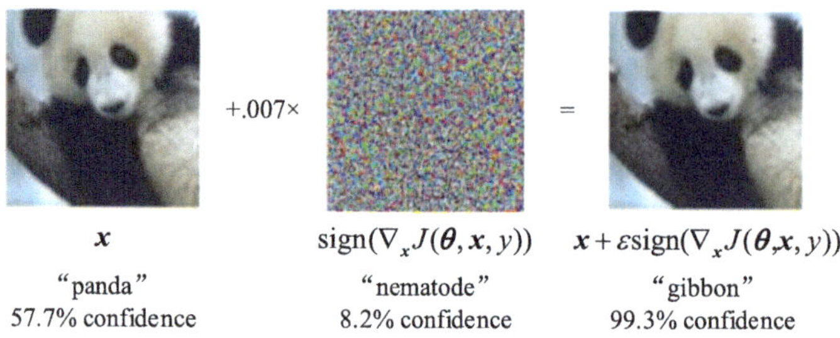

Fig. 4.10 Schematic diagram of adversarial sample attack

various means or directly attack the machine learning model to achieve the intended countermeasures.

The adversarial sample attack refers to the attacker designing a targeted numerical vector by slightly modifying the data source. The user cannot perceive it, causing the machine learning model to make a misjudgment after receiving the data. For example, through the machine learning model, the leftmost picture in the figure below (Fig. 4.10) can be successfully recognized as a panda. However, by adding subtle noise to this picture, the picture on the right is obtained, and this picture will be incorrectly recognized. It is a gibbon, and the confidence level is still very high. Although for humans, such subtle disturbances are indistinguishable by the human eye, this disturbance is enough to change the prediction results of the model.

The adversarial attack first constructs an adversarial example, then inputs the adversarial example into the machine learning model like normal data, and finally obtains the deception recognition result. The construction of adversarial examples can be divided into two categories, white-box attacks and black-box attacks, based on the amount of information the attacker has access to within the machine learning model.

A white-box attack means that the attacker analyzes the target model and obtains the model's parameters to construct a adversarial example. In contrast, a black-box attack occurs when the attacker lacks knowledge of the machine learning system's algorithms and parameters. Consequently, the entire target model is a black box for the attacker. The attacker initially provides input and observation output, after which they train an infinitely similar model to the target model based on the input and the observed output. Finally, the attacker uses the approximate model to generate an adversarial example that makes the target model classify incorrectly.

D. **Advanced Persistent Threats Attack**

An Advanced Persistent Threat (APT) attack, also known as a directed threat attack, is a specific type of cyberattack. It refers to various advanced attack methods to carry out organized and long-term continuous network attacks on high-value targets [5]. An APT attack requires long-term management and planning to be successful, so it is highly concealed and targeted. It tends not to pursue short-term economic gains and pure system destruction but instead focuses on system intrusion step by step.

(1) APT Attack Method

APT attacks typically gain access to a target network via one of the following six methods:

- Break through the web server facing the external network through attacks such as SQL injection;
- Use the intruded web server as a springboard to scan other servers or desktop terminals on the intranet and prepare for further intrusion;
- Obtain the administrator account through password cracking or sending fraudulent emails, and finally, break through the AD server or core development environment;
- The private mailbox of the attacked person automatically sends a copy of the email to the attacker;
- Malicious software, such as Trojan horse, backdoor, Downloader, and other malicious software, is then implanted, returning a large number of sensitive files (e.g., WORD, PPT, PDF, CAD files);
- Attachments with malicious programs are sent through senior managers' emails to trick employees into clicking on them and hacking intranet terminals.

The APT attack method can be divided into five distinct stages:

1. Intelligence gathering stage. Attackers use technology and social engineering methods to collect a significant amount of key information about the system's business processes and usage. Through sorting and analysis, it is possible to identify potential security weaknesses in the system. Concurrently, it collects vulnerabilities, creates Trojan horse programs, and develops attack plans to prepare for subsequent stages of attacks.
2. Enter stage. An APT attack uses a large amount of information collected through a 0-day vulnerability to attack a predetermined target host, send a malicious program to the target host, and trick users into running the malicious program. Another method is to send an email containing a malicious URL to the intended target. When the email is opened, the malicious program downloads by itself and executes, achieving its goal.
3. Command and control stage. APT attack determines whether it is an important computer by searching for sensitive information. After finding the target computer, it will establish communication with it through the network communication protocol, establish a secret channel through the internal firewall between the controlled computer and the C&C server, and confirm the computer and the C&C server's successful intrusion to maintain communication.
4. Horizontal expansion stage. Use the channel to find important information and establish the target system. APT determines whether it is an important computer by searching for sensitive information on the target network. After finding the target computer, it uses techniques and tools that contain specific algorithms to increase the attacker's use of authority at a high level, allowing attackers to easily access and control the target computer.

5. Data mining and transmission stage. Attackers use advanced evasion techniques to transmit data to specified machines or external networks. There are numerous methods for data transmission, including the obfuscation of trusted data and DNS traffic.

(2) APT Attack Threat

The primary concern regarding the security of big data is the potential for APT attacks, which employ a range of techniques to achieve their objectives. By leveraging sophisticated techniques and social engineering, attackers can gain internal access and authority to critical national infrastructure and enterprises, leading to big data leakage. This could have far-reaching consequences, impacting sectors such as energy, electricity, finance, and national defense, and national defense, and potentially undermining the stability and security of the nation. At present, information systems are facing significant challenges to their security, with a proliferation of cyberattacks targeting big data. In April 2011, Sony's PSN network platform was hacked by hackers, resulting in the exposure of the personal data of over 70 million users. In April 2012, the CIA website was hacked and forced to go down for several hours. A number of internal court documents were made public. In the era of big data, big data in connected devices such as mobile phones and social networks can effectively double-lock individuals from the two dimensions of time, space, and social context, thereby posing significant risks to user privacy and security.

Advanced persistent threats targeting big data tend to exhibit a set of distinctive technical characteristics, including:

1. Attackers Often Use Malicious Websites to Lure Targets into the Bait by Means of Phishing.
2. Malicious emails are a common method of attack, with attackers often disguising themselves as legitimate senders. Furthermore, the malicious code embedded in email attachments frequently exploits 0 day vulnerabilities, and traditional email content analysis is also ineffective in detecting such attacks.
3. Some attacks are implemented directly through SQL injection into the target public website.
4. When attackers control the victim machine, they frequently use SSL connections. Consequently, most existing content detection systems cannot analyze the transmitted content, and they are unable to analyze suspicious connections.

4.5.2 Big Data Security Protection Technology

In response to the growing concern surrounding the potential risks of big data in cyberspace, various technologies have emerged with the aim of enhancing security and privacy protection [6]. These technologies include a variety of approaches, including the development of trusted computing technology, access control mechanisms, anonymization privacy protection technology, federated learning algorithms, and identity authentication solutions.

A. **Trusted Computing**

Trusted computing technology is an innovative security paradigm that ensures the integrity and confidentiality of computer operations. It guarantees that the outcomes of computations align with expectations and that the processes themselves can be monitored and regulated without external interference. In essence, it represents an advanced computing model that integrates active defense mechanisms with the computational process, offering features such as authentication, state assessment, and secure data storage.

The core of Trusted Platform Module (TPM) is a processor that incorporates security functions. The main system architecture is presented in Fig. 4.11.

The application of trusted computing technology is currently more advanced in the context of trusted payment terminals, such as UnionPay electronic payment. The Trusted Cryptography Module (TCM) is embedded in the system terminal. Once authentication with the back-end server is complete, users can use UnionPay cards to conduct online payment, access account information, and perform other functions, with the assistance of the TCM to ensure the security and dependability of the transaction [7]. After the system was unveiled at the 2008 "Financial Electronics Show," the system attracted considerable attention.

B. **Access Control**

Access control technology restricts the user's access to specific information items or restricts the use of certain control functions according to the user's identity and a certain defined group to which it belongs [8]. The technology in question is comprised of three fundamental elements: the subject, the object, and the control strategy. The technical principle is presented in Fig. 4.12.

Access control technology can be divided into three categories: discretionary access control, mandatory access control, and role-based access control.

Discretionary access control is based on the access control list. This method involves attaching a subject list to the object, where the resource owner determines the resource access rights of other subjects. This approach offers flexibility and convenience, but it also has disadvantages, including lower security performance and greater management difficulty.

In contrast, mandatory access control is a multi-level access control strategy that implements mandatory access control on subjects and objects. This access control method has high security performance, but it is challenging to manage [9].

Role-based access control is a method of linking permissions with roles, and users obtain permissions for the role by acting as members of the appropriate role, as displayed in Fig. 4.13. This approach streamlines the complexity of authorization management, minimizes management overhead, and effectively improves the flexibility of enterprise security policies.

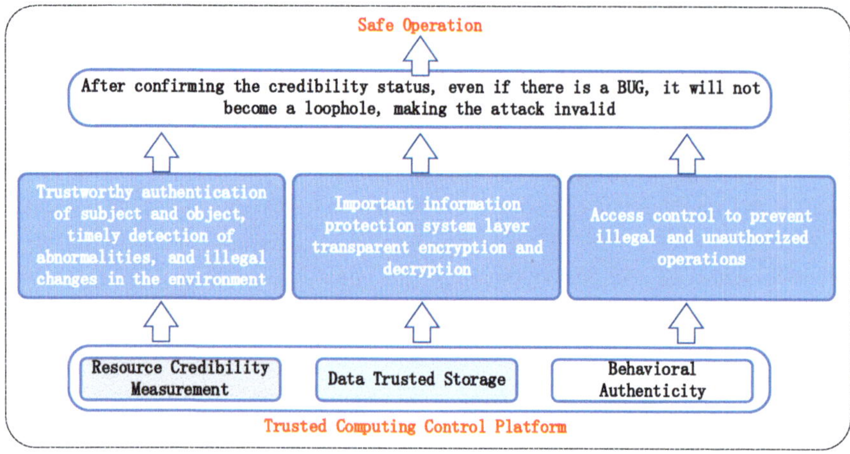

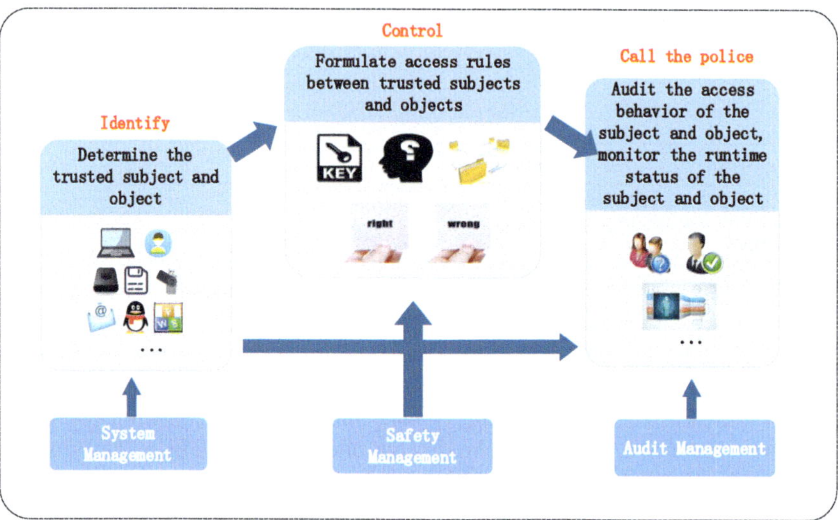

Fig. 4.11 Trusted platform module system architecture diagram

C. Anonymized Privacy Protection

Privacy protection is used to describe the use of anonymization technology, which is employed to prevent entities, such as individuals or collectives, from being identified by external parties and thus ensuring that they receive the appropriate level of protection. Data releases often contain a data table file containing multiple records. Each record has individual attributes. The aggregation of these attributes constitutes a microdata set [10]. The specific information is shown in Table 4.2.

The majority of cybercriminals employ link attacks to steal microdata sets. Link attacks are an attack method that acquires and integrates information through multiple

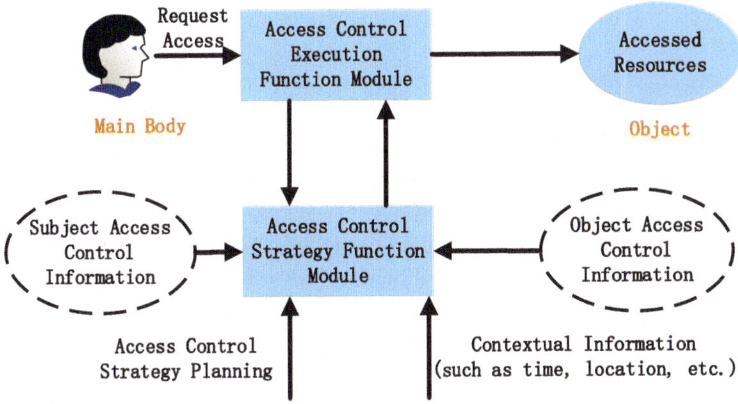

Fig. 4.12 Schematic diagram of access control technology principle

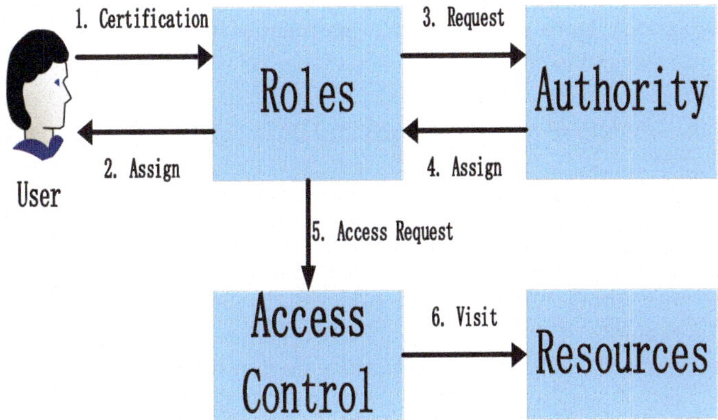

Fig. 4.13 Schematic diagram of the role-based access control model

Table 4.2 Table of microdata set attribute

Attributes	Content
Identifier	Unique identification of the individual, such as ID number and name
Quasi-identifier	Link with other data sheets to mark individual identity, such as gender and date of birth
Sensitive attributes	The attributes that need to be kept secret when publishing include salary and beliefs
Non-sensitive attributes	The attributes that can be disclosed are also called ordinary attributes

channels to obtain private data. The concept of big data is a powerful catalyst for this attack method. Anonymization is one of the main technologies to prevent privacy leakage caused by link attacks. Anonymized privacy protection methods include data encryption-based methods, restricted release methods, and data distortion-based methods.

The data encryption method uses encryption technology to hide sensitive data in the data mining process, including SMC (Secure Multi-party Calculation). Even if two or more sites complete the calculation through a certain protocol, each party only knows its input data and all the final results after the data calculation; it also includes distributed anonymization that ensures site data privacy and collects enough information to achieve data anonymity with the largest utilization rate.

Based on the restricted release method, the original data is selectively released, not released, or sensitive data with low accuracy is released to achieve privacy protection. Current research on this type of technology focuses on "data anonymization" to ensure that the risk of revealing sensitive data and privacy is within a tolerable range, including K-anonymity, L-diversity, and T-closeness.

The more typical privacy protection model based on the restricted release method is k-anonymity, proposed by Samarati and Sweeney in 2002. In response to deanonymization attacks, each record in the data released by k-anonymity must be indistinguishable from at least $k - 1$ other records (called an equivalence class). When the attacker obtains the k-anonymous data, he will get the records of at least k different people, and thus cannot make an accurate judgment. The parameter k represents the strength of privacy protection. The larger the value of k, the stronger the privacy protection, but the more information is lost and the lower the data availability. However, Machanavajjhala of Cornell University in the United States and others discovered the defect of K-anonymity in 2006. It relies heavily on the predefined generalization layer or the total order relationship on the attribute domain, making the anonymous result produce high information loss, resulting in privacy leakage. For example, if the K-anonymity data obtained by the attacker is all AIDS (Acquired Immune Deficiency Syndrome) patients in the equivalence category, the attacker can easily make a judgment that the attacked person definitely suffers from AIDS. To prevent consistency attacks, the new privacy protection model L-diversity improves K-anonymity to ensure that the sensitive attributes in any equivalence class have at least l different values. Experimental results show that this method can reduce information loss and improve the availability of published data. Based on L-diversity, T-closeness requires the distribution of sensitive attributes in all equivalence classes to be as close as possible to the attribute's global distribution.

However, privacy protection methods based on restricted releases are constantly improving. However, at the same time, new attack methods appear, making the traditional privacy protection model based on K-anonymity fall into such an endless cycle. Basically, the shortcoming of the traditional privacy protection model is that it makes too many assumptions about the attacker's background knowledge and attack model. However, these assumptions are often not completely true in reality, so attackers can always find a variety of attack methods to use. This problem was not solved until the emergence of differential privacy technology based on data distortion.

In data distortion-based methods, sensitive data is altered by the introduction of noise and other methods while keeping certain data or data attributes unchanged and maintaining certain statistical properties. The aforementioned methods can be classified into three categories: randomization, blocking, and cohesion, and differential privacy protection. Randomization means adding random noise to the original data and then publishing the disturbed data. The blocking method entails the exclusion of specific data from publication. The cohesion method involves the aggregation of statistical information within groups of raw data records. Differential privacy protection refers to adding noise to ensure that deleting or adding records in a dataset will not affect the outcomes of the subsequent analysis. Accordingly, even if the attacker obtains two datasets that differ by only one record, the results produced by analyzing the two are the same. They cannot infer the information of the hidden record. This method is a new privacy protection model proposed by Dwork of Microsoft Research in 2006. This method addresses two major limitations of the conventional privacy protection model: (1) It defines a fairly strict attack model that does not account for the background knowledge of the attacker has, even if the attacker has mastered all the record information except a specific record (the maximum background knowledge assumption). This ensures that the privacy of the record cannot be revealed; (2) it provides a comprehensive definition and a quantitative evaluation method for the level of privacy protection. The numerous advantages of differential privacy led to its rapid replacement of the traditional privacy protection model. It has since become a prominent topic in contemporary privacy research, attracting attention from diverse fields such as theoretical computer science, database management, data mining, and machine learning.

D. **Federated Learning**

Federated learning is a machine learning setting where multiple clients (such as mobile devices or the entire organization) train models collaboratively under a central server (such as a service provider). Furthermore, this setting ensures the decentralization of training data and uses local data collection. The minimization principle represents a solution for large-scale deployment to protect privacy, as it serves to reduce some of the systemic privacy risks and costs brought by traditional centralized machine learning methods. The typical workflow of federated learning is as follows:

- Problem identification: model engineers identify the problems to be solved using federated learning;
- If necessary, client settings set the client (such as an app running on a mobile phone) to store the necessary training data locally (although there are limitations on time and quantity). In many cases, the app already stores data (for example, a text message app must store text messages and a photo management app must store photos). However, in other cases, additional data or metadata, such as user interaction data, must be retained to provide labels for supervised learning tasks;
- Simulation prototype development (optional): model engineers may develop prototypes for model architecture and use proxy datasets to test and learn hyperparameters in a federated learning simulation environment;

- Federated model training: start multiple federation training tasks to train different variants of the model or use different optimized hyperparameters;
- Federated model evaluation: after the task is fully trained (usually several days), analyze the model, and select excellent candidate models. The analysis may include metrics calculated on the standard dataset in the data center, or the model may evaluate the results of a federated evaluation of local client data on the client-side;
- Deployment: finally, once the model has been selected, it is necessary to initiate the standard model deployment process. This includes conducting manual quality assurance, implementing real-time A/B testing (using the new model on some models and then deploying it on other models), and comparing its performance with that of the staged rollout. The staged rollout allows for the rollback of the process when poor behavior is identified, thus minimizing the impact on a significant number of users. The particular methodology employed during the installation of the model is set by the owner of the application and is typically distinct from the model training process. In other words, this step can be used in a similar manner for models that have been trained using either federated learning or traditional data center methods.

E. **Identity Authentication**

Identity authentication is the process of verifying the identity of an operator engaged in interaction with an information system or network. The conventional methods to authentication rely on the use of passwords or other forms of credentials that are known to the user. However, modern authentication technologies, which are based on big data, go beyond this by collecting and analyzing user and device behavior data. By identifying patterns and characteristics in this data, these advanced systems can more accurately ascertain and confirm the identity of the user with greater accuracy, thereby providing a more robust and dynamic approach to authentication.

The authentication technology is characterized by robust security features. It is challenging for an attacker to simulate user behavior to pass the authentication. Concurrently, the user is not unduly burdened by the implementation of the method, which can also facilitate the unification of each system's authentication mechanism. However, this method is not without its limitations. The absence of a large amount of data in the initial stage of the authentication process may result in an inaccurate analysis, with the potential for user privacy to be compromised through the collection of information.

In the real world, the fundamental techniques used for user authentication can be divided into the following three categories:

1. Identity authentication based on information secrets, whereby an individual's identity is proven based on the information they possess, such as a static password;
2. Identity authentication based on trusted objects, whereby an individual's identity is proven based on the items they possess, such as dynamic tokens, smart cards;

3. Identity authentication based on biometrics, whereby an individual's identity is proven based on their unique physical characteristics, such as fingerprints, iris, voice, or face.

In the online world, the methods are analogous to those in the real world. In order to achieve enhanced identity authentication security, certain scenarios will select two of the above three types to combine and use, a process known as two-factor authentication. Two-factor identity authentication has a wide range of applications, including banking, securities, online games, e-commerce, e-government, online education, enterprise informatization, and other fields. Two-factor identity authentication can be employed to protect many types of application systems, including host computers, diverse network equipment, and various application systems that use computers, mobile phones, telephones, and digital TVs as operating terminals.

Two-factor identity authentication can be implemented in the user's daily website and forum login. When two-factor authentication is adopted by a user's daily website and forum, the original login password and forum can be used as an authentication factor, along with dynamic tokens and other forms of authentication. This configuration forms a two-factor authentication mechanism, providing a robust security layer for user access to these online platforms.

The following introduces several typical identity authentication technologies as examples to understand common big data security protection technologies based on identity recognition.

(1) Static Password

Static password represents the most conventional form of authentication technology. In the "account + password" authentication method, the password is a fixed string chosen by the user. Once established, this static password remains constant unless the user proactively alters it. Upon entering the correct password when logging into the network, the system recognizes the individual as an authorized user. However, to prevent forgetting their passwords, many users opt for easily guessable combinations, such as birthdates or phone numbers, or they may write them down and store them in a secure location, inadvertently risking exposure. Since static passwords are fixed data, they must be stored in the computer's memory during authentication, making them susceptible to interception by Trojan horse programs or network eavesdropping during transmission. While the static password mechanism is straightforward to implement and use, it has significant security vulnerabilities. Static passwords are prone to a range of threats, including shoulder surfing, guessing, dictionary attacks, brute force cracking, theft, monitoring, replay attacks, and Trojan infiltration. These vulnerabilities highlight the need for more sophisticated and secure authentication methods to protect against the evolving landscape of cyber threats.

One method of improving the security of static passwords is to require users to change them on a regular basis. However, this approach had caused challenges in the use and management of static passwords, particularly when a user is required to manage numerous passwords, such as in the case of individuals with dozens of passwords to process. The necessity of frequently changing these passwords can lead to issues such as wrong or forgotten passwords. It is challenging to mandate that all users perform regular password modifications. Notwithstanding regular password modifications, the password will remain fixed for a considerable period of time. In essence, the shortcomings and deficiencies of static passwords are mainly evident in the following key areas:

- The ease of use and security of static passwords are mutually exclusive, and neither can be taken into account. Simple and easy-to-remember passwords are weak, whereas complex static passwords provide a high level of security, albeit at the cost of usability;
- Static passwords are susceptible to a range of security threats and present a significant risk to the user;
- The risk and cost of static passwords are high. In the event of a leak, the greatest loss may be incurred, and it is often the case that the static password has been leaked before it is lost;
- The use of static passwords is inconvenient, particularly when users are required to maintain more than a dozen static passwords. The process of reporting the loss of a static password and subsequently resetting it can be time-consuming, and it can negatively impact user experience;
- Therefore, despite the simplicity of using and deploying static password mechanisms, in terms of security, static password is a single-factor authentication method that no longer meets the Internet's demand for authentication security.

(2) Dynamic Token

A dynamic token is a sophisticated identity authentication terminal device used to generate dynamic passwords. It is imperative that this device be used in conjunction with a back-end authentication system. Dynamic passwords, also known as one-time passwords, represent a transformation in the traditional password paradigm. It is a modified version of the password. The change in the dynamic password is attributed to alterations in the operational factors that generate it, including time, frequency, transaction amount, counterparty account number, transaction serial number, and other information. The generation of dynamic passwords typically employs a combination of operational factors. One such factor is the seed key of the token, which serves as a fixed identification code representing the user's identity. The second factor is the change factor, which represents a continuous change in these change factors. Changes have produced dynamic passwords that are constantly changing.

The dynamic token is a straightforward and accessible mechanism. The use of a constantly changing password ensures that the security of the system is greatly improved compared with a static password. It has a variety of forms, including hardware tokens, software tokens, mobile phone tokens, SMS tokens, etc., and has a wide range of applications. It can be widely used in banking, securities, online games, e-commerce, e-government, and online education. In the fields of enterprise informatization, it can serve to safeguard various application systems, including host computers, various network devices, and diverse systems that use computers, mobile phones, telephones, and digital TVs as operational terminals.

(3) Digital Signature

Digital signatures, also known as public-key digital signatures, are cryptographic constructs that either accompany or are intrinsic to a data unit. They enable the recipient to authenticate the origin of the data and to ensure its integrity. Furthermore, they serve to safeguard the data from being forged by third parties, including the recipient. Essentially, they are a means of digitally endorsing messages, which can be disseminated over communication networks. Digital signatures typically involve a pair of complementary operations: one for signing and the other for verifying.

Both public-key and private-key cryptosystems can generate digital signatures; however, the current standard predominantly relies on public-key cryptosystems. This includes both ordinary digital signatures and special digital signatures. Commonly used digital signature algorithms include RSA, ElGamal, Fiat-Shamir, Guillou-Quisquater, Schnorr, the Ong-Schnorr-Shamir digital signature algorithm, Des/DSA, the elliptic curve digital signature algorithm, and the finite automata digital signature algorithm. Special digital signatures include, but are not limited to, blind signatures, proxy signatures, group signatures, undeniable signatures, fair blind signatures, threshold signatures, signatures with message recovery functions, etc. These are closely related to specific application environments. The implementation of digital signatures involves a number of legal considerations. The U.S. federal government has established its own Digital Signature Standard (DSS), which is predicated on the discrete logarithm problem within a finite field.

Digital signatures play an important role in the network environment:

- Anti-counterfeiting. Only the signer knows the private key, thereby preventing others from constructing the correct one.
- Identifiable identity. In a network environment, it is imperative that the receiver be able to authenticate the identity claimed by the sender.
- Anti-tampering (anti-corruption of the integrity of the information). In a digital signature, the signature and the original file are combined to form a single, indivisible dataset that cannot be altered, thereby ensuring the integrity of the data.
- Anti-replay. If technologies such as adding serial numbers and timestamps to signed messages are used, replay attacks can be prevented in digital signatures.
- Non-repudiation. In a digital signature system, the receiver is required to return a self-signed message indicating that it has received it. This is referred to as non-repudiation, which can be achieved through the involvement of a third party or the introduction of a third-party mechanism.
- Confidentiality. The digital signature can encrypt the message to be signed.

4.5.3 Big Data Security Construction Case

A. **Big Data Security Practices in 360 Company**

In order to help enterprises quickly implement applications based on big data, the 360 Enterprise Security Group has developed a 360 SecGate security big data platform,

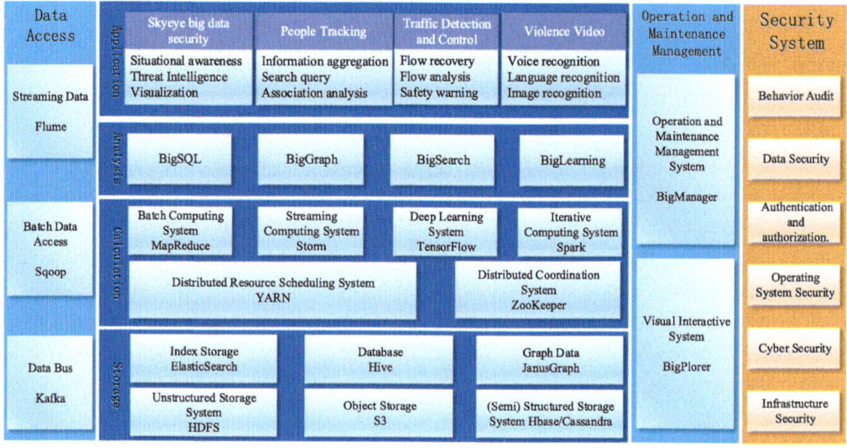

Fig. 4.14 A diagrammatic representation of the 360 SecGate security big data platform system architecture

which supports PB to EB-level big data applications. To ensure the platform's big data security, the company makes use of a variety of standards and specifications.

The 360 SecGate Security Big Data Platform is an open platform that integrates multiple open-source systems, including data access, data storage, data calculation, data analysis, data application, data operation, maintenance management, security protection, and other functional modules (Fig. 4.14). The platform's independent and other security system construction can help companies rapidly build specialized products for massive data analysis applications, thereby tapping the value of big data while safeguarding data security across all aspects.

The 360 SecGate security big data platform integrates a suite of security measures across key domains—system, network, data, and application—to ensure robust big data security. It includes essential components such as behavior audits, data protection, authentication and authorization protocols, operating system safeguards, network defenses, and technical facility security, thereby offering a well-rounded approach to data security.

1. Behavior audit. All actions undertaken by entities on the platform will be documented in audit records and stored in encrypted form. The analysis of audit data can facilitate the realization of security functions, including entity behavior mining and the discovery of abnormal behavior. These works refer to standards such as GB/T 18794.7-2003 and GB/T 17143.8-1997.
2. Data security. All data stored on this platform is encrypted and stored in accordance with the specific characteristics of the data. The utilization of diverse encryption methods is contingent upon the nature of the data. When the encrypted data is used within this platform, transparent encryption and decryption can be realized without affecting data analysis and data mining. This aspect of the

work complies with the regulations and technical requirements of relevant state departments, including those set forth by the Secrecy Administration.

3. Authentication and authorization. The authentication and authorization processes are embedded throughout the entire functionality of this platform, including but not limited to data acquisition, data storage, data calculation, user operations, etc. The authentication and authorization processes are based on the internationally accepted RBAC (Role-Based Access Control) model, which enables the implementation of fine-grained control over user groups, users, roles, and permissions. The Kerberos protocol is used for account information security authentication. Furthermore, a single sign-on functionality is available, which enables the consolidation of user management and authentication processes.
4. Operating system security. This mainly includes operating kernel security reinforcement, operating system patch update, operating system authority control, operating system port management, the operating system running program detection, etc. For further details, please refer to GB/T 20272-2006 "Information Security Technology Operating System Security Technical Requirements."
5. Cyber security. Before deployment, a network plan is developed according to the specific requirements and circumstances of the customer. This plan is designed to ensure that the network possesses the essential attributes of isolation, confidentiality, and stability. The platform's network is divided into two levels: the business level and the management level. Physical isolation is maintained between the two levels. In the operational process, network security incidents were promptly addressed in accordance with the standards set forth in GB/T 28517-2012, GB/T 32924-2016, and GB/T 25068.
6. Infrastructure security. A comprehensive approach is employed to safeguard the various equipment on the platform, utilizing anti-virus, border protection, intrusion protection, situational awareness, threat intelligence, and other measures. These works refer to standards such as GB/T 20281-2015 and GB/T 31505-2015, GB/T 20275-2013.
7. The 360 Enterprise Security Group has played an active role in the development and revision of big data security standards, including GB/T 35274-2017 "Information Security Technology Big Data Service Security Capability Requirements," and multiple upcoming big data security standards. The preparation unit and platform developers ensure the timely implementation of the relevant requirements of these standards into the product development and operational management work.

B. **Big Data Security Practices in Alibaba**

In order to ensure the compliance and security of the entire data business link, Alibaba provides a big data platform for the e-commerce industry. The platform ensures the protection of consumer privacy and trade secrets at the three levels of business, data, and ecology. As illustrated in Fig. 4.15, there are a number of security risks and challenges associated with these aspects.

Firstly, with regard to the design of the business model, the big data security platform is established in accordance with the specific characteristics of e-commerce

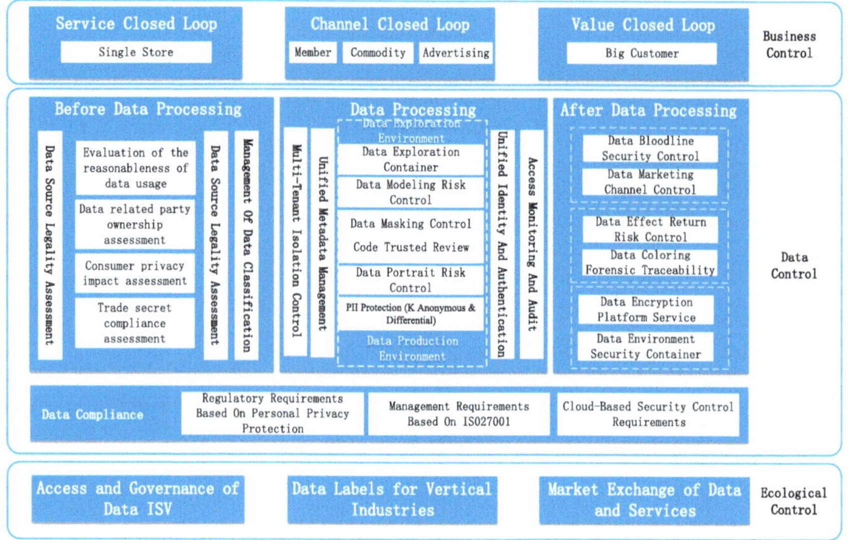

Fig. 4.15 Alibaba big data security system architecture diagram

and the boundary of data ownership. This is done in a manner that encompasses three distinct loops: an in-store service closed-loop based on private domain data, an in-platform channel closed-loop, and a value closed-loop based on public domain data. The internal channel and value closed-loop ensure that the business's authorization boundary as a whole is both reasonable and clear. The data processing logic is based on the security principle of availability and invisibility. The utilization of data is not merely the presentation of the data itself, but rather the application of its intrinsic value.

Secondly, this big data security platform establishes a comprehensive data management and control system based on data business links, including data security management and control before data processing, during data processing, after data processing, and data compliance. In the data compliance layer, the relevant specifications are referenced as follows: "GB/T 35273-2017 Personal Information Security Specification," "GB/T 35274-2017 Information Security Technology Big Data Service Security Capability Requirements," "GB/T 31168-2014 Information Security Technology Cloud Computing Service Security Capability Requirements." In addition, the ISO 27001 series standards are implemented to guarantee the protection of personal privacy information, the security control of cloud services, the security of big data services, and compliance with the national regulatory requirements [11].

Finally, the implementation of a data labeling system based on vertical industries and the establishment of a market management mechanism for data ecology, facilitated by the entry and exit of data ISVs (independent software vendors), ensures an effective balance between business and security.

Through continuous practice and self-improvement, Alibaba has developed a distinctive big data security management concept and a big data security capability maturity model.

With regard to the domain of big data security management, Alibaba has established a set of standardized architectural frameworks for the collection, computation, storage, service, and application of big data. The construction of a big data system architecture has enabled Alibaba to gradually form a data-centric architecture and big data security management concept. The organization's data security is guaranteed throughout the entire data lifecycle. By complying with standards, data security risks are effectively controlled. This not only enhances the data security capabilities of the company itself and its ecological partners but also facilitates the circulation and sharing of data resources within the ecosystem. Data is the soil from which value grows.

Besides, leveraging the extensive big data security expertise cultivated within its internal operations and gleaned from external engagements, Alibaba has honed a Big Data Security Capability Maturity Model, as illustrated in Fig. 4.16. This model serves as a valuable tool for assessing current big data security strengths and delineates a strategic path for enhancing Alibaba's future security initiatives.

Currently, the Big Data Security Task Force of the National Information Security Standardization Technical Committee is in the process of developing national standards for the "Data Security Capability Maturity Model." Alibaba's existing big data

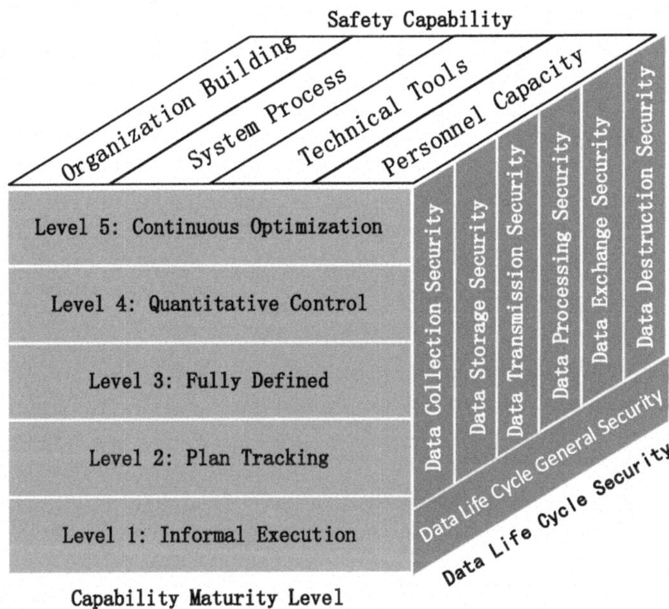

Fig. 4.16 Schematic diagram of Alibaba's big data security capability maturity model

security practices have been instrumental in shaping these standards. The company has successfully implemented its model across various domains, including internal business operations, the broader ecosystem, and multiple industry sectors. Alibaba's pioneering efforts have led to the model's adoption in over 30 enterprises across different industries, significantly bolstering their data security defenses. This has not only elevated the maturity of data security capabilities within these organizations but also contributed to the advancement of the national data security framework.

C. **Big Data Security Practices in China Mobile**

China Mobile has accumulated a substantial corpus of customer, production data, and management data over the course of its long-term business operations. These data exhibit the characteristics of a large scale, multiple types, and high accuracy [12]. In addition, due to the characteristics of new technologies and new formats, such as centralized data management, data opening, and equipment virtualization, new security challenges have emerged in the development of big data services.

Since 2015, China Mobile has progressively enhanced the velocity of its construction of a comprehensive big data security assurance system. The system framework covers six systems, including security strategy, security management, security operations, security technology, compliance evaluation, and service support. The system involves the collection, transmission, and storage of big data, as well as the process of use, sharing, and destruction. Furthermore, the advancement of "big data security protection" measures and the implementation of pilot research on "big data security applications" ensure the comprehensive protection of big data confidentiality, integrity, availability, and traceability. The environment in which big data is stored, processed and transmitted is secure, manageable, controllable, and reliable.

For more than two years of operation, the big data security assurance system has been an invaluable asset in facilitating the open sharing of company data, optimizing data usage, and ensuring data security. Related achievements have been promoted to become international, national, and industry standards, and efforts are being made to become domestic and international best practices. The framework of its big data security assurance system is shown in Fig. 4.17.

The following seven aspects will present an overview of the functions and characteristics of China Mobile's big data security framework.

1. The security strategy system clarifies the overall requirements and direction of security assurance work from the top-level design level, fully grasps the objects of big data security work, proposes the fundamental principles and core strategies of security management and control, establishes and improves the internal management process of big data security, and continuously strengthens the management of big data external cooperation. Corresponding management systems include: "Guidelines for the Prevention and Control of Big Data Security Risks," and "Overall Strategy for Big Data Security Assurance."

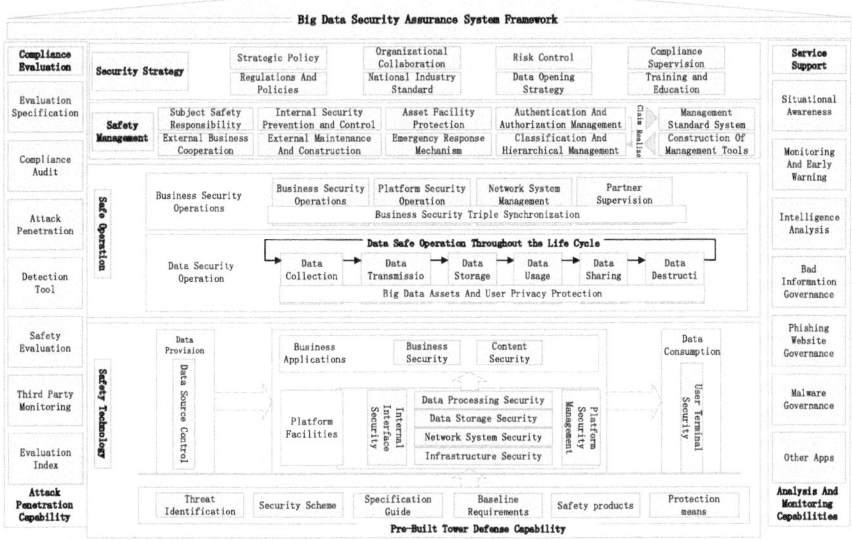

Fig. 4.17 Framework diagram of China mobile's big data security system

2. The safety management system establishes a management system to clarify the primary security responsibility of the operator and implement safety management measures. The relevant systems include third-party cooperative management, internal safety management, data classification, classification management, emergency response mechanisms, asset facility protection, certification authorization management, and other safety management requirements. The corresponding management systems include: "Big Data Security Management Requirements," "Big Data Security Management and Control Classification and Hierarchical Implementation Guide," and other related documents are primarily aimed at guiding market, business, support, and other departments in the implementation of data and platform security management.
3. The security operation system is designed to facilitate comprehensive security management of big data and data-driven business operations. This is achieved by defining operational roles and clarifying the security responsibilities of operating organizations. The system design includes business security operations, data security operations, and the management and control of security emergencies. The aforementioned management systems include "Big Data Security Operation Requirements," etc. and are mainly targeted at business departments, system construction, and support departments, with the objective of implementing standardized procedures.
4. The security technology system encapsulates the company's requirements and strategies for establishing robust big data security safeguards. This system design covers the general technical specifications for data security at every stage of the data lifecycle. It also includes baseline security configuration standards for the

various infrastructure and application components within the big data platform. Furthermore, it features a security protection system architecture for the big data platform, which is designed and implemented based on the general specifications for data protection and platform baseline specifications. Additionally, it provides technical guidelines for the implementation of sensitive data security desensitization. The corresponding enterprise standards include: "General Technical Requirements for Big Data Security Protection," "Big Data Platform Security Baseline Requirements," "Big Data Security Protection Technology Implementation Guide," "Big Data Security Desensitization Implementation Guide." These standards are primarily intended for system construction, operation, and maintenance departments, offering regulatory guidance for the application in scenarios that include system planning, construction, operation, and acceptance process.

5. The safety compliance evaluation system includes safety operation management compliance evaluation and safety technology compliance evaluation methods, and evaluation procedures. The objective of its construction is to achieve continuous optimization of the security assessment capabilities, achieve a comprehensive assessment of the risk points of each link of the big data business, and ensure the effective implementation of the security management system and technical requirements. The corresponding corporate standards include: "Big Data Security Technology Compliance Evaluation Method," etc., which are mainly aimed at security management departments and system construction departments in scenarios involving security acceptance, inspection, and self-inspection, as well as other related activities.
6. The big data service support system is based on big data resources and is designed to provide support services for information security. These include research and promotional activities related to the application of big data in the security field, the provision of novel technical means for the company's information security governance, and offering assistance to external security services in achieving Data Value-Added.
7. A review of the security standards practice experience of China Mobile indicates that national standards lack adequate security technology, especially in terms of technical security requirements for big data-related products. For example, big data basic software products represent a fundamental component of the construction of big data platforms. This can be defined as a "big data operating system." It is imperative that unified security requirements be established at each stage of the development process, including model selection, procurement, construction, and other links. It is recommended that the security functional requirements related to big data basic software be promoted in accordance with national standards. Furthermore, it is necessary to develop supplementary national standards for big data security evaluation and big data security applications.

4.6 Summary

This chapter introduces the cognitive foundation of cyberspace security, including an overview of the relevant big data resources, an examination of the associated big data analysis technology, and a discussion of the latest developments in cyberspace big data security protection technology.

The study of big data and cyberspace security requires an understanding of four fundamental concepts: complex networks, information networks, cyberspace, and cyberspace security. The focus is on information security confrontation, that is, multi-level attack and defense at the basic and system layers.

Basic resource data constitutes the fundamental basis for research into the domain of security attacks and defense. A public dataset may include resources related to network security, including personal data, industry data, security traffic data, security log data, and security public opinion data. The public application dataset includes KDD CUP, ISOT, ECML-PKDD, CSIC, CTU-13, DREBIN, AMD, etc., which can be used in the fields of computer network intrusion detection, botnet detection, malware detection, and other fields.

The commonly used cyberspace big data analysis techniques include security event correlation analysis, network anomaly detection and analysis, data content security analysis, and security situation awareness analysis. Network security situation awareness technology is a complex system of various security analysis technologies, with the primary objective being situation understanding. The implementation of situational assessment, situational prediction, and situational visualization allows users to gain a comprehensive understanding of the pertinent information, thereby facilitating informed decision-making.

In the contemporary academic and industrial contexts, big data occupies a pivotal position. It is regarded as a substantial shift in the way individuals live, work, and think. The security of big data and the technology used to protect it are of great significance in the field of cybersecurity research.

Although big data analysis technology is extensively deployed in fortifying cyberspace security, the security of big data itself presents a significant concern. Cyberspace and big data security issues have repercussions that extend beyond individuals and enterprises; the proliferation of false information can severely skew the outcomes of analyses reliant on big data security techniques. Currently, leading Internet companies have initiated the development of data security protection frameworks. However, the efficacy of these systems is variable. The development of a systematic, comprehensive, cutting-edge, and efficient cyberspace data security protection system is emerging as a pivotal area of research and innovation for the future, with the objective of addressing and surmounting the multi-faceted challenges inherent in big data security.

4.7 Exercises

1. State the three main characteristics of complex networks.
2. Briefly describe the basic concepts of cyberspace and cyberspace security.
3. What are the main types of information security issues associated with information systems?
4. State the connotation of the six basic principles of information security and confrontation theory, the six system-level principles, and the relationship between the two.
5. List the common big data resources in cyberspace, and briefly describe their main content and characteristics.
6. What is anomaly detection? What are the main application scenarios of anomaly detection technology in network security issues?
7. Please design an analysis system to detect illegal text content (assuming that there are enough normal text and illegal text corpora with labels), mainly to explain the method's framework principle and the key technologies involved.
8. Briefly describe the definition and connotation of cybersecurity situational awareness and outline the core steps in the process of cybersecurity situational awareness.
9. List three common attack techniques against big data and briefly explain their attack principles.
10. What is federated learning? Briefly describe the general process of federated learning.
11. For big data security, briefly describe three common security protection technologies and their protection principles.
12. Please start from the basic theory of information security confrontation, analyze the relationship between cyberspace big data security analysis technology and its security issues.

References

1. China Academy of Information and Communications Technology (CAICT). Big data white paper;2016. http://www.cac.gov.cn/files/pdf/baipishu/dashuju2016.pdf.
2. Lee W, Stolfo S. Data mining approaches for intrusion detection;1998.
3. The Great Wind. Israel: turn big data into a weapon against terrorism;2017. http://wemedia.ifeng.com/12643194/wemedia.shtml.
4. National Information Security Standardization Technical Committee Big Data Security Standards Special Working Group. Big data security standardization white paper;2018. https://www.tc260.org.cn/upload/2018-04-16/1523808293220001658.pdf.
5. Lin M. APT attack method and detection method based on big data. Comput CD Softw Appl. 2014;(10):91.
6. Dengguo F, Min Z, Hao L. Big data security and privacy protection. Chin J Comput. 2014;37(1):246–58.

7. InfoQ. Big Data Analysis in the Security Field;2014. http://www.infoq.com/cn/articles/big data-analytics-for-security.
8. The Open Web Application Security Project. Web attack source tracing based on big data;2015. http://123.207.31.216/OWASP-CHINA/OWASP_Conference/owasp-2015fh/files/06WEBV2.pdf.
9. Zuoning C, Guangyi W, Sutai H et al. Big data security and autonomous controllability. Chin Sci Bull. 2015;(z1):427–32.
10. Pingshui W, Jiandong W. Research progress of anonymization privacy protection technology. Comput Appl Res. 2010;27(6):2016–9.
11. Xin L, Xiaolu H. Research on improving big data security system. Inf Secur Res. 2015;1(3):211–6.
12. Yang W, Peiyue G. Research on value operation of mobile communication big data resources. J Wuhan Univ Technol (Inf Manag Eng Ed). 2016;38(3):347–50.

Chapter 5
Big Data and Natural Language Processing (NLP)

5.1 Introduction

In recent years, the exponential growth in big data has primarily originated from individuals' daily activities, particularly in the context of Internet-based services. As indicated by the Digital Universe study conducted by the International Data Corporation (IDC), the aggregate volume of data generated and replicated globally in 2011 surpassed 1.8 Zettabytes, with 75% of this total originating from individual sources. Text data constitutes the largest category of these data. For example, chat or moment messages are generated by users on WeChat on a daily basis. Additionally, topic or comment messages are generated by social media represented by Sina Weibo. The daily data volume of the original text content generated by users is 100 million posts. In order to mine valuable information from these text data and obtain political and economic value, it is necessary to utilize natural language processing technology.

In the context of the current era of big data, text data is typically characterized by the following attributes: 1. The acquisition of massive text data is now more readily achievable than was previously the case. For example, the downloading of basic public resources and the crawling of Twitter data can result in the acquisition of a substantial corpus of text data. 2. The diversity of text data is more abundant when it is massive. The majority of the text data is presented in an unstructured format. Furthermore, the form and style are subject to constant change. The same information described by different individuals will form a corpus of text data that is not repetitive. 3. The information contained in the text data is updated with greater frequency. When new phenomena emerge, Weibo and WeChat Moments frequently serve as the initial platforms to document these developments. The text data in these locations frequently reflects the prevailing social circumstances in real time.

The characteristics of text data in the era of big data have promoted the development of natural language processing technology; however, this has also brought about a number of challenges. 1. The quantity of data does not necessarily correlate with its value. Conversely, it frequently results in the dissemination of information

S. Luo and L. Pan, *Big Data Analysis*, Big Data Management,
https://doi.org/10.1007/978-981-95-5725-7_5

spam 2. The diversification of text data also implies a lack of prior knowledge. The primary focus of traditional natural language processing is on structured or semi-structured data. The availability of structured knowledge allows for the preliminary organization of data prior to its analysis. 3. The advent of real-time data generation has precipitated a shift in the landscape of data analysis, moving it from an offline to an online domain. This has created a need for the adaptation of new natural language processing technologies to big data real-time processing frameworks. In conclusion, the vast quantity of text data provides a comprehensive data foundation for natural language processing technology, yet it also presents novel problems and challenges. It is therefore of great value to study NLP technology and combine it with big data analysis technology to adapt to the era of big data.

This chapter introduces and discusses four significant aspects of natural language processing: basic cognition, basic resources, analysis technology, and application cases. The chapter on basic cognition provides an overview of the background and significance of natural language processing, an analysis of the current research status, an examination of the basic theoretical methods employed, and a discussion of the challenges currently facing the field. The following section presents a statistical description of the existing corpus, language knowledge base, knowledge graph, and other resources. The analysis technology chapter also addresses the current fundamental tasks of natural language processing and their implementation methods, as well as how each task is achieved in the era of big data. In the following chapter, a number of representative cases will be provided that illustrate how big data and natural language processing are currently being combined. In conclusion, the critical conclusions of each section will be summarized at the end of this chapter.

5.2 Foundations of NLP

5.2.1 Brief History of Research

The initial research on natural language processing is grounded in an empiricist approach.

In 1913, the Russian scientist Markov (1856–1922) used the method of manual frequency checking to count the occurrence frequencies of vowels and consonants in the literary work "Eugen Onegin," a lengthy poem by Alexander Pushkin. He proposed the theory of the Markov random process and established the Markov model. His research is based on the statistical data of Russian vowels and consonants, and the methods used are predominantly statistical and empirical.

In 1948, the American scientist Shannon applied the discrete Markov process's probabilistic model to the description of automata. He drew a parallel between the act of transmitting language through media such as communication channels or acoustic speech and the process of transmitting information through a "noisy channel" or "decoding." Additionally, he adopted the concept of "entropy" from thermodynamics to quantify the channel's information capacity, or the language's information content. Moreover, he used a manual method to quantify the probability of English letters, and used probability technology to determine the zero-order of English letters for the first time; the entropy value is 4.03 bits.

This statistically based empiricism represents a significant departure from the approach advocated by Noam Chomsky. In 1956, Chomsky incorporated the concept of the finite-state Markov process from Shannon's work, used finite-state automata as a means of describing language grammar, defined finite-state language as a product of finite-state grammar, and established a finite-state model of natural language. This early research work led to the establishment of a new research field, "formal language theory," which attempted to provide a unified mathematical description of natural and formal language. Formal language theory has also become the most important theoretical cornerstone of computer science.

From the late 1960s to the 1970s, the transformational generative grammar served as a formal description method of natural language, offering a powerful tool for computer processing of natural language and greatly promoting the development of natural language research and language processing. This approach, to a certain extent, addressed some of the limitations inherent in traditional linguistics, propelling the evolution of linguistic theory and methodology. However, it is argued that statistical methods are only capable of explaining the surface phenomena of language, and are unable to account for the internal rules or generation mechanisms by which it is governed. It is a methodology that is distinct from the empirical approach that was characteristic of early natural language processing. The research approach of transformational generative grammar is informed by the philosophical tradition of rationalism.

From the late 1950s to the mid-1960s, empiricism in natural language processing became progressively more prevalent.

In the late 1950s, the Bayesian method was first employed to solve optical character recognition problems. In 1964, Mosteller and Wallace used the Bayesian approach to successfully resolve the distribution of the original authors of anonymous articles' distribution problem in the Federalist, thereby demonstrating the superiority of the empirical method.

In the 1950s, the first online Brown corpus was established. The advent of canons has led to the development of statistical methods for the automatic acquisition of language knowledge from corpora, which have become an important area of research within the field of natural language processing.

In the 1960s, statistical methods were successfully employed in the development of speech recognition algorithms. The hidden Markov model, the noisy channel, and the decoding models are of critical importance. However, during this period between the 1960s and the early 1980s, the prevailing approach in natural language processing remained rule-based rationalism, and the empirical method did not receive special attention. This situation underwent a transformation in the early 1980s. A review of the history of natural language processing researchers has revealed that the finite-state model and empirical methods, which were previously overlooked, still have merit. It has become increasingly evident that the rule-based rationalist method is flawed.

In the final five years of the 1990s (1994–1999), research and development in the field of natural language processing has experienced a period of unprecedented prosperity. Probability and data-driven methods have almost become the standard methods of natural language processing. Statistical methods have become integral to a range of natural language processing applications, including machine translation, text classification, information retrieval, question and answer systems, information extraction, and language knowledge mining. The empirical method based on statistics has gradually become the dominant approach in natural language processing research. Since 2008, inspired by the success of deep learning in image and speech recognition, researchers have gradually introduced deep learning into natural language processing research. Initially, this involved the use of word vectors, which were later replaced by word2vec in 2013. This new approach combined deep learning with natural language processing, leading to significant advancements in the field. It has reached a high point and has achieved particular success in the field of machine translation, question and answer systems, and reading comprehension. Subsequently, Recurrent Neural Network (RNN), Gated Recurrent Unit (GRU), and Long Short-Term Memory (LSTM) became the most commonly used natural language processing methods. In 2018, Google's release of the Bi-directional Encoder Representations from Transformers (BERT) model triggered a new peak [1]. It broke 11 records in the field of natural language processing and attracted a lot of attention from industry and academia.

5.2.2 *Basic Concepts*

Natural Language Processing (NLP) is also referred to as Natural Language Understanding (NLU). Since the beginning of artificial intelligence (AI) research, the investigation of the fundamental aspects of human comprehension of natural language has constituted a pivotal research area.

As stated by Zong Chengqing in the 1996 publication "Computer Processing of Natural Language," "natural language processing is a technology that uses computers as tools to process various types of human-specific written and spoken language." [2].

American computer scientist Bill Manaris defined natural language processing as follows: "Natural language processing requires the development of models that represent language capabilities and language applications, and the establishment of computational frameworks to achieve this. These models should be developed in such a way that they can be continuously improved, and they should inform the design of various practical systems. Finally, it is important to discuss the evaluation techniques of these practical systems."

5.2.3 Basic Method

There are three principal methodological approaches in the field of NLP: empirical methods based on statistics, rationalist methods based on rules, and deep learning methods. In order to study language, the statistic-based approach uses probabilistic or random methods to establish a probabilistic language model. This approach is well-suited to the examination of superficial linguistic phenomena and their interdependence. Induction is a frequently used method, whereas deduction is a less common approach.

The development of NLP techniques has benefited considerably from statistical methodology. Among the key contributions in this regard have been those of the hidden Markov model, maximum entropy model, n-grams, probabilistic context-free grammar, noise channel theory, the Bayesian method, the minimum editing distance algorithm, the Viterbi algorithm, a search algorithm, a two-way search algorithm, weighted automata, and support vector machines.

The rule-based approach is also referred to as the symbolic approach. The fundamental premise is the "physical symbol system hypothesis." This hypothesis claims that a physical symbol system is capable of simulating intelligent human behavior. The physical symbol system comprises a set of physical symbol patterns that can be used to construct various symbolic expressions, which in turn represent the structure of the symbols. This approach is well-suited to the analysis of deep-level linguistic phenomena and long-distance dependencies. The process of deduction is used with greater frequency, while the use of induction is less common.

Furthermore, rule-based methods that facilitate the development of NLP techniques include a range of approaches, including the finite-state transfer network, finite-state transcription machine, recursive transfer network, expanded transfer network, phrase structure grammar, bottom-up analysis, top-down analysis, left corner analysis method, Earley algorithm, CYK algorithm, Tomita algorithm, complex feature analysis method, unification operation, dependency grammar, first-order predicate calculus, semantic network, frame network, etc.

The deep learning approach uses neural networks to solve NLP problems. An embedding layer is used to map discrete symbols into relatively low-dimensional continuous vectors, thereby enabling the use of neural networks in natural language processing.

On the other hand, technologies developed based on deep learning methods in NLP contain convolutional neural networks (CNN), recurrent neural networks (RNN), attention mechanisms, reinforcement learning, deep generative models, and memory enhancement models.

In many NLP tasks, the combination of deep learning with rule-based and statistical methods is often adopted as a means of leveraging the strengths of each approach to achieve optimal results.

5.2.4 *Facing Difficulties*

At present, NLP is confronted with two significant challenges. The first challenge is the construction of large-scale corpus data, and the second is the improvement of semantic analysis technology.

In large-scale corpus construction, the necessity for a substantial corpus is a consequence of the statistical machine learning foundation of NLP technology. However, the manual construction of a corpus is a time-consuming and laborious process, and therefore, data sharing is an essential initial step in promoting research and development. Moreover, it is inherently challenging to include all instances within a given field in any given corpus. Additionally, the level of detail in the annotation of the corpus is challenging to grasp. The classification is insufficiently detailed to provide a comprehensive and accurate description of the language. The classification of detailed rules results in the annotation information becoming overly extensive, which in turn reduces annotation efficiency and introduces data sparseness issues. Consequently, due to the inherent difficulties associated with manual corpus labeling, some researchers are currently engaged in utilizing a substantial quantity of unlabeled data or partially labeled data to solve specific problems, a process known as semi-supervised learning. Nevertheless, the research in this area remains in an immature state.

In terms of perfecting semantic analysis technology, two main semantic analysis methods have emerged: the semantic analysis method based on knowledge or semantic rules, and the other based on statistics. However, the aforementioned two approaches remain challenging in accurately representing the semantics of natural language. Semantic analysis methods based on knowledge or semantic rules are unable to include all linguistic phenomena, and their reasoning process is complex, unable to deal with uncertain events. It is difficult to establish accurate knowledge and semantic rules. In contrast, statistics-based methods necessitate a vast corpus, and the advantages and disadvantages of the corpus significantly impact the model performance.

5.3 NLP Data Resources

5.3.1 *Basic Corpus*

In statistical NLP methods, it is generally necessary to collect a corpus of texts as the basis for the establishment of statistical models. Such texts are collectively referred to as a "corpus." A corpus base is a database comprising extensive collections of processed and annotated corpora, which have undergone screening and other forms of preprocessing. Statistical methods are typically founded upon a substantial corpus, hence they are also designated as corpus-based NLP methods.

A. **LDC Chinese Tree Bank**

The LDC Chinese Tree Bank (CTB)[1] is a Chinese syntactic treebank developed by the University of Pennsylvania (UPenn) and released through the Linguistic Data Consortium (LDC). The corpus collected by the treebank is based on material from the Xinhua News Agency and Hong Kong News. The seventh edition consists of 2,400 text files, containing 45,000 sentences, 1.1 million words, and 1.65 million Chinese characters. The files are stored in two encoding formats: GBK and UTF-8.

B. **BFS-CTC Corpus**

The Beijing Forest Studio-Chinese Tagged Corpus (BFS-CTC) [3] is a Chinese sentence sense structure tagging corpus constructed by Beijing Forest Studio (BFS).

In regard to the labeling of content, the sentence meaning structure model provides a definition of the elements and their combination relations at the sentence pattern level, description level, object level, and detail level. This includes the type of sentence meaning, the identification of predicates and their respective tenses, semantic case type, and other information. Furthermore, it furnishes lexical and phrase structure syntax data, which is conducive to comparative analysis and research of morphology, syntax, and sentence meaning. In terms of corpus organization structure, the corpus includes four parts: the original sentence database, the lexical tagging database, the syntactic annotation library, and the sentence meaning structure annotation library. According to the requirements of research, the data can be subjected to in-depth processing based on the aforementioned annotations, with the addition of more targeted extended annotation libraries based on the core annotation library. This is achieved through the use of a unique ID number assigned to each sentence for identification purposes. In regard to the corpus's source and scope, the original data is derived from the news corpus. Following manual collection and sorting, the corpus comprises a reasonable coverage of the six main sentence types, including subject-predicate sentences, non-subject-predicate sentences, and sentences with the "ba" construction, with a scale of 50,000 sentences.

Based on modern Chinese semantics, BFS-CTC provides multi-level sentence meaning structure labeling data and performs lexical and grammatical annotation in a manner that is consistent with existing annotation specifications. The fundamental structure is shown in Fig. 5.1. The morphology, syntax, and sentence meaning marked by BFS can be used as standalone elements or in a comprehensive manner, facilitating a range of research endeavors in NLP.

C. **Lancaster Chinese Corpus**

The Lancaster Corpus of Mandarin Chinese (LCMC) is a balanced corpus of modern Chinese, constructed in June 2003 by Dr. Xiao Zhonghua, a student of Professor Tony McEnery, over a period of more than half a year.

This project is undertaken by the Department of Linguistics at Lancaster University. The project was initially established by the Contrasting Tense and Aspect

[1] http://www.cis.upenn.edu/~chinese/ctb.html.

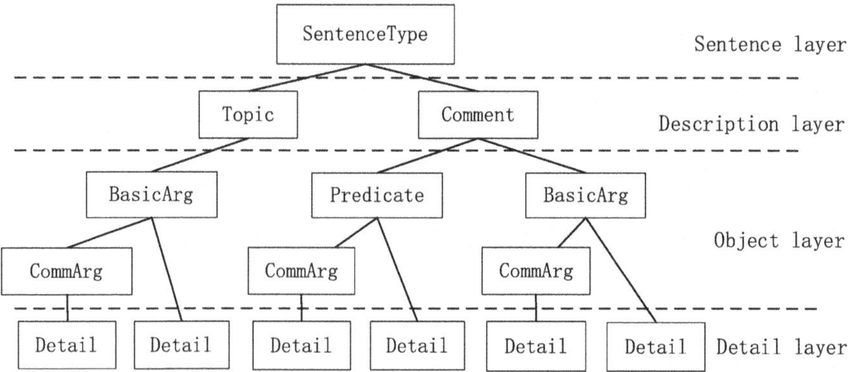

Fig. 5.1 Basic form of sentence meaning structure

in English and Chinese project, which was funded by the British Economic and Social Research Council. A comprehensive Chinese corpus that is accessible to the public free of charge. Accordingly, the objective is to construct a modern Chinese corpus that is analogous to the Freiburg-LOB Corpus of British English and the Freiburg BROWN Corpus of American English. This will facilitate engagement in corpus-based Chinese monolingual or Chinese-English (English-Chinese) bilingual comparative research.

The LCMC corpus is a Chinese written corpus compiled strictly following the Freiburg-LOB Corpus of British English (FLOB) model. However, it has been refined in two key aspects.

Firstly, within the FLOB sampling category, Xiao Zhonghua changed the "Western and Adventure Novels" of the corresponding sample in the FLOB to "Martial Arts Novels." Table 5.1 represents the sampling category.

Secondly, due to the constraints of the source text, Xiao Zhonghua extended the publication date of the sample to encompass a two-year window centered around

Table 5.1 LCMC Sampling category table

Code	Sampling type	Code	Sampling type
A	News report	J	Academics, technology
B	Editorial	K	General novel
C	News review	L	Detective fiction
D	Religious	M	Science fiction
E	Technology, business	N	Martial arts novels
F	Popular social life	P	Romance novel
G	Biography and essays	R	Humor
H	Other: reports and official documents, etc.		

1991 (that is, from 1989 to 1993). He believes that the two-year range will not impact the homogeneity of the entire corpus.

The LCMC corpus requires users to employ retrieval tools that are capable of reading XML, such as Xaira or WordSmith Tools version 4.0.

D. **Taiwan Academia Sinica Corpus**

In the early 1990s, Taiwan's Academia Sinica established two key linguistic resources: the Chinese Balanced Corpus (Sinica Corpus[2]) and the Chinese Treebank (Sinica Treebank).[3]

The Sinica Corpus was constructed in accordance with the word segmentation standards established by the Taiwan Computational Linguistics Association. The corpus comprises approximately 5.2 million words, corresponding to 7.89 million Chinese characters. The corpus was selected from a range of disciplines, including philosophy, art, science, life, society, and texts in the field of literature, published between 1990 and 1996. In 2003, two supplementary Chinese corpora were incorporated. One is a Chinese-English parallel corpus containing 2,373 Chinese-English parallel texts published between 1976 and 2000. The corpus comprises approximately 103,000 Chinese-English sentence pairs, 320 thousand English words, and 5.3 million Chinese words. The other attached corpus is the modern Chinese corpus developed by the Institute of Computational Linguistics of Peking University. It comprises approximately 85 million Chinese characters, and the collected chapters were published after 1919. The latest version of the Sinica Treebank, version 3.0, comprises 61,087 sentence trees and more than 360,000 Chinese words. A total of 1,000 sentence trees can be accessed via the public domain for purposes of scientific investigation. The underlying structural framework of SinicaTreebank is based on the head-driven principle, which posits that a given sentence or phrase can be understood as consisting of a central component and a number of parameters or attachments. The head in this context defines the phrase category, as well as its relationship to other components.

E. **BYU-BNC Corpus**

The British National Corpus (BYU-BNC: British National Corpus, hereinafter referred to as "BNC Corpus") is one of the most representative contemporary English corpora on a global scale. The Oxford Press, Longman Publishing Company, and Chambers have developed a large-scale corpus in collaboration with the Harlow Publishing Company, Oxford University Computing Service Center, Lancaster University English Computer Center, and British Library, which was completed in 1994.

The corpus is designed to encompass both written and oral language. Its CD-ROM version comprises over 100 million words, including more than 90 million words from the written corpus and over 10 million words from the oral corpus. In terms of application, the corpus can be used not only with its latest iteration of SaRA search

[2] http://www.sinica.edu.tw/SinicaCorpus/s.

[3] http://godel.iis.sinica.edu.tw/CKIP/engversion/treebank.htm.

software but also with a range of general search software, and it can be searched directly online.

F. **Proposition Bank**

PropBank, NomBank,[4] and the Penn Discourse Tree Bank (PDTB)[5] are all extensions of the Penn Tree Bank.

The proposition bank was initially constructed based on the Penn English Treebank (Penn English Treebank) with the incorporation of semantic data. The fundamental premise is that the treebank offers insight solely into the syntactic structure of sentences, which is insufficient for computers to fully comprehend human language. The initial objective of PropBank was to facilitate the labeling of verb senses and the identification of argument information related to each word sense within the Pennsylvania English treebank. The labeling scheme was developed by a joint effort of the University of Pennsylvania (UPenn)'s relevant research groups, Broadcast News Navigator (BNN), MITRE,[6] and New York University in 2000. The following section will present a discussion and development of the topic. Subsequently, PropBank expanded the labeling of Chinese propositions. The Chinese PropBank version 2.0 selected a 500,000-word tree structure of sentences from the Chinese treebank version 6.0 for predicate-argument labeling, including 81,009 verb instances (11,171 verbs) and 14,525 name instances (1,421 nouns).

G. **Prague Dependency Tree Bank**

The Prague Dependency Treebank (PDT)[7] is a project of the Institute of Formal and Applied Linguistics (Faculty of Mathematics and Physics)[8] at Charles University in Prague. The corpus currently comprises three corpora: the Czech Dependency Treebank, the Czech-English Dependency Treebank, and the Arabic Dependency Treebank.

The Prague Dependency Treebank contains three levels:

1. Morphological layer: The lowest layer of PDT, which contains all morphological information labels;
2. Analytic layer: The middle layer of PDT is mainly the surface syntactic information labeling in the sequential relationship. The level is conceptually close to the syntactic labeling in Penn Treebank;
3. Deep grammatical layer (tectogrammatical layer): The highest layer of PDT represents the deep grammatical structure of sentences. The tectogrammatical tree structures (TGTS) contain only those words (subjective words) (autosemantic word) nodes in the sentence that have actual meanings (e.g., no preposition nodes), and meet the condition of projectivity. No cross edges are present, and each node is assigned an operator, such as ACTOR, PATIENT, and ORIGIN.

---

[4] http://nlp.cs.nyu.edu/meyers/NomBank.html.

[5] http://www.seas.upenn.edu/~pdtb.

[6] http://www.mitre.org/.

[7] http://www.elsnet.org/nps/0040.html.

[8] http://ufal.mff.cuni.cz/.

The Czech dependency tree bank's corpus was mainly derived from the Czech National Corpus newspaper news field, with a corpus size of approximately 1.5 million words.

5.3.2 Language Knowledge Base

Language knowledge bases have important applications in the fields of natural processing and linguistics research. These include, for example, vocabulary knowledge bases, syntactic rule bases, grammatical information bases, and semantic concept bases, as well as other varieties of language knowledge resources. All of these are constructed by means of NLP systems. Such systems represent an essential and, indeed, an indispensable foundation for the study of linguistics. For an extended period, numerous experts and linguists in NLP have dedicated significant effort to the establishment of language knowledge bases, resulting in notable achievements. This section provides a concise overview of select research findings.

A language knowledge base contains a more extensive range of content than a corpus. In general terms, language knowledge bases can be divided into two distinct categories. The first category comprises explicit representations of language knowledge, such as dictionaries, rule bases, and semantic concept bases. These representations are formalized in a way that allows for precise description and analysis. The second category, which is more closely related to the corpus, encompasses the category, meaning, and usage of each language unit. The primary component of the corpus is text, which is a compilation of sentences. Each sentence is a linear, unstructured text sequence, and the knowledge it contains is implicit. The objective of corpus processing is to make tacit knowledge explicit, thereby facilitating machine learning and citation.

A. **WordNet**

WordNet is an English machine-readable vocabulary knowledge base developed by a research group led by George A. Miller of the Cognitive Science Laboratory of Princeton University in the United States.[9] It is a synthesis of traditional dictionary information, computer technology, and mental language, representing an organic combination of scientific research results. Since 1985, the WordNet project has been in full swing and has developed into a very influential English vocabulary knowledge resource database.

WordNet is a vast lexical database structured according to a semantic relation network. A variety of lexical relations and semantic relations are used to represent the organization of vocabulary knowledge. The WordNet source files comprise two

[9] http://wordnet.princeton.edu/man/wnstats.7WN.

basic components: the word form and the meaning. A standardized word form represents the form of a word, while a synset represents the word's meaning. Lexical relationships pertain to the interconnections between two word forms, whereas semantic relationships concern the interlinks between two word meanings.

WordNet classifies nouns, verbs, adjectives, and adverbs into a synset, which is a group of words that share a common semantic relationship. These synsets are then distributed across different source files based on syntactic categories and other organizational principles. Adverbs are stored in a single file, while nouns and verbs are organized into different files according to their semantic categories. Adjectives are divided into two files, one comprising descriptive adjectives and the other relational adjectives. In WordNet version 2.0, there are approximately 114,648 nouns and 79,689 synsets, a significant proportion of which are collocation words. Additionally, there are 21,436 adjectives, 18,563 adjective synonym sets; 11,306 verbs, 13,508 verb synonym sets; and 4,669 adverbs, 3,664 adverb synonym sets.[10]

B. **FrameNet**

FrameNet is an emerging online knowledge base that was established based on Frame Semantics [4]. The frame network dictionary offers definitions of words, annotated examples of sentences in which they are used, a variety of indexing methods, and explanations of the headwords. The syntactic features of the headwords indicate their valence structure.

In FrameNet, words are classified into Lexical Units (LU) based on their distinct meanings. In theory, each meaning of a polysemous word is associated with a distinct semantic frame, serving as the essential means of organizing the vocabulary's semantic knowledge [5]. Lexical units with identical semantics are allocated in the same frame, and the meaning of the lexical unit is described by the frame to which it belongs. The semantic frame is analogous to the conceptual structure of a script. It comprises multiple frame elements to describe a specific situation type, objects, events, event participants, and props.

The evolution of FrameNet has practical significance in lexicography, NLP, and semantic research.

C. **EDR Electronic Dictionary**

The EDR Electronic Dictionary is constructed by the Japan Electronic Dictionary Research Institute (EDR) [6] comprised of 11 sub-dictionaries, including conceptual dictionaries, dictionaries, and bilingual dictionaries. The project's development was funded by the Japan Key Technology Center and eight Japanese computer manufacturers, including Fujitsu, NEC, Toshiba, Hitachi, Sharp, Oki, Panasonic. The project spanned nine fiscal years (1986–1994). The Japan Electronic Dictionary Research Institute was disbanded on March 31, 2002, and the EDR dictionary was subsequently continued to be provided by the Communications Research Laboratory (CRL) of Japan.

---

[10] http://wordnet.princeton.edu/man/wnstats.7WN.

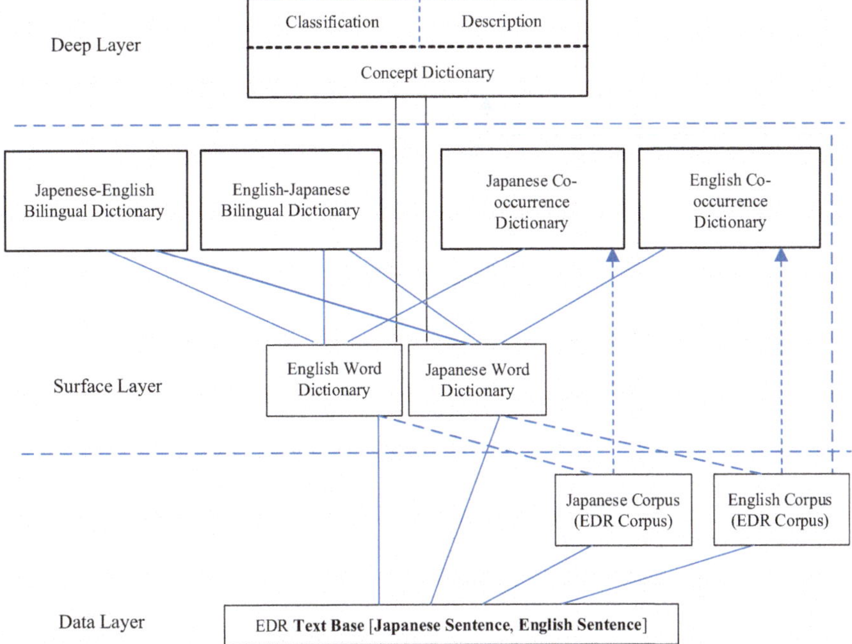

Fig. 5.2 The level of EDR and its sub-dictionaries

The EDR comprises five types of dictionaries: the Word Dictionary, the Japanese-English Bilingual Dictionary, the Concept Dictionary, the Japanese and English Co-occurrence Dictionary, and the Technical Term Dictionary, Terminology Dictionary. Additionally, it includes the EDR Corpus. The hierarchy of the EDR dictionary and the relationship between its sub-dictionaries are illustrated in Fig. 5.2.

D. **HowNet**

HowNett [7], established by machine translation specialists Dong Zhendong and Dong Qiang, is a common sense knowledge base that employs Chinese and English words as the basis for describing concepts and reveals the interrelationships between concepts and their attributes as the basic content.

As a knowledge system, HowNet strives to capture the shared and distinctive characteristics of concepts. By categorizing diverse conceptual relationships, this knowledge network system is trained on computers to facilitate the computational representation of knowledge.

In HowNet, Yiyuan is a particularly significant concept, representing the most basic and the smallest unit of meaning. Its inherent simplicity makes it challenging to subdivide. One of the key features of the construction method of the knowledge network is to adopt a bottom-up induction method. This involves the observation and analysis of all the fundamental semantic elements, leading to the formation of an initial annotation set representing the original meaning. This set is then evaluated with the incorporation of additional concepts, ultimately leading to the establishment of a perfect annotation set. HowNet is a language knowledge system with a substantial corpus of content and a logical structure. It can be applied in a variety of contexts, including the calculation of vocabulary semantic similarity, the disambiguation of vocabulary semantics, the recognition of noun entities, and the classification of text. Further details on HowNet can be found on its homepage[11] and in other relevant literature.

E. **Peking University Comprehensive Language Knowledge Base**

The Comprehensive Language Knowledge Base (CLKB) of Peking University was established by Professor Yu Shiwen of the Institute of Computational Linguistics (ICL/PKU) at Peking University. The database encompasses a comprehensive range of linguistic elements, including words, phrases, sentences, text units, morphology, syntax, and semantics levels, which extend from Chinese to multilingual and from general to professional domains.

CLKB is currently the largest and most widely recognized Chinese language knowledge resource in the world. The database includes a modern Chinese grammar information dictionary, a Chinese phrase structure rule database, a modern Chinese multi-level processing corpus, a multilingual concept dictionary, a parallel corpus, and a multi-domain corpus. The modern Chinese Grammatical Knowledge Base (GKB) is a comprehensive digital dictionary designed for language information processing. It contains 80,000 Chinese words, classified according to their grammatical functions (advantages) and detailed grammatical attributes.

It represents a significant milestone in China's artificial intelligence and information processing research, extending over five decades. It provides substantial support for the theoretical research and the advancement of information processing technology in China. For detailed information on each sub-library of CLKB, interested readers are encouraged to refer to the relevant literature.

5.3.3 Knowledge Graph

The Knowledge Graph, also referred to as the Scientific Knowledge Graph, is a semantic network that exhibits both robust expressive capacity and modeling flexibility [8]. It is capable of modeling entities, concepts, attributes, and their relationships in the real world. Additionally, the Knowledge Graph serves as a "protocol" of data modeling and a data exchange standard for derivative technologies. Its related technologies cover knowledge extraction, knowledge integration, knowledge management, and knowledge application.

[11] http://www.keenage.com.

A. DBpedia

DBpedia [9] is a multilingual, comprehensive knowledge base created by researchers from the University of Leipzig in Germany. The project extracts structured data from Wikipedia entries in multiple languages. DBpedia publishes its data on the Internet in the form of linked data, which allows other datasets to be connected with Wikipedia on data nodes. The involvement of this semantic technology has enabled Wikipedia to easily and quickly derive many innovative applications with ease and speed, including a mobile phone version, map integration, multi-faceted search, relational query, document classification, and labeling. DBpedia employs a relatively strict ontology, which includes definitions of people, places, music, movies, organizations, species, diseases, etc.

Additionally, it has also established data links with multiple datasets, including Freebase, OpenCYC, and Bio2RDF. DBpedia uses the Resource Description Framework semantic data model. By the end of 2014, DBpedia had accumulated in excess of 3 billion fact triples, establishing it as one of the largest multi-domain knowledge ontologies globally.

DBpedia supports three general methods of data acquisition, which can be obtained through the Pubby-related data interface, a similar data browser, and various crawlers. Additionally, DBpedia supports the SPARQL (SPARQL Protocol and RDF Query Language) service and RDF file download, and provides a comprehensive user interface and a range of functions to facilitate the utilization of DBpedia's diversified data.

B. Freebase

Freebase is a project with notable similarities to DBpedia. Both projects extract structured data from Wikipedia and publish RDF. However, Freebase is a fully structured database, with a data source that extends beyond Wikipedia. Additionally, Freebase imports numerous professional datasets and provides advanced data query and entry mechanisms that are more comprehensive and powerful than those of DBpedia. The import and use of Freebase data are subject to the Creative Commons Attribution License.[12]

The main data sources of Freebase include Wikipedia, the global celebrity database NNDB, the open music database MusicBrainz, and community users' contributions. Furthermore, individual users manually edit the pages one by one to add data. As of the end of 2014, Freebase contained 68 million entities, 1 billion pieces of relationship data, and over 2.4 billion pieces of factual triad information [9]. In June 2015, Freebase was integrated into Wikidata as a comprehensive entity. In 2016, Google announced that it would migrate Freebase's data and API services to Wikidata, subsequently leading to the official closure of Freebase.

C. YAGO

YAGO is a linked database developed by the Max Planck Institute in Germany in 2007. In response to the situation where the application at that time only

[12] http://wiki.freebase.com/wiki/Freebase_license.

used a single source of background knowledge, it has established a high-quality, high-coverage knowledge base of multi-source background knowledge. Although WordNet, constructed by experts, has an extremely high degree of accuracy in terms of ontology knowledge, its scope of coverage remains limited to certain common concepts or entities. In contrast, Wikipedia contains a substantial corpus of entity-specific knowledge. However, the hierarchical structure of the concepts it provides, the lack of precision in the label structure, and its unsuitability for direct use in ontology construction represent significant limitations. YAGO mainly incorporates data from Wikipedia, WordNet, and GeoNames, integrating WordNet's vocabulary definitions and Wikipedia's classification system to enhance YAGO's entity classification system. Furthermore, YAGO incorporates temporal and spatial considerations, as exemplified by the inclusion of temporal and spatial attribute descriptions in numerous knowledge items.

By 2018, YAGO contained over 10 million entities and over 120 million triples of knowledge. Notably, it serves as a back-end knowledge base for IBM Watson [9].

D. **Knowledge Vault**

On May 16, 2012, Google introduced Knowledge Vault, a cutting-edge technology that boosts search engine performance and elevates result quality. By meticulously structuring topic-related information, it delivers a more intuitive and insightful search experience.

Based on machine learning, Knowledge Vault can extract data from a multitude of sources, including text, tabular data, page structure, and manual annotation. Additionally, it can infer facts and relationships based on all available data. However, due to the prevalence of erroneous data within the network, the framework relies on an existing knowledge base, such as Freebase, to ensure the accuracy of its operations [10].

E. **Sogou Knowledge Cube**

Sogou Knowledge Cube is a novel knowledge base search engine recently launched by Sogou with the objective of facilitating user access to information. It is the pioneering knowledge base search product in the domestic search engine industry.

Knowledge Cube mainly re-optimizes search results through the integration of a vast quantity of disparate Internet data, thereby facilitating the presentation of the most core information to users. In contrast to the conventional "keyword search," Zhilifang does not merely extract data from web pages; rather, it employs "semantic understanding" technology to understand the user's search intent and deliver accurate search results.

The KnowledgeCube is capable of intelligently analyzing the user's query intention, providing the desired answer based on reasoning and computing capabilities. By analyzing and mining the entire web page, the accuracy of the knowledge base data is ensured. Furthermore, it offers a comprehensive knowledge system, facilitating a more thorough understanding of knowledge points.

5.4 NLP Techniques in Big Data

The era of big data has ushered in a paradigm shift within traditional data science, which was previously based on mathematical statistics. This has led to the emergence of novel data analysis methods. Deep learning represents a significant advancement in machine learning, with multi-layer neural networks emerging as a central focus in the field of big data processing and analysis. From machine learning to deep learning, we have experienced not only early symbolic induction, machine learning, statistical machine learning, and neural networks, but also the research and practice of data mining since the end of the 20th century. It has been demonstrated that deep learning can identify the potential value of big data. The field of deep learning draws upon multiple disciplines, including computer science, engineering technology, and statistics. It is applicable to a wide range of disciplines, including politics, finance, astronomy, geography, and social sciences. The advantage of deep learning is that it is capable of expressing a high degree of complexity and can process data with high-dimensional and sparse features. It is therefore imperative that the challenges currently being faced by NLP be introduced into the development of deep learning ideas, methods, and technologies in order to facilitate the implementation of timely and effective solutions.

The application of deep learning to NLP and the discovery of the potential value inherent in big data have emerged as a significant area of interest and research within the industry. As an essential research direction in the intersection of computer science and artificial intelligence, NLP synthesizes the insights and achievements of linguistics, computer science, logic, psychology, and artificial intelligence.

The modern Chinese sentence semantics-based meaning model, rooted in linguistics, constitutes fundamental applied research. Its technical roadmap includes planning of both the fundamental and practical application parts, as illustrated in Fig. 5.3. The direct application of this sentence meaning model involves extracting the essence of sentence meaning by analyzing the sentence meaning, capturing its informational depth to the fullest extent, and subsequently applying this model's expression format directly. Scenarios that necessitate the direct deployment of this model primarily revolve around simulating human communication processes, encompassing question answering systems, human-computer dialogues, machine translation, etc. In these contexts, reliance solely on specific linguistic facets proves insufficient to fulfill the application's requirements. Instead, it becomes imperative to harness a directional model to allow the computer to simulate the intricate manner in which humans comprehend language, thereby enhancing the accuracy and naturalness of interactions.

Considering the rapid development of the Internet, there have been significant shifts in the manner in which public opinion is disseminated. The Internet has become a crucial medium through which the government can gain insight into the lived experiences and public opinion of citizens. Furthermore, the Internet has become the main platform for individuals to express their opinions. The Internet environment is characterized by a high degree of complexity. It is of particular importance to

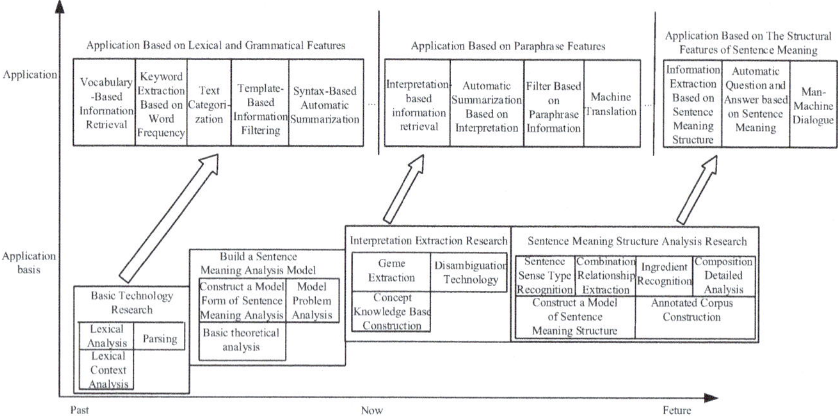

Fig. 5.3 Schematic diagram of the technical route of sentence meaning model research

develop effective strategies for guiding speech, monitoring and evaluating the risks faced, and taking timely countermeasures. Consequently, NLP applications for public opinion situation awareness and clue discovery directions have become a prominent area of interest, encompassing techniques such as named entity recognition, hot topic extraction, entity-relation extraction, event extraction, event summary, and sentiment analysis. The block diagram of NLP technology is presented in Fig. 5.4.

The following section will provide a comprehensive overview of the current big data analysis technology and its applications in various domains, including named entity recognition, entity-relation extraction, sentiment classification, text summarization, machine translation, automatic question and answer.

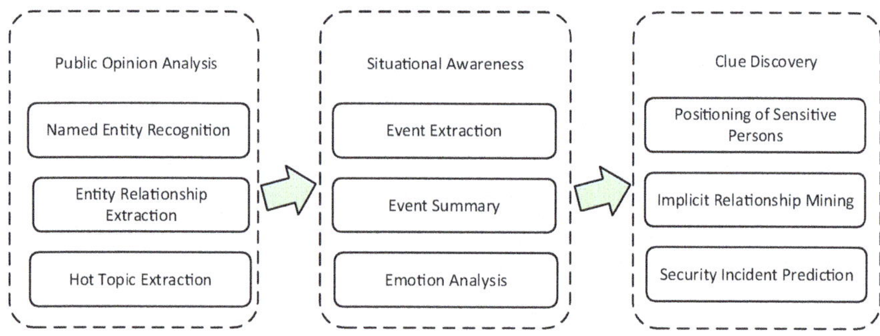

Fig. 5.4 Block diagram of natural language processing technology

5.4.1 *Entity-Relation Extraction*

A. **Task definition**

Entity-relation extraction represents a significant area of research within the domain of information extraction. Its primary objective is to extract the semantic relationship between entity pairs within sentences, with the aim of categorizing the relation category between entity pairs in the unstructured text based on entity recognition. This process ultimately leads to the formation of three tuples, namely <entity 1, relation, entity 2>. For example, if the sentence were to be "Tsinghua University is located in the nearest neighbor of Beijing," and the entities were "Tsinghua University" and "Beijing," the model could obtain the relation of "location" through semantics, and then extract the triplet <Tsinghua University, located in, Beijing>.

In 1998, the final Message Understanding Conference (MUC), which was funded by the Defense Advanced Research Project Agency (DARPA), introduced the entity-relation extraction task for the first time. The template relation described in the MUC is the earliest description of an entity relation.

In 1999, the National Institute of Standards and Technology (NIST) organized the Automatic Content Extraction (ACE) evaluation, during which one of the key evaluation tasks was entity-relation recognition.

The entity-relation extraction that is undertaken as part of MUC and ACE evaluation meetings is limited to a few types of entity relations between named entities (including names of persons, places, organizations). These include relations pertaining to employment, geographic location, and people-society organizational relations. Another significant evaluation meeting in the field of information extraction is SemEval (semantic evaluation), which follows MUC and ACE. A considerable number of academic institutions have participated in this evaluation meeting. In SemEval-2007, evaluation task 4 defines seven entity relations between collective nouns or noun phrases, yet it offers a limited English corpus. Subsequently, the evaluation task 8 of SemEval-2010 constituted an improvement and expansion of the aforementioned entity-relation types, which were extended to encompass the following nine types: component-whole, instrument-agency, member-collection, cause-effect, entity-destination, content-container, message-topic, product-producer, and entity-origin. Taking into account the sequence of the entity pairs within sentence instances, an "Other" category has been introduced to describe instances that do not align with the aforementioned relation types. This has resulted in the generation of 19 entity relations. The SemEval-2010 evaluation has prompted a renewed surge of interest in the study of entity-relation extraction between common nouns or noun phrases.

B. **Method overview**

Entity-relation extraction tasks include both distant supervision-based relation extraction methods and entity-relation joint extraction methods based on a variety of usage approaches.

(1) Distant supervision-based relation extraction

In the distant supervision-based entity-relation extraction method, the training of a model necessitates the availability of a substantial quantity of labeled training data. However, the process of labeling is very time-consuming and challenging. In 2009, Mike M proposed a distant supervision-based method to generate training data automatically. This method employs a two-step process to automatically build a large amount of training data. First, the method aligns the knowledge base with unstructured text. Second, it reduces the model's dependence on manually labeled data and enhances the model's cross-domain adaptability.

The distant-based relation extraction method assumes that if a relation exists between two entities within the knowledge base, then all sentences containing these two entities will use the specified relation expression. Although the distant-based method mitigates the model's dependence on artificially labeled data to some extent, it inevitably introduces a considerable number of noise samples into the dataset. For example, Microsoft founder Bill Gates is a relational triplet in the knowledge base. The sentence "Bill Gates' turn to philanthropy is related to Microsoft's antitrust issues in the United States and the European Union." It contains two entities, "Microsoft" and "Bill Gates." According to the distant supervision method, the relation type between the entity pairs in this sentence will be represented as "founder." Nevertheless, this sentence does not indicate that these two entities have this type of relation.

In the realm of Multi-Instance Learning (MI), many researchers address the challenge of noisy samples, an issue often introduced by distant supervision methods. The essence of the multi-instance learning approach can be summarized as follows: each data entry in the training set is considered a "bag," which contains multiple examples (instances). These bags are assigned an aggregate label, whereas the individual instances within them are not labeled. A bag is deemed positive if it contains at least one instance that is positively marked; conversely, a bag with a negative label implies that all of its instances are negatively marked.

The processing of noise data in distant supervision is a challenging task due to the lack of a clear supervisory signal to capture noise. In 2015, Zeng Daojian et al. developed a solution to the noise problem caused by distant supervision. They regarded the problem of distant supervision relation extraction as a multi-instance learning problem. To address the issue of data labeling errors, this approach employs multi-instance learning to extract training examples with high confidence from the training set, which are then used to train the model. Although multi-instance learning has the potential to reduce the impact of noise effectively, this approach is not without its shortcomings: (1) Multi-instance learning selects the sample with the highest confidence in each package as a positive sample for training. Consequently, a considerable amount of valuable supervisory data is discarded; (2) In multi-instance learning based on the "at least one" assumption (at least one example in the package is a positive example), the relation type of the package may be erroneous. In 2017, He Dengchao et al. also regarded distant supervision relation extraction as a multi-instance learning problem, and the aforementioned two issues persist.

In 2016, Lin Yankai et al. addressed the initial issue identified in Zeng Daojian's method, proposing a model based on the attention mechanism. Firstly, PCNN is used to encode each sentence within the packet, thereby generating a sentence vector. Secondly, the weight of the sentence vector within the packet is determined through the calculation of the similarity between each sentence vector and the packet relation vector. Finally, a weighted sum of each sentence in the packet is carried out, with the objective of fully utilizing the effective examples within the packet while simultaneously reducing the noise.

In 2017, Ji Guoliang et al. proposed a sentence-level attention model for multi-instance learning problems. The primary distinction between the method proposed by Lin Yankai lies in the manner of calculating the attention weight. This method initially uses the "entity-entity=relation" method to represent the relationship between entities. Subsequently, a convolutional neural network is used to capture the features of an entity description page, thereby enhancing the entity representation. Finally, different weights are given to sentences by calculating the relation between entities and the sentence similarities. This enables the effective utilization of pertinent examples within the packet while minimizing the influence of irrelevant data.

In 2017, Luo Bingfeng et al. proposed a novel approach wherein a noise matrix is used to align the noise distribution, model the noise, and mitigate its impact through matrix multiplication. Two kinds of transition matrices are constructed: "dynamic transition matrix" and "global transition matrix." The "dynamic transition matrix" is demonstrated to be superior to the "global transition matrix." In the training process, the method of course learning is introduced, through which some prior knowledge can be incorporated into the training model.

In 2018, Du J et al. discovered that the conventional one-dimensional attention model was insufficient for learning different contexts and predicting the relationship between entity pairs when selecting effective examples. The traditional attention model is only capable of learning the probability distribution of the weight of a single aspect of the sentence; it is unable to learn the distribution of multiple aspects. A novel multi-layer self-attention mechanism is proposed for relation extraction in the MIL framework. Nevertheless, this method ignores the possibility that all examples within the package may be classified as noise samples.

In 2018, Qin P et al. attempted to use a deep reinforcement learning framework and a generative adversarial network to automatically identify noisy samples within each relation type's instance set, operating without any supervision information.

In 2018, Yuan C et al. used cosine similarity to calculate the similarity between each sentence and the best sentence. The higher the similarity, the greater the weight of the sentence, and vice versa. Firstly, the best sentence is obtained according to at least one hypothesis of MIL. Subsequently, the weight of each sentence is obtained by calculating the similarity between each sentence and S. Finally, the feature representation of the package is obtained through a weighted sum. In addition, it is observed that as the sentence grows, it will contain some words that are irrelevant to the target relation and are distant from the entity. To address this issue, the method assumes that the relation between entities and words changes with the distance between entities and words. This assumption can reduce the word's weight as the distance between the entity and the word increases.

In 2018, Yuan Y et al. proposed a cross-relation, cross-package selection attention model, which is still under construction as a sentence package under the MIL framework. This method improves sentence-level attention by considering the correlation between relations and selects at the package level with an attention layer. The first attention layer attempts to reduce the influence of noisy or mismatched sentences, while the second attention layer is focused on high-quality bags.

(B) Entity-relation joint extraction

The conventional entity-relation extraction method employs a pipelined approach. This method initially engages in named entity recognition and subsequently classifies the identified entities through pairwise combinations, culminating in the extraction of triples that denote entity relationships as the end product. However, the methodology encounters three primary issues: (1) error propagation, whereby the error of the entity recognition module can affect the performance of the relation classification; (2) the neglect of relational interdependencies between the subtasks; and (3) the generation of unnecessary redundancy, as the process of pairing entities followed by relation classification can introduce unrelated entities, thereby increasing the volume of redundant information and increasing the error rate.

In order to address the shortcomings of the pipeline method, the method of joint learning has been introduced into the field of entity-relation extraction. The method can be broadly classified into two categories: (1) joint extraction methods based on parameter sharing; (2) other joint extraction methods.

In 2016, Miwa M et al. used the underlying expression of a shared neural network for joint learning. First, the word embedding layer reads the input sentence; then, the Bi-directional Long Short-Term Memory (Bi-LSTM) layer is used to encode the input sentence. Finally, a neural network is used for Named Entity Recognition (NER) and a Bi-LSTM for Relation Classification (RC). Compared with the prevailing NER model and the Bi-LSTM+CRF model, this method incorporates label embedding with respect to the previous prediction label. Subsequently, it is passed into the current decoding process, where it replaces the conditional random field layer, thus addressing the issue of label dependency in NER. During the training, the shared parameters are updated through the backward propagation algorithm in order to establish the interdependence between the two subtasks. In 2017, Zheng Suncong et al. used parameter sharing for joint learning as well. The distinction lies in the decoding models used for NER and RC. The NER model uses an LSTM for decoding, considers the distance between entities, and uses a CNN for relation classification. In the context of relation classification, it is necessary to pair entities according to the NER prediction results. Subsequently, a CNN is used to classify the relation between the text and the entities. The underlying model parameters are primarily shared across the model. During training, both tasks will update the shared parameters through the backward propagation algorithm to achieve the two subtasks' dependence. In 2017, Li Fei et al. also applied similar ideas to the task of relation extraction in the field of biomedicine.

The above three methods all use parameter-sharing methods to achieve joint learning. However, this method presents two problems: (1) The parameter-sharing method has two subtasks, yet the interaction between these subtasks is realized through parameter sharing; (2) during training, NER must be performed first. Subsequently, pairwise matching is performed according to the NER's prediction data to classify the relation, which may still generate redundant information from unrelated entities.

In 2017, Zheng Suncong et al. proposed a novel tagging scheme that effectively transforms the relation extraction process, which typically involves sequence tagging and classification tasks, into a sequence tagging problem. By employing an end-to-end neural network model, the relation entity triples can be obtained directly. However, this method cannot handle the overlapping of entity relations. In 2018, Zhou et al. modeled a sequence tagging model proposed by Zheng et al. [11]. This model introduced pre-trained entity features and used an attention mechanism to select important information from the output of the coding layer.

In 2017, Katiyar A et al. proposed a joint model that outputs both entity tags and relation tags while parsing each word in a sentence. In this approach, only word embedding is used in the feature part, with other features such as location and dependency tree not used. The network architecture comprises one input layer and two output layers: One output layer is responsible for outputting entity labels, while the other is tasked with outputting relation labels.

In 2018, Zeng X et al. proposed an end-to-end model with a copy mechanism to complete the joint extraction of entities and relations. The model includes two modules: encoder and decoder. Two different decoding strategies are used: (1) a unique decoder is used to generate all triples; (2) multiple separate decoders are used, with each decoder generating a triple. This model allows for the effective handling of overlapping relations.

In 2019, Takanobu R et al. used a hierarchical reinforcement learning framework to implement the joint extraction of entities and relations. The framework was decomposed into a two-level reinforcement learning strategy hierarchy, which was used for relation extraction and entity recognition.

C. **An end-to-end neural network model for extracting entities and relationships**

In 2016, Miwa M et al. proposed an end-to-end neural network model for the extraction of entities and relationships. The method considered both the word sequence information of the sentence and the dependency tree's substructure information. These two parts of information are modeled by a bidirectional LSTM network, and the two resulting models are stacked to facilitate the extraction of relations and use information about entities throughout the process. In the training process, the entity pretraining and scheduled sampling methods are used to further improve relation extraction performance.

The task of entity-relation extraction is traditionally regarded as a pipeline task comprising two subtasks: entity extraction and entity-relation extraction. The process of identifying entities is referred to as named entity recognition, whereas entity-relation extraction is primarily concerned with relation classification. Entity recognition is a relatively traditional task. The most advanced method is LSTM-RNN. The most advanced method is to add a CRF layer based on Bi-LSTM-RNN, which can perform both Part-of-Speech (POS) tagging and NER simultaneously. The traditional relation classification method is primarily based on the feature/kernel approach. In addition, there are several approaches based on neural network models, including: (1) embedding methods; (2) CNN-based methods; (3) RNN-based methods. However, the word sequence and dependency tree information were not simultaneously considered in the RNN-based method.

In general, the model includes three principal types of presentation layers. The first type is the embedding layer, which is used to represent words, POS tagging, dependencies, and entities. The second type is the sequence layer, which is used to represent word sequences. This layer is a Bi-LSTM-RNN. The third type is the dependency layer, which represents the relationship between two target words (entities) on the dependency tree. The structure of the neural network model is presented in Fig. 5.5.

The embedding layer encompasses several distinct elements. The embedding for words and their POS tags serves as input to the RNN. The entity embeddings are produced as the output from the sequence layer. The dependency type embeddings are fed into the dependency layer as the inputs. The sequence layer mainly describes the

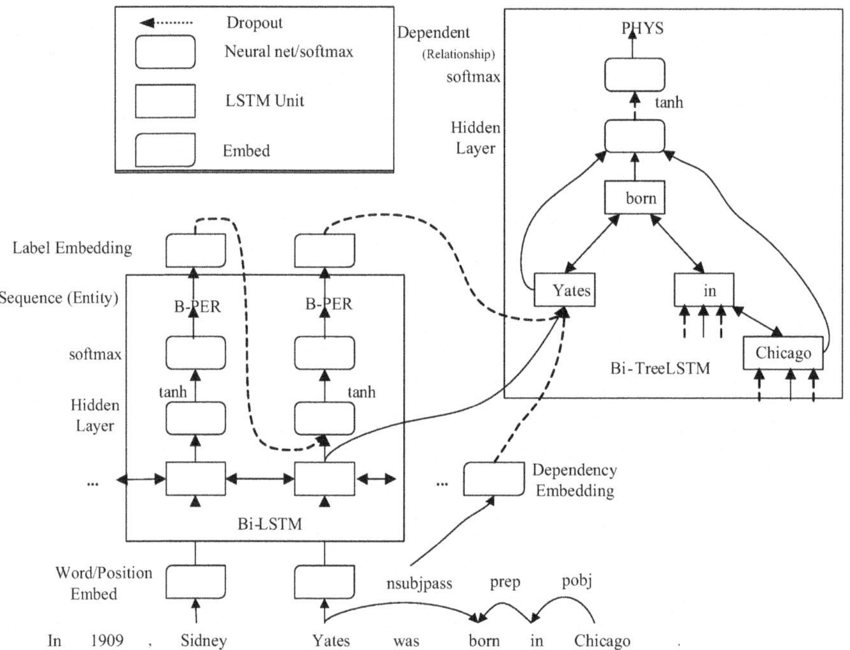

Fig. 5.5 End-to-end entity-relation extraction model structure

context information and entities of a sentence, and conveys the sequential information of the sentence through a Bi-LSTM-RNN.

The entity recognition layer is essentially a sequence tagging task. The tags of the entities in the text adopt the Begin, Inside, Last, Outside, and Unit (BILOU) coding mode, wherein each tag represents both the entity type and the position of a word within that entity. Entity recognition is achieved through a hidden layer and a Softmax output layer. In the decoding process, the label of a predicted word is used to predict the subsequent word's label. The input to the decoding layer is the connection between the output of the sequence layer and the embedded vector result of the previous tag.

The dependency layer is used to describe the relation between a pair of target words in the dependency tree. The primary objective of this layer is to identify the shortest path between the pair of target words on the dependency tree. This method uses a bidirectional tree structure of LSTM-RNN to capture the dependent structure around the target word pair. Information may be transmitted from the root node to the leaf node, or conversely, from the leaf node to the root node.

Relation classification layer: (1) For two words with L or U labels (BILOU mode), a candidate relation can be constructed; (2) NN predicts a relation label for each candidate relation, with a direction (except for negative relation).

The model training process mainly adopts the following optimization methods: the model parameters are primarily updated through BPTT (Backpropagation through time) and Adam, including weights, deviations, and embeddings. Additionally, gradient clipping, parameter averaging, and L2 regularization were used. Dropout is also applied to the embedded layer and the final hidden layer for the purposes of entity detection and relation classification. The model also uses two enhancement techniques: scheduled sampling and entity pretraining.

The experiment uses two datasets, ACE04 and ACE05, for end-to-end relation extraction. Each dataset contains seven entity types and 6–7 relation types. The SemEval-2010 Task 8 dataset is used for the purpose of relation classification. The dataset contains relations between 9 types of nouns. The training set comprises 8,000 sentences, while the testing set consists of 2,717 sentences.

The experimental results show that this method is comparable to the existing feature-based methods, exhibiting better recall and F1 values while exhibiting slightly weaker accuracy. The effects of entity pretraining, scheduling sampling, tag embedding, and sharing parameters on the model's performance are analyzed. The experimental results show that entity pretraining has a significant impact on performance. This is due to the fact that if the entity's information is not adequately trained, it will have a more pronounced effect on the subsequent relation extraction module. Furthermore, the impact of the structure of the LSTM-RNN of the dependency layer on the performance of relation extraction is investigated. The experiment revealed that the discrepancy between the tree selection structure and the sequence structure is not substantial. The structure of the tree itself has a large impact on performance, as evidenced by the following tree structures: SPTree (shortest-path tree), SubTree (subtree), FullTree (whole dependent tree), SubTree-SP (no special consideration for the shortest path), and FullTree-SP (no special consideration for the shortest path).

5.4.2 Named Entity Recognition

A. **Task definition**

According to the interpretation of the Automatic Content Extraction (ACE)[13] evaluation program of NIST in the United States, there are three forms of entity mention in the text (entity mention, or "referential item"): named reference, nominal reference, and pronoun reference. For example, in the sentence "men's table tennis team attended the meeting pointed out the focus of current economic work." There are three referent items for the entity concept "Liu Guoliang," of which "Head coach of the Chinese Men's Table Tennis Team" is a nominal reference, "Liu Guoliang" is a naming reference, and "he" is a pronoun reference

The term "named entity" was first used in MUC-6. At that time, the focus was on the information extraction related to company activities and defense-related activities from unstructured texts, such as those found in newspapers. The key elements of structured information are person names, place names, organization names, time, and digital expressions (including time, date, currency amount, and percentage). Therefore, an evaluation task organized by MUC-6 is the recognition of these entity references and their categorization, which is referred to as the Named Entity Recognition and Classification (NERC) task. This is, of course, to identify the boundaries and categories to which these entities refer [1].

B. **Method overview**

It is notable that one of the earliest works on named entity recognition was the research conducted by Rau LF, which aimed to identify and extract company names from text.. Subsequently, further research was conducted on the topic of proper noun recognition, though it failed to garner significant interest. It was not until the named entity recognition and classification evaluation tasks were organized by MUC-6 for the first time, and the subsequent series of evaluations, that this technology experienced a period of rapid development. In addition to the MUC conference, other related evaluation conferences include the Conference on Computational Natural Language Learning (CoNLL), ACE, and Information Extraction-Entity Recognition Evaluation (IEER), etc.

Prior to the organization of the NERC mission by MUC-6, the primary focus was on the identification of three principal types of proper nouns: person names, place names, and organization names. Since the MUC-6, place names have been further refined into cities, states, and countries. Subsequently, the classification of personal names has also been expanded, with some researchers further differentiating individuals into categories such as politicians, entertainers, and others.

In the evaluation task of the CoNLL organization, proper nouns have been expanded to include the identification of product names. In some other research work, film titles, book titles, project names, research field names, email addresses,

[13] http://www.itl.nist.gov/iad/mig//tests/ace/.

and telephone numbers have also been involved. It is worth noting that a considerable body of research has been conducted on the recognition of specific nouns (e.g., protein, deoxyribonucleic acid (DNA), ribonucleic acid (RNA), etc.) and their relationship in the field of bioinformatics. Some studies do not limit the type of "entity," regarding it as an open-domain NERC. They divide the category of "named entity" into a hierarchical structure, and attempt to cover names and names that appear more frequently in newspapers. Strictly referential words are grouped into approximately 200 categories. This section is primarily concerned with the identification of the three principal categories of proper nouns: person names, place names, and organization names.

As with other tasks in NLP research, early approaches to named entity recognition methods are predominantly rule-based. The implementation cost of the system is relatively high, and its portability is subject to certain restrictions.

Since the late 1990s, and particularly following the advent of the twenty-first century, statistical methods based on large-scale corpora have gradually become the dominant approach within the field of NLP. A large number of machine learning methods have been effectively deployed across the full spectrum of NLP. The various machine learning methods can be roughly grouped into the following categories of named entity recognition based on machine learning: supervised learning methods, semi-supervised learning methods/weakly supervised learning methods, unsupervised learning methods, and hybrid methods. A brief overview of these methods is presented in Table 5.2.

Some named entity recognition tools have been publicly released on the Internet to download and use freely. Stanford NER is a named entity recognition system (Stanford Named Entity Recognizer) based on a conditional random field model developed by the NLP Research Group of Stanford University. The system parameters are based on CoNLL, MUC-6, MUC-7, and ACE named entities, and the corpus is trained.[14]

BANNER is an English named entity recognition system for the biomedical field developed by Arizona State University. Its basic model is also a conditional random field.

MALLET is an open-source statistical NLP software package developed by the University of Massachusetts (UMASS) Amherst. It includes functions such as text classification, clustering, topic modeling, and information extraction. In the application of its sequence annotation tool, named entity recognition can be realized. The system is a finite-state transition machine realized by a hidden Markov model, maximum entropy, and conditional random field models.

As previously stated, all the aforementioned recognition and extraction systems are based on English text. Zhao Hai from Microsoft Research Asia and Shanghai Jiaotong University has developed Chinese named entity recognition systems and made them available online. The S-MSRSeg Chinese word segmentation and named entity recognition tool, released by Microsoft Research Asia, is based on the work of Gao Jianfeng et al. The named entity recognition tool for BaseNER unsegmented

[14] https://nlp.stanford.edu/software/CRF-NER.shtml.

Table 5.2 Overview of named entity recognition methods based on statistical models

Type	Model or method used	Representative works
Supervised learning method	Hidden Markov model or language model	Liu Feifen et al.; Zhang Huaping et al.; Sun Jian et al.; Zhou Guodong et al; Bikel DM et al.
	Maximum entropy model	Tsai TH et al.; Borthwick A; Mikheev A et al.
	Support vector machines	Yi E et al.; Asahara M et al.
	Conditional random field	Leaman R et al.; Finkel JR et al.; McCallum A et al.
	Decision tree	Isozaki [61]; Paliouras G et al.; Sekine S et al.
Semi-supervised learning method/weakly supervised learning method	Bootstrapping learning using labeled small datasets (seed data)	Singh S et al.; Nadeau D; Collins M; Collins M et al.
Unsupervised learning method	Use vocabulary resources (such as WordNet) for context clustering	Etzioni O et al.; Shinyama Y et al.
Hybrid method	Combining several models or using statistical methods and an artificially summarized knowledge base	Liu Xiaohua et al.; Finkel SD et al.; Zhou Guodong Wu Youzheng et al; Jansche M et al.

Chinese text developed by Zhao Hai is based on the CRF++ model, and uses n-gram feature settings for parameter training.

Statistical methods for named entity recognition, which leverage extensive training data from large-scale corpora, depend heavily on a multitude of manually labeled texts. Theoretically, increasing the volume of training data should enhance the effectiveness of the recognition process. However, when the data scale reaches a certain level, a series of automation capabilities, including data storage, transmission, and calculation, must be enhanced in parallel. Consequently, a named entity recognition method driven by big data was developed. The advent of big data has led to the integration of novel technologies into existing named entity recognition methods, including massively parallel processing databases, data mining, distributed file systems, distributed databases, cloud computing platforms, the Internet, and scalable storage systems. The integration of these technologies allows named entity recognition to surpass the confines of laboratory settings, meet the demands of practical applications, and make a meaningful contribution to societal value.

C. **Sequence labeling method based on deep neural network**

Generally, the task of named entity recognition involves identifying the boundaries and categorizing the types of named entities within a given unstructured text. Depending on the specific language, unstructured text corresponds to a sequence of words for languages like Chinese and Japanese, or to a sequence of word tokens for languages that use spaces to delimit words, such as English and German. In the context of named entity recognition, each word or token in the corpus is typically assigned a label corresponding to its entity category. Consequently, the task of named entity recognition is transformed into a sequence labeling problem, where each element in the sequence is assigned a label from a pre-defined category set.

At present, sequence labeling methods based on deep neural networks have become the mainstream research direction. Like traditional machine learning methods, the sequence labeling method based on deep neural networks implements automatic feature extraction and selection processes, and then inputs feature vectors to the classifier to determine the final category (label). The implementation of traditional machine learning methods requires researchers to perform feature engineering on data. This approach is often time-consuming and labor-intensive. The emergence of deep neural networks has addressed this challenge. However, the current deep neural network exhibits limited interpretability, rendering it challenging for humans to derive meaningful insights from the parameters.

In 2015, Huang Zhiheng et al. used a series of models based on LSTM 2015 to process sequence labeling tasks. The models include LSTM networks, bidirectional LSTM (BI-LSTM) networks, LSTM-CRF models that use Conditional Random Field (CRF), and BI-LSTM-CRF models. In a pioneering application of the BI-LSTM-CRF model to the sequence labeling task, the authors employed this approach for the first time. The structure of BI-LSTM allows the entire BI-LSTM-CRF model to effectively use past and future input features at the same time. Furthermore, the CRF model enables the incorporation of sentence-level tag information. The results of the experiments demonstrate that the BI-LSTM-CRF model is effective not only in named entity recognition tasks, but also in tasks such as POS tagging and chunking analysis.

Huang et al. have proposed a model including three fundamental components: LSTM, BI-LSTM, and CRF. Among them, LSTM is an advanced form of RNN. Human thought processes are characterized by a tendency to draw upon existing knowledge and prior experiences when addressing current challenges. For example, when reading an article, one must comprehend the text currently being read based on what has been previously read. The assembly line production method may be employed as an analogy here. The input raw materials are subjected to a series of continuous processing steps prior to becoming products for sale. Each step depends on the previous step. If these individual steps are considered to be identical, they can be reduced to units of action that must be repeated. This represents the fundamental RNN neuron within a neural network model, expressed in mathematical form as follows:

$$\boldsymbol{h}_t = f\left(\boldsymbol{W}\left[\boldsymbol{x}_t, \boldsymbol{h}_{t-1}\right]\right)$$

where $\boldsymbol{x}_t$ is the input from the outside to the RNN neuron at time, $\boldsymbol{h}_{t-1}$ is the internal state of the RNN neuron at the previous time, $\boldsymbol{W}$ is the parameter matrix, and $f(\cdot)$ is a nonlinear function, called an activation function. $\boldsymbol{h}_t$, which is obtained after the calculation, is the internal state of the RNN neuron at time t.

The basic idea of LSTM is the same as that of RNN, but its structure is more complex and can learn long-term dependence information. There are three gates in an LSTM neuron; in addition to the internal state, similar to the RNN neuron, there is also a cell state. These three gates are the input gate, forget gate, and output gate, which are used to protect and control the cell state. Input gates, forget gates, and output gates are all nonlinear functions. The input gate determines what new information is stored in the cell state. The forget gate determines what information is discarded from the cell state, and the output gate updates the old cell state. Based on the current cell state, it is determined which value is to be output (the internal state of the LSTM neuron).

Considering the LSTM processing sequence input as a one-way work from beginning to end, if the content of the entire sequence are known, can the reverse processing sequence input achieve similar effects? BI-LSTM answers this question very well. It includes two LSTM models, one forward processing sequence, and the other reverse processing sequence to get two output sequences. The outputs of the corresponding positions can be connected in series to get the final output.

CRF is a discriminant undirected graph model that models the conditional probability of multiple variables given observations. If $\boldsymbol{x} = \{x_1, x_2, \ldots, x_n\}$ is the observation sequence, and $\boldsymbol{y} = \{y_1, y_2, \ldots, y_n\}$ is the corresponding label sequence, then the goal of the conditional random field is to construct a conditional probability model $P(\boldsymbol{y} \mid \boldsymbol{x})$. The tag sequence $\boldsymbol{y}$ can have a certain structure, which is a certain correlation between its elements. In the sequence labeling task, the observation sequence is text (word or word sequence), and the labeling sequence is the corresponding class label sequence, which has a linear sequence structure [12].

Let $G = \langle V, E \rangle$ denote an undirected graph with a one-to-one correspondence between a node and an element in the tag sequence $\boldsymbol{y}$, y_v denotes the tag variable corresponding to the node v, and $n(v)$ denotes the adjacent node of the node v. If the graph of each variable y_v of G satisfies the Markov property, namely

$$P\left(y_v \mid \boldsymbol{x}, \boldsymbol{y}_{V \setminus \{v\}}\right) = P\left(y_v \mid \boldsymbol{x}, \boldsymbol{y}_{n(v)}\right)$$

then $(\boldsymbol{y}, \boldsymbol{x})$ constitutes a conditional random field [12].

Considering that named entity recognition requires a large amount of feature engineering knowledge and dictionaries to obtain better results, Chiu and Nichols proposed a hybrid BI-LSTM and CNN structure in 2015. Character features are extracted from each English word. The BI-LSTM converts the word features, including word embedding, additional word features, and character features extracted by CNN, into named entity tag scores.

CNN is a feed-forward neural network comprising artificial neurons that are capable of responding to surrounding units within a part of the coverage area. CNN comprises a hierarchical structure including the data input layer, convolutional calculation layer, excitation layer, pooling layer, and fully connected layer. The data input layer is primarily responsible for preprocessing the original data. The neurons in the convolutional calculation layer can be regarded as a set of filters. The inner product is calculated using fixed weights and local data in different positions, thereby achieving the convolution operation and obtaining a set of features. The excitation layer performs a nonlinear mapping operation on the output results of the convolutional calculation layer. The pooling layer compresses the amount of data and parameters to reduce over-fitting. The fully connected layer is typically located at the end of the CNN. The connection method is identical to that employed in traditional neural network neurons, ensuring that the form of the output data meets the application requirements.

In order to leverage the flexibility of the neural network model and alleviate its inexplicability, it is necessary to integrate deep neural networks with structured logic rules. In 2016, Hu Zhiting et al. proposed a comprehensive framework that uses declarative first-order logic rules to enhance the performance of diverse neural networks. As illustrated in Fig. 5.6, in each round of iteration, the posterior regularization rules are adjusted in a logically constrained environment (the posterior regularization method involves constraining the model posterior in an unsupervised environment), and the student network is mapped to the regularized subspace of this rule, thereby obtaining the teacher network (the specific method provides a closed solution). Subsequently, the student network is updated in order to achieve a balance between its capability to imitate the teacher network output and its ability to predict the true mark. During the subsequent training phase, these two network models undergo iterative evolution. The aforementioned process excludes the neural network framework, thereby enabling its application to diverse neural networks, including CNNs and RNNs.

Traditional sequence labeling systems require a large amount of task-specific knowledge, which is typically presented in the form of manual features and data preprocessing. To this end, Ma and Hovy combined Huang et al.'s method and Chiu et al.'s method to introduce a new neural network structure in 2016. This new structure employed a combination of BI-LSTM, CNN, and CRF, while extracting words benefited from English text benefit from the representation of levels and character

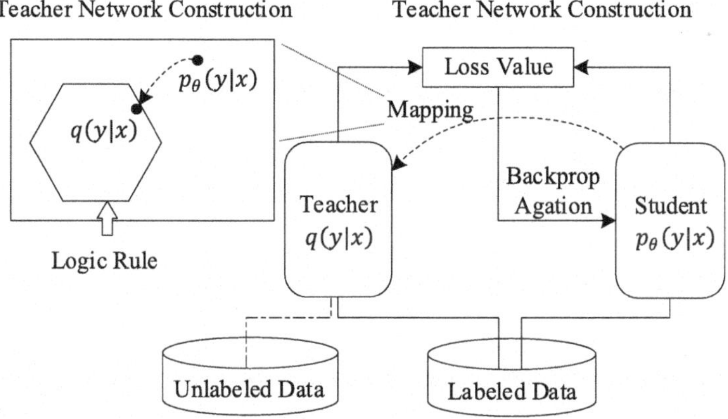

Fig. 5.6 General framework proposed by Hu Zhiting et al.

levels. The system is designed as an end-to-end solution, obviating the need for feature engineering and data preprocessing. This makes it suitable for deployment in a wide range of sequence labeling tasks.

Traditional character-based methods often fall short in capturing the nuances of word sequences and lexical information, while word-based methods can be impeded by segmentation errors. In 2018, Zhang Yue and Yang Jie introduced a pioneering lattice-structured LSTM model that addresses these limitations. This model encodes input character sequences and potential matching words from a dictionary, employing a gate loop cell mechanism to select the most pertinent characters and words within a sentence. This approach effectively sidesteps the initial word segmentation errors, leveraging the contextual understanding provided by the LSTM architecture to enhance the accuracy of language processing.

Due to the absence of natural word boundaries in Chinese, the efficacy of word embedding and dictionary features depends on the accuracy of word segmentation. Errors in word segmentation and the presence of out-of-set words can have a negative impact on named entity recognition. In 2019, Zhu Yuying and Wang Guoxin proposed a Convolutional Attention Network (CAN), which primarily includes a character-level CNN and a local attention layer, as well as a Gate Recurrent Unit (GRU) and a global self-attention layer. This network is capable of capturing Information about adjacent characters and sentence context.

RNN is often used in named entity recognition tasks; however, the nature of the cycle limits its computational efficiency. Some studies aim to use CNN instead of RNN to fully leverage the capabilities of GPU parallel computing. Nevertheless, CNN is challenging to capture long-term contextual information in the sequence. In response to this challenge, Chen Hui et al. proposed in 2019 the Gated Relation Network (GRN), which is designed to capture long-term contextual information. The proposed method involves three main steps: first, using CNN to explore the local context features of each word; second, modeling the relationship between words, as a gate to fuse the local context features into the global context features; and finally, predicting the label.

5.4.3 Emotion Classification

A. **Task definition**

Sentiment classification is the process of categorizing textual content into distinct groups, such as positive or negative, based on the conveyed sentiment and emotional undertones. This technique assesses the author's inclination, viewpoints, and attitudes, providing a nuanced understanding of the text's emotional landscape. It is a process of analyzing, processing, summarizing, and reasoning the subjective text with emotional color. In some instances, it is also referred to as tendency analysis or opinion mining.

Sentiment classification refers to the attitude of individuals with regard to entities such as products, services, and organizations, encompassing their views, emotions, and assessments. The rapid start and development of this field were propelled by the advent of social media on the Internet, including the proliferation of product reviews, forum discussions, and the rapid development of Weibo and WeChat. This is the first time in human history that such a vast digital record has been created.

This technology has a wide range of applications. Companies may utilize it to understand users' evaluations of products. Government departments can understand their attitudes in real time by analyzing their comments on specific events, policies, regulations, or social phenomena. Consequently, sentiment classification has emerged as a fundamental technology that facilitates public opinion analysis [1].

B. **Method overview**

The field of text sentiment classification draws upon a multitude of disciplines, including NLP, data mining, machine learning, statistical analysis, and probability theory. These diverse areas of expertise collectively form the foundation of research in this domain. Text sentiment analysis can be divided into the following three principal categories, distinguished by the level of textual granularity: word-level, sentence-level, and text-level. The distinction between subjective and objective text is based on the presence or absence of emotional content. Related research is primarily concerned with discerning the positive and negative sentiment in subjective text. The proliferation of e-commerce and online social platforms has resulted in a considerable increase in the number of Internet users sharing their comments and opinions online. This has led to the creation of a vast repository of digital data, offering scholars invaluable insight into the attitudes, perceptions, and opinions of Internet users with regard to a diverse range of products and services. For businesses, this data provides a valuable source of consumer feedback, allowing them to gain insights into their customers' preferences and experiences. Over the past few years, research in the field of text sentiment classification has witnessed a surge of interest from both academic and industrial circles, with significant advancements being achieved in this area of study.

Traditional text sentiment classification research can be divided into sentiment dictionary-based methods and traditional machine learning-based methods. The sentiment dictionary-based method needs to manually construct the sentiment dictionary and set a scoring rule for the language expression, recognize the sentiment words in the sentence, record the sentiment intensity, and calculate the sentiment positive and negative tendency score of each sentence through rule matching. Then, the score of the sentence is accumulated to obtain the emotional score of the text. This method is the simplest method of sentiment analysis, but it is easily affected by the domain of the sentiment dictionary and the completeness of the artificial rules. Due to the rapid development of the Internet, new sentiment vocabulary emerges endlessly, and sentiment dictionaries and manual rules cannot be updated in time, and new data cannot be processed in time. The sentiment analysis method based on machine learning selects the positive and negative sentiment characteristics, uses a large number of artificially labeled corpora to train the classifier, analyzes and recognizes the test set data, automatically judges the sentiment trend of the text, and has good generalization ability.

Initially, Pang Bo et al. used movie reviews as a corpus for emotional analysis technology research, applied machine learning methods to emotional positive and negative tendency analysis, and used Support vector machine (SVM), Naive Bayes (NB), and Maximum Entropy (ME) three classifiers to conduct experiments. The experiment shows that when using SVM, unigram word feature, and BOOL feature weight, the classification accuracy is relatively high. It is found that in most cases, the number of times the emotional feature appears is not important. What is important is whether it appears and in which category it appears. Since the method of machine learning transforms the entire text into feature vectors and analyzes the feature vectors to obtain the sentiment tendency, choosing the feature vector is an important factor affecting the accuracy of sentiment analysis. The expression of online comment text is different from that of written text. It contains all kinds of new online words and is full of emojis, which can directly express the tendency of the text. Therefore, selecting the characteristic elements that express the sentiment tendency according to the characteristics of the review text has become the key to the sentiment analysis technology of the review text. Liu Quanchao et al. conducted sentiment analysis on the Weibo corpus by collecting online emoticons, new online vocabulary, etc., adding new text sentiment features, and achieving better results.

The study of sentiment analysis of Chinese comment text started late. Despite the lack of domestic research work, it has been a gradual increase in interest and investment from researchers. Zhao Yanyan et al. at Harbin Institute of Technology summarized and analyzed the research status and development direction of text sentiment classification, divided text sentiment analysis into three steps: sentiment information extraction, information analysis, and information induction retrieval, and made a detailed introduction for sentiment analysis evaluation tasks and resources. Zhang Ziqiong et al. of the Harbin Institute of Technology conducted sentiment analysis research on Internet product reviews, explained the existing problems, elaborated research methods such as emotional analysis of praise and criticism, subjective content identification, and economic value mining of online reviews, and explained the achievements of each research method. From the perspective of text granularity, Yang Ligong et al. of Beijing Institute of Technology, from the construction of corpus, the construction of sentiment dictionary, the extraction of sentiment words, the judgment of sentiment sentences, the analysis of text-level sentiment, and the analysis of evaluation objects and opinion holders. The review identified shortcomings in the current sentiment analysis system and proposed research goals for advancing the field. These studies indicate the objectives and directions for future development and inquiry in the field of sentiment analysis.

In recent years, deep learning has been gradually applied to the field of NLP. In the field of text sentiment analysis, whether deep learning models can be used to automatically extract sentiment features from text is the key to whether deep learning can achieve a higher accuracy rate in sentiment analysis. Socher R et al. use an RNN to perform sentiment analysis based on a syntactic tree, and experiments on movie review texts. The results show that the model can identify negative words' influence on sentiment orientation analysis, but the method relies on syntactic analysis. Kim Y et al. use CNN to perform emotion classification, obtain word embedding vectors through word2vec training, map the text into a feature matrix in a low-dimensional space, and use a convolutional layer, a down-sampling layer to extract features, and realize text emotion classification. This method provides new ideas for future research, and many researchers put forward innovations on this basis. Tai K S et al. constructed a parse tree of LSTM networks, developed the standard LSTM time series chain structure into a syntax tree structure, and achieved better text sentiment classification results. Zhang Yangsen et al. used an RNN to conduct sentiment analysis research on Weibo. The word vector sequence is input to the input layer, and the sentence vector is obtained by the mathematical transformation of the hidden layer, and then enters the output layer.

The experimental results demonstrate that computing the sentence vector representation facilitates an understanding of the deep structure of the sentence and contributes to the advancement of text sentiment analysis in diverse fields of research. Guan Pengfei et al. proposed a two-way LSTM model with two-way attention enhancement, which directly learns the weight distribution of each word's sentiment tendency based on the word vector, with the objective of identifying words that can enhance the classification effect. Based on the end-to-end sentiment analysis of the review target, Li Xin et al defined the End to End Aspect-Based Sentiment Analysis (E2E-ABSA) task as a one-stage sequence labeling problem and introduced a unified tagging schema that integrates the location information and sentiment information of the aspect simultaneously within a single tag. The model is more powerful in its downstream applications, and the representation of words at the semantic level is more effective. Li Zheng et al. used a stacked bidirectional LSTM as the fundamental model. The upper LSTM was tasked with predicting the unified label (i.e., the evaluation object + emotional polarity), while the lower LSTM was responsible for identifying the boundaries of the review object. Peng Haiyun et al. introduced the Aspect Sentiment Triplet Extraction (ASTE) task. The objective of the Triplet Extraction task is to identify all aspects, the associated sentiment, and the corresponding opinion term within a given comment. Subsequently, the three aspects are matched to form triples comprising an aspect, sentiment, and opinion term.

C. **Sentiment classification method based on deep neural network**

RNN, as the name implies, is a neural network with a cyclic structure. The chain structure of RNN enables it to solve the problem of sequence labeling. However, issues such as gradient explosion and gradient disappearance often exist in the training of RNNs, making it difficult for RNNs to maintain long-term memory.

The LSTM network is an extension of the RNN. The original intention of the design is to solve the problem of the long-term dependence of RNN. LSTM is also a special RNN, so it also has a chain structure. But compared with the repetitive module of the cyclic neural network, it has a different structure. It has four neural network layers, and each network layer interacts especially, not a single simple neural network layer. At the same time, LSTM has three thresholds, namely the input gate, the forget gate, and the output gate. These three thresholds are used to protect and control the unit state.

The complete input of the LSTM network at the current time is the hidden layer state $\boldsymbol{h}_{t-1}$ at the previous time, the input $\boldsymbol{x}_t$ at the current time, and the cell state $\boldsymbol{c}_{t-1}$ at the previous time. The input gate determines the information to be input into the cell state $\boldsymbol{c}_t$ at the moment, and its specific calculation method is shown in Formula (5.1). In addition, Formula (5.2) represents the original input information at this moment.

$$\boldsymbol{i}_t = \sigma\left(\boldsymbol{W}_i\left[\boldsymbol{h}_{t-1}, \boldsymbol{x}_t\right] + \boldsymbol{b}_i\right) \tag{5.1}$$

$$\boldsymbol{c}_t' = \tanh\left(\boldsymbol{W}_c\left[\boldsymbol{h}_{t-1}, \boldsymbol{x}_t\right] + \boldsymbol{b}_c\right) \tag{5.2}$$

where $\boldsymbol{h}_{t-1}$ is the hidden layer state at the previous time, $\boldsymbol{x}_t$ is the input at the current time, $\boldsymbol{W}_i$, $\boldsymbol{W}_c$ are the hidden layer state at the previous time and the weight parameters of the current time input, $\boldsymbol{b}_i$, $\boldsymbol{b}_f$ are bias terms, σ represents the sigmoid function, and tanh is the hyperbolic tangent function.

The forget gate of LSTM is responsible for deciding what information needs to be discarded, and its specific calculation method is shown in Formula (5.3).

$$\boldsymbol{f}_t = \sigma\left(\boldsymbol{W}_f\left[\boldsymbol{h}_{t-1}, \boldsymbol{x}_t\right] + \boldsymbol{b}_f\right) \tag{5.3}$$

where $\boldsymbol{h}_{t-1}$ is the hidden layer state at the previous moment, $\boldsymbol{x}_t$ is the input at the current moment, $\boldsymbol{W}_f$ is the hidden layer state at the previous moment and the weight parameter of the current moment input, $\boldsymbol{b}_f$ is the bias term of the forget gate, σ represents the sigmoid function.

Next, the cell state needs to be updated. The specific calculation method is shown in Formula (5.4) as follows:

$$\boldsymbol{c}_t = \boldsymbol{f}_t * \boldsymbol{c}_{t-1} + \boldsymbol{i}_t * \boldsymbol{c}_t' \tag{5.4}$$

where $\boldsymbol{f}_t$ is the state of the forget gate at the current moment, $\boldsymbol{c}_{t-1}$ is the cell state at the previous moment, $\boldsymbol{i}_t$ is the state of the input gate at the current moment, and $\boldsymbol{c}_t'$ is the input information at the current moment.

Moreover, the output gate determines what information is output. The specific calculation method is shown in Formula (5.5):

$$\boldsymbol{o}_t = \sigma\left(\boldsymbol{W}_o\left[\boldsymbol{h}_{t-1}, \boldsymbol{x}_t\right] + \boldsymbol{b}_o\right) \tag{5.5}$$

where $\boldsymbol{h}_{t-1}$ is the hidden layer state at the previous time, $\boldsymbol{x}_t$ is the input at the current time, $\boldsymbol{W}_o$ is the weight parameter of the hidden layer state at the previous time and the current time input, $\boldsymbol{b}_o$ is the bias term of the output gate, σ represents the sigmoid function.

LSTM is based on the improvement of a recurrent neural network, which has the output of the corresponding hidden layer unit at each moment. The specific calculation method is as follows:

$$\boldsymbol{h}_t = \boldsymbol{o}_t * \tanh(\boldsymbol{c}_t) \tag{5.6}$$

where $\boldsymbol{o}_t$ is the state of the output gate, $\boldsymbol{c}_t$ is the current cell state, and tanh is the hyperbolic tangent function.

The sentiment classification method based on LSTM relies on the word2vec algorithm to generate word embedding representations. Word2vec is an open-source toolkit for obtaining word vectors launched by Google in 2013. Its input is a large corpus, and the corresponding language model is trained by an artificial neural network. The real goal of word2vec is to preserve the word vector representation obtained in the process of generating a language model. word2vec contains two models: the continuous bag of words (CBOW) model and the Skip-gram model. The methods used to implement these two models are Hierarchical Softmax and Negative Sampling. Figure 5.7 shows the CBOW model and the Skip-gram model, where $w(t)$ represents the position t of the current word in the sentence.

Both the CBOW model and the Skip-gram model are three-layer models, with an input layer, a projection layer, and an output layer. The CBOW model predicts the probability of the current word based on the context. Its input is the word vector in the window. If the window size is 5, the input is $v(w(t-2)),\ v(w(t-1)),\ v(w(t+1)),\ v(w(t+2))$. The projection layer adds the input word vectors. The output layer continuously performs logistic regression classification by constructing a Huffman tree, and continuously corrects the intermediate variables and word vector representation. The Skip-gram model predicts the probability of the context based on the current word. Its input is the word vector of the current word, and its output is the word vector of the surrounding words.

The structure of the LSTM-based sentiment classification method is depicted in Fig. 5.8. Firstly, the positive and negative samples of the labeled training data are divided. Secondly, each original sentence in the sample set is vectorized using the

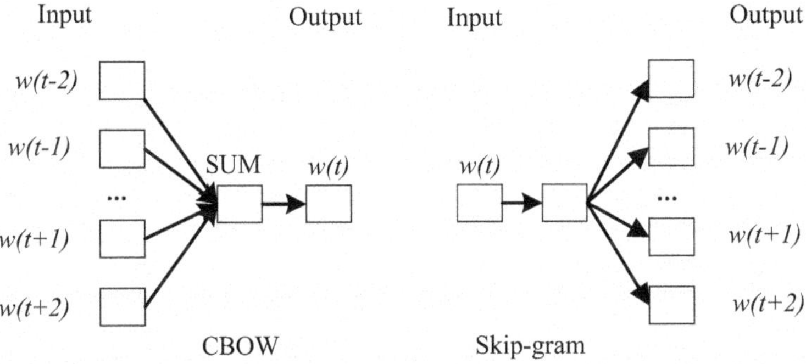

Fig. 5.7 CBOW and skip-gram model

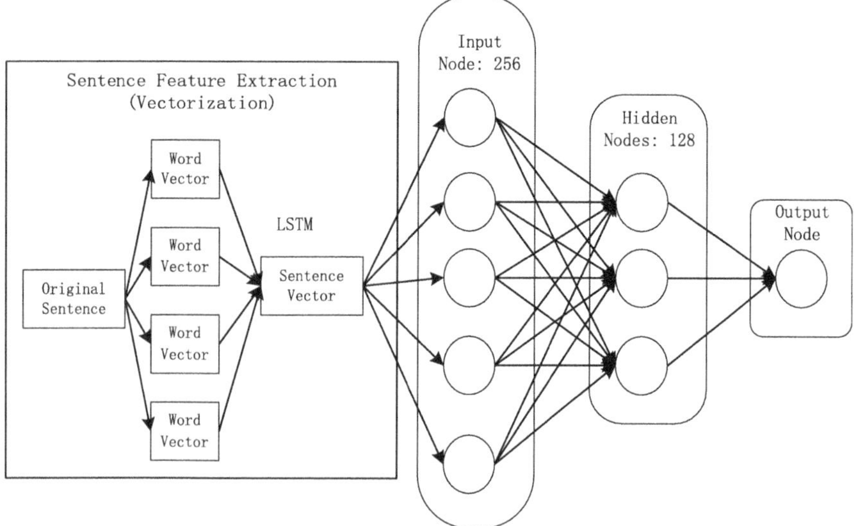

Fig. 5.8 Structure diagram of sentiment classification based on LSTM

word embedding method. Consequently, the generated words are the vector representation of the LSTM neural network. Finally, the output is passed through the hidden layer and the fully connected layer, and the LSTM model is trained according to the true emotional tendency of the text. The method can automatically extract the emotional features of the text, save the process of manual feature selection, make full use of the semantic information of the text, and improve the accuracy of emotion classification. The LSTM model is trained according to the true sentiment tendency of the text.

5.4.4 Text Summary

A. **Task definition**

The objective of a text summary is to provide users with a concise and accurate text description by compressing and refining the original text. According to the number of documents processed, text summaries can be divided into single-document text summaries for a single document and multi-document text summaries for document sets. According to whether the context environment is provided, text summarization can be divided into topic- or query-related text summary and normal text summary. Text abstracts can be classified according to their intended use and the context in which they are to be applied. For example, they may be used as biographical abstracts, opinion abstracts, or for the generation of academic literature reviews. These abstracts are typically designed to meet particular application requirements.

In the era of big data, the primary objective of text summarization has shifted from its traditional document summarization for news documents to Internet text summarization for heterogeneous Internet texts. The main research objects are news, social media, academic literature, etc. It has the characteristics of large scale, diverse types, strong timeliness, multi-language, and viewpoint. Consequently, the new abstraction technology for text big data includes incremental abstraction, abstraction for specific text types, update abstraction, cross-language abstraction, opinion abstraction, comparative summary, etc.

Text summarization can be defined as an information compression process whereby one or more input documents are compressed into a short summary. It is inevitable that this process will lead to information loss, but it is required to retain as much important information as possible. The automatic summarization system typically involves three main stages: understanding the input document, key screening points, and abstract synthesis. The depth of document understanding required varies considerably. Most text summarization systems necessitate only a relatively shallow document understanding, such as paragraph division, sentence segmentation, and lexical analysis. There are also abstract systems that rely on syntactic analysis, semantic role labeling, and reference digestion, and even deep semantic analysis techniques.

B. **Method overview**

The methods used in text summarization can be classified into two primary categories: extractive summarization and abstractive summarization. These categories are distinguished based on their implementation approaches. The extraction method is relatively simple. It typically employs a variety of methods to evaluate the structural units (sentences, paragraphs, etc.) within a document, assigning a specific weight to each structural unit, and ultimately selecting the most pivotal structural unit to construct an abstract. The generative method typically necessitates the employment of natural language understanding technology for the analysis of the grammar and semantics of the text, fusion of information, and the use of natural language generation technology for the generation of novel abstract sentences.

The current mainstream text abstract research work roughly follows the following technical framework:

$$\text{Content Representation} \rightarrow \text{Weight Calculation}$$
$$\rightarrow \text{Content Selection} \rightarrow \text{Content Organization}$$

Firstly, the original text is expressed in a manner that is conducive to subsequent processing. Subsequently, the model calculates the significance of distinct syntactic or semantic units. Subsequently, units are selected according to their importance, and the final summary is formed through the content abstract. Existing research work is carried out in response to different settings and scene requirements, offering a range of design solutions for each technical point in the above framework. Additionally, numerous related studies have endeavored to integrate multiple technical points within a unified framework.

(1) Content representation and weight calculation

Each sentence in the original document is composed of multiple vocabulary or units. In the subsequent processing stage, the basic unit of analysis is the vocabulary and other elements, and a comprehensive evaluation score is assigned to the sentence. Taking the extraction method based on sentence selection as an example, the importance score of a sentence is measured by the importance of its components. Given that the frequency of a word in a document can partially reflect its importance, we can use the probability of a word in each sentence as the score for that word. Consequently, the sentence score can be obtained by summing the probabilities of all the words present within it [13, 14]. Additionally, some works that consider more details and use the Bayesian topic model with robust expansibility to model the topic relevance probability of vocabulary itself [15–17].

In some methods, each sentence is represented as a vector, with the dimension equal to the total vocabulary size. The weighted frequency [18, 19] is typically used to value the sentence vector's corresponding dimension. The weighted frequency can be defined in many ways, including the term frequency-inverse document frequency (TF-IDF) weight, which is commonly used in information retrieval. Additionally, research has been conducted on the use of implicit semantic analysis or other matrix factorization techniques to obtain low-dimensional implicit semantic representations and use them [20]. Subsequently, a similarity measure, such as cosine similarity, is calculated on the obtained vector representation. Subsequently, a weighted graph is constructed in accordance with the calculated similarity. In this model, each node in the graph corresponds to a single sentence. In a multi-document summarization task, sentences that are important may be similar to other sentences, and thus can be connected by a similarity edge weight between nodes and the importance score of the sentence [19, 21, 22]. There is also a lot of work trying to capture the concepts described in each sentence, such as the named entities or verbs contained in the sentence. For the sake of simplification, bigrams are more commonly used as concepts in existing work [23, 24]. Additionally, numerous efforts have been made to employ sequence labeling, using bidirectional GRU (gated recurrent unit) to model word-level and sentence-level representations, respectively. For instance, extractive summaries can be modeled as a sequence labeling task for processing. The fundamental concept is to assign a two-category label (0 or 1) to each sentence in the original text. 0 indicates that the sentence is not included in the summary, and 1 represents that the sentence belongs to the summary [25, 26].

Conversely, many abstract tasks have already been subjected to a certain number of public datasets used for the training of supervised scoring models. For example, for extractive summaries, the artificially constructed summaries include sentences or concepts present in the original document. This allows for the gathering of data regarding the selection of summary sentences, with different units being considered. Subsequently, some features are extracted manually from each unit, and supervised learning is conducted by using a regression model [27, 28] or a ranking learning model [29, 30] to obtain the score of sentence or concept correspondence. The description of document content is structured, so it is also used for supervised training of abstracts by using common sequence tagging or general structural prediction models such as a hidden Markov model (HMM), a conditional random field (CRF), or a structured support vector machine (SVM) [31–33]. The extracted features include location, containing vocabulary, the similarity with neighboring sentences, etc. Generally, features related to specific settings are introduced for specific summary tasks. For example, in query-related summary tasks, the matching or similarity with the query must be considered.

(2) Content selection

From either an affective or a practical standpoint, the final abstract is typically constrained in length. Once the importance score of a sentence or other unit has been obtained, it is necessary to consider how to contain as much important information as possible in the shortest possible length, and then select the original content on this basis.

– Greedy selection

A greedy selection can be made based on the importance scores of sentences or other units. In order to ensure the integrity of the final abstract, it is essential to consider the similarities between units and to avoid including duplicate information. The most straightforward and prevalent mechanism for removing redundancy is Maximal Marginal Relevance (MMR). In each selection process, the greedy selection is the most relevant to the query, or the content is the most important, simultaneous, and existing. This is achieved by choosing the result with the least overlap of information. Some methods directly consider the importance and diversity of content selection in the same probability model framework. This approximates the likelihood function based on greedy selection and achieves good results.

Subsequently, the research group with a discrete optimization direction has intervened in automatic summarization research, noting that numerous greedy selection objective functions, including the MMR method, exhibit sub-modularity. Note that the objective function selected for the content is $F(S)$, and its independent variable S is the set of units to be selected. The submodular function requires that, for any given unit u, the following properties are satisfied: this property is known as diminishing returns, which is consistent with the intuition of greedy selection of summary content. As a consequence of the instantaneous optimality of each step, an additional sentence is selected with each iteration, and the quantity of information is limited to a maximum increase relative to the preceding step. Using a specific greedy method to approximate the submodular functions' optimization problem generally has the theoretical guarantee of the worst-case approximation ratio. In practical applications, research has found that the greedy method can often find an optimal solution. The simplicity and efficiency of the greedy algorithm have contributed to the proliferation of content selection based on submodular function optimization, which has been expanded a lot in recent years. A variety of submodular function optimization or partial submodular function optimization problems and corresponding greedy solutions have been proposed for the selection of specific sentences or syntactic units. Some studies integrate sequence labeling with Seq2Seq and reinforcement learning. A sentence compression model can be learned based on sequence labeling using Seq2Seq. The quality of selected sentences can then be measured using the model, and reinforcement learning can be employed to complete the training of the model.

– Global optimization

The content selection method based on global optimization also aims to maximize summary coverage information and minimize redundancy. Additionally, a variety of constraints derived from the task's nature and the method can be considered in the optimization problem. The most commonly used formal framework is integer linear programming based on 0–1 binary variables. It is assumed that in the result of solving the optimization problem, the variable has a value of 1. In that case, it means that the unit corresponding to the variable should be selected in the final summary. Since integer linear programming is generally an NP-hard problem in terms of computational complexity, the solution process of this approach can be slow in practice and is not suitable for high real-time application scenarios. Some research works simplify the problem and use dynamic programming strategies to design more efficient approximate solutions. There are also many research works that attempt to convert the problem into a minimum segmentation problem and solve it quickly under some special cases or use dual decomposition techniques to convert the problem into multiple simple subproblems in an attempt to find a better approximate solution. More general global optimization acceleration schemes remain an open issue.

(3) Content organization

– **Content simplification and integration**

The sentences obtained through sentence extraction lack sufficient concision, necessitating the application of techniques such as sentence compression, simplification, and rewriting to address this issue. Among these technologies, sentence compression gained particular prominence as a means of simplifying abstract content. The current main practice is based on syntactic rules or text rules. For example, if a phrase is of significant importance and is to be selected to construct an abstract, then the modified headword phrase should also be selected to ensure that the resulting text is grammatically correct. These rules can be directly used in the post-processing stages that are to be conducted after the selection of the content. Such rules may be applied to the optimization model as constraints. Once the optimization problem has been solved, the resulting simplified structure will naturally conform to the established rules. Local rules can be easily expressed as linear inequality constraints between variables, rendering them particularly well-suited for incorporating the integer mentioned above into the linear programming framework. There is a substantial body of research on sentence simplification and rewriting, mainly using machine translation models to transliterate sentence strings or syntax trees. However, due to the high cost of training and the efficiency and performance of phrase structure syntax analysis, there has been minimal direct integration and application of related modules in abstract systems.

Some non-extractive summarization methods focus on fusing the original sentence information to generate a new summary sentence. Based on syntactic analysis and alignment technology, the final sentence can be directly generated from the merged word map. Alternatively, the merged information can be introduced into the optimization model in the form of constraints. Moreover, Seq2Seq-based models are frequently unsuitable for the generation of lengthy texts. For summaries, it is more like a sentence compression rather than a summary. Accordingly, some studies have proposed using real abstracts to guide the generation of text abstracts. The fundamental concept is that abstracts of comparable sentences exhibit a degree of similarity, and these abstracts serve as soft templates to facilitate external knowledge. The Sum model contains three components: Retrieve, Rerank, and Rewrite.

Some researchers try to express the original document as a deep semantic form (such as a deep semantic map) by first understanding the original document's semantics, then analyzing the deep semantic representation of the abstract (such as a deep semantic subgraph), and finally getting the deep semantic representation of the summary. A recent attempt is to generate summaries based on Abstract Meaning Representation (AMR). The summary sentences obtained by such methods are not based on the original sentences; rather, they are directly generated from semantic expressions using semantic analysis and natural language generation technology. This method is relatively complex, and it remains an open question whether natural language understanding and natural language generation can be fully resolved. The current generative summarization method is still in its exploratory phase, and its performance is not yet satisfactory.

Additionally, some studies have improved the conventional method to generate vocabulary chains, employing natural language analysis and processing to generate abstracts. In contrast to the conventional approach of analyzing a single word, this method employs WordNet and part-of-speech tagging tools to examine the meaning of words and to gather those that are related to a specific topic within the original document, thereby forming a vocabulary chain. Another study considered that the traditional vocabulary chains exclusively consider nouns, thereby neglecting the information conveyed by other grammatical parts. This results in a reduction in accuracy. Consequently, further improvements have been made, with the introduction of predicates and adjectives, respectively (i.e., the lexical chain of adverbs). The three together form a holographic lexical chain, which is used for Chinese text summarization, and the accuracy rate has been significantly improved.

- **Sorting of the content**

With regard to the sorting of the selected content, it is acknowledged that related research is still at a relatively preliminary stage. In the context of single-document summarization, the presentation order of the selected content in the original document can be taken to reflect the correct organization order between the content. Consequently, the order of the selected content in the original text is usually maintained directly. For multi-document summarization tasks, the selected content comes from a range of documents. Consequently, it is of greater importance to consider the cohesion and continuity of the content. Early entity-based methods calculated the coherence between sentences by modeling the probability of entity description transitions. Based on this, finding a set of optimal ranking problems can be reduced to the NP-complete traveling salesman problem, which is a challenging problem to solve with high accuracy. Consequently, a variety of approximate algorithms have been applied to content ranking. In recent years, deep learning technology has been used in sentence coherence modeling and ranking tasks. Li Jiwei et al. proposed a discriminative model and a generative model based on LSTM, which can achieve ideal ranking results. There are also ideas that the content ranking can be improved by combining sentence ranking with a new scoring method. Since previous models scored sentences based on their representation of sentences, this led to a separation of scoring and selection. The initial step is to score the sentences, which is then followed by the selection process, which does not leverage the inter-sentential relationships. A new scoring method is proposed, which uses the benefit of sentences as the scoring method, taking into account the relationship between sentences. The scoring and extraction of some of the sentences are completed by a one-way GRU and a double-layer MLP. As text analysis and reference resolution technology continue to advance, the potential for more effective solutions to the problem of sentence ordering in multi-document summaries will grow.

(4) End-to-end summary

As deep learning technology has achieved a series of significant breakthroughs in tasks such as distributed semantics, language models, and machine translation, there has also been a notable increase in the research and development of related methods for the purpose of addressing more complex, abstract tasks. At present, sequence-to-sequence learning based on an encoder-decoder architecture is the most popular due to its capacity to circumvent the arduous manual extraction of features and the modularization of technical points, including the assessment of importance and the selection of content. Training can be initiated with minimal input and output data. However, these methods necessitate a considerably larger training corpus than traditional methods. Moreover, the current mainstream neural framework is not yet able to effectively encode long documents. Consequently, the majority of contemporary research in this field is limited to sentence-level simplification and title generation. Generally, only the first sentence of the document is used as the input, and a short sentence is used as the output [34, 35]. It is notable that few recent works have considered the simultaneous integration of sentence selection and summary generation in a neural network framework. Concurrently, efforts have been made to encode sentence levels and introduce hierarchical attention mechanisms, yet the resulting improvement in performance has not reached the levels achieved by traditional methods.

C. **Generative summary method based on the end-to-end models**

Extractive abstracts are a relatively mature field of study. This method typically uses a sorting algorithm to categorize the processed sentences. Nevertheless, extractive abstracts are less conducive to semantic understanding and are unable to establish comprehensive semantic information within text paragraphs. In contrast, generative technology necessitates that the model comprehend the article's semantics and then summarize the abstract, which is more similar to human practices. However, this technology requires machine learning, and it has not yet matured for a long time. In recent years, researchers have discovered that using end-to-end models is more effective than other methods for generating abstracts, which are inherently challenging tasks. The following section will present a typical end-to-end generative summary model and discuss proposed improvements to address issues related to OOV (out-of-vocabulary) words, low-frequency words, and repetition in the generated summary.

For the first time, in a study by Rush et al. [34], a deep neural network was used to create a generative summary. This work was inspired by the success of machine translation based on deep neural networks, which combine a neural network-based language model with context-based encoders. The encoder is modeled based on the attention mechanism proposed by Dzmitry B et al., which employs an attention mechanism. The model can learn important information from the input text, thereby facilitating the generation of a comprehensive summary. The encoder and the generation model are trained concurrently in the sentence summary task, using the cluster search method to decode and generate the summary.

Nallapati R et al. have implemented numerous improvements to address problems in Rush's work, including excessive vocabulary, OOV, and generating repeated words. Firstly, they proposed the introduction of a Large Vocabulary Trick (LVT) as a means of solving the issue of overly large decoder vocabulary. The LVT's principle uses negative sampling to estimate the denominator of the softmax, thereby greatly reducing the computational and training time requirements. Secondly, the traditional POS tagging, NER, and TF-IDF (Term Frequency-Inverse Document Frequency) features are added to facilitate the identification of the sentence's pivotal elements. In the coding phase, the extracted POS, NER, and TF-IDF features are quantized and then spliced with the word vector. If TF-IDF is a continuous value, it must first be discretized before being spliced into the word vector. Moreover, the Generator-Pointer is introduced to solve non-vocabulary (OOV) and low-frequency words. Some OOV or unusual words in the abstract can be found from the input sequence, so this work is to design a switch in the decoding part $P(s_i == 1) = f(h_i,\ y_{i-1},\ c_i)$. In the event that the switch is activated, the default prediction vocabulary will be initiated. If the switch is turned off, it is necessary to indicate a position within the original text as the desired output. In accordance with Eq. (5.7), the loss function is calculated as follows:

$$\log P(y|x) = \sum_i \left(g_i \log\left\{P(y_i|y_{-i}, x)P(s_i)\right\} + (1 - g_i)\log\{P(p(i)|y_{-i}, x)(1 - P(s_i))\}\right) \tag{5.7}$$

When $g_i = 0$, the model is a copy that is point to a certain position $p(i)$ from the original text, specifically the probability distribution of direct reuse of attention $P_i^a(\mathrm{j})$, which is the probability of pointing to the jth position of the original text at the ith moment; when $g_i = 1$, it is the generation probability that is the probability of calculating the vocabulary. Moreover, the work also introduces two novel techniques: Hierarchical Attention, which is designed to capture the important information of sentences, and Temporal Attention, which is intended to address the issue of repeated words in continuous generation.

As proposed by A et al., a Pointer Network represents an effective method for addressing the generation of OOV words and low-frequency terms. The approach proposed by Generator-Pointer can be reduced to one of two possibilities, without considering the possibility of fusion. The approach of this work is to directly use a network to learn the weight between these two probabilities, which is at the tth time of the decoding part, the probability $p_{\text{gen}} \in [0, 1]$ is generated through the context vector h_t^*, the decoding state s_t, and the decoding input x_t, as shown in Eq. (5.8):

$$p_{\text{gen}} = \sigma\left(\mathrm{w}_{h^*}^T h_t^* + w_s^T s_t + w_x^T x_t + b_{ptr}\right) \tag{5.8}$$

Vector w_{h^*}, w_s, w_x, and scalar b_{ptr} are the parameters obtained through learning, and σ is the sigmoid functions. p_{gen} is used as a soft switch to choose whether to generate a word from the dictionary by calculation P_{vocab}, or copy a word from the input sequence by calculating the attention distribution a^l. For each text, let the expanded vocabulary represent the combination of vocabularies, and all words appear in the source text. The following probability distribution is obtained through the expanded vocabulary, as shown in Eq. (5.9):

$$P(w) = p_{\text{gen}} P_{\text{vocab}}(w) + (1 - p_{\text{gen}}) \sum_{i:\mathrm{w}_i = w} a_i^t \tag{5.9}$$

Note that if w is a superclass word, then $P_{\text{vocab}}(w)$ is 0; similarly, if w does not appear in the source text, then $\sum_{i:\mathrm{w}_i=w} a_i^t$ is 0. One of the primary benefits of the pointer generation model is its capacity to generate superclass words. In contrast, the baseline model is constrained by the pre-defined vocabulary.

As proposed by A et al., the coverage mechanism is to solve the problem of generating consecutive repeated words. In the coverage model, a coverage vector c^t is proposed, which is the sum of the attention distributions at all previous decoding time steps, as given in Eq. (5.10):

$$c^t = \sum_{t'}^{t-1} a^{t'} \tag{5.10}$$

In essence, vector c^t represents the cumulated sum of the attention distribution calculated by each preceding prediction step. This vector records which elements of the original text have been the object of the model's attention, and its influence is reflected in the calculation of the attention weight of the current step. The coverage vector thus serves as an additional input for the attention mechanism, as illustrated in Eq. (5.11):

$$e_i^t = v^T \tanh(W_h h_i + W_s s_t + w_c c_i^t + b_{attn}) \tag{5.11}$$

The purpose of this method is to tell the model the words that it has previously focused on when calculating the level of attention attributed to the current step. This approach is designed to prevent the occurrence of sustained attention on specific words.

Moreover, the coverage mechanism incorporates an additional coverage loss, as shown in Eq. (5.12), to penalize repeated attention weights.

$$\text{cov loss}_t = \sum_i \min(a_i^t, c_i^t) \tag{5.12}$$

It is worth noting that this loss will only penalize repeated attention weights and will not force the model to pay attention to every word in the original text.

Consequently, the overall loss function of the model is shown in Eq. (5.13):

$$\text{loss}_t = -\log P(w_t^*) + \lambda \sum_i \min(a_i^t, c_i^t) \tag{5.13}$$

5.4.5 Machine Translation

A. **Task definition**

Machine Translation (MT) [12] is defined as the use of computers to translate between different languages. The language being translated is typically referred to as the source language, while the resulting language of the translation is known as the target language. Machine translation can be defined as the process of transforming text from the source language to the target language.

B. **Method overview**

The study of machine translation, which concerns the use of computers to facilitate automatic conversion between natural languages, represents a significant research avenue within the broader domains of artificial intelligence and NLP. The implementation of machine translation has been identified as a significant factor for the advancement of national unity, reinforcement of cultural exchange, and the facilitation of international trade. It is regarded as a key technology that can effectively address the communication challenges posed by linguistic differences between diverse countries and nations.

Since the end of the 1940s, machine translation research has generally gone through two stages of development: the period when rationalist methods dominate (1949–1992) and the period when empirical methods dominate (1993–2020). In the early stages of machine translation, rationalist methods were the predominant approach. This entailed the observation of the laws governing the conversion between different natural languages, with the aim of expressing translation knowledge in the form of rules. While this approach allows for the analysis, conversion, and generation of natural language at the deep level of syntax and semantics, it encounters challenges in acquiring translation knowledge, long development cycles, and high labor costs.

In consequence of the expansion of the Internet, especially the booming development of big data and cloud computing over the recent years, the empirical methods have become the dominant paradigm within the field of machine translation. The empirical method places an emphasis on data-driven analysis, employing mathematical models to describe the conversion process of natural language. These models are trained automatically on large-scale multilingual text data, facilitating the development of machine translation systems. A notable example of this type of method is statistical machine translation. The fundamental concept is to describe the translation process through the implicit structure (e.g., word alignment, phrase segmentation, phrase adjustment, synchronous grammar, etc.), use features to describe translation rules, and use feature locality. A dynamic programming algorithm is used to translate polynomial time complexity in an exponential search space efficiently. In 2006, the launch of Google Translate marked the advent of data-driven statistical machine translation methods as a mainstream commercial machine translation system. However, statistical machine translation still encounters challenges in translation performance, particularly in relation to hidden structure and feature design. Local features are often challenging to capture global dependencies, and the log-linear model faces difficulties in addressing linear inseparability in the translation process.

Since 2014, end-to-end neural machine translation has developed rapidly, with a notable enhancement in translation quality when compared to statistical machine translation. In 2015, Baidu released the world's first Internet neural network translation system, while in 2016 Google enabled neural translation for nine languages. The Google Neural Machine Translation (GNMT) system was developed. In 2017, the Facebook Artificial Intelligence Research Institute announced its use of CNN to achieve Neural Machine Translation (NMT), which can achieve performance levels similar to RNN-based NMT, while offering a processing speed that is nine times faster. In June 2017, Google released an NMT model that was entirely based on the attention mechanism. Neural machine translation has replaced statistical machine translation as the core technology of commercial online machine translation systems, including those offered by Google, Microsoft, Baidu, and Sogou.

C. **Machine translation method based on deep neural network**

Neural Machine Translation (NMT) is a machine translation method proposed in 2014. Compared with traditional Statistical Machine Translation (SMT), NMT has the capacity to train a neural network that can map from one sequence to another, and the output can be a variable-length sequence, which is used in translation, dialogue, and text summary aspects, and can achieve excellent performance. The majority of NMT is based on the encoder-decoder structure. The encoder encodes the source language sequence and extracts the information in the source language. The decoder then converts this information into another language, namely the target language, thus completing the language translation.

As the name implies, sequence-to-sequence learning involves the training of a sequence pair. In the case of a Chinese sentence "我也爱你" and a corresponding English sentence "I love you too," the input of the sequence is "我也爱你," and the output of the sequence is "I love you too." In the context of deep learning, one of the primary challenges arises from the inherent variability in sequence length, which contrasts with the fixed input and output dimensions characteristic of this technique. However, the RNN structure provides a solution to this problem. In general, the encoder and decoder use an LSTM or GRU structure in practical applications.

As illustrated in Fig. 5.9, inputting a sentence "ABC," along with the terminal symbol, <EOS>, of the sentence, results in the output "XYZ," and the terminal symbol, <EOS>. In the encoder, a word is an input at each time step until a terminal is reached, at which point the last hidden layer h_t of the encoder becomes the input for the decoder. In the decoder, the initial input is the last hidden layer of the encoder, and the output is the target sequence word X. Subsequently, the hidden layer and its output X are employed as the input or the subsequent time step, thereby generating the second word Y in the target sequence. This process is repeated until the final time step, at which point the terminal symbol <EOS> is output. Given an input sequence $(x_1, \ldots, x_T)$, the output sequence $(y_1, \ldots, y_{T'})$ is generated iteratively through the following equation:

$$\begin{aligned} h_t &= f\left(W^{hx} x_t + W^{hh} h_{t-1}\right) \\ y_t &= W^{yh} h_t \end{aligned} \tag{5.14}$$

where W^{hx} is the weight input to the hidden layer, W^{hh} is the weight from the hidden layer to the hidden layer, h_t is the hidden node, and W^{yh} is the weight from the hidden layer to the output. In this structure, the objective is to estimate the conditional probability $p(y_1, \ldots, y_{T'} | x_1, \ldots, x_T)$, first obtain the fixation of $(x_1, \ldots, x_T)$ through the last hidden layer of the encoder, the vector of dimensions represents v, and then

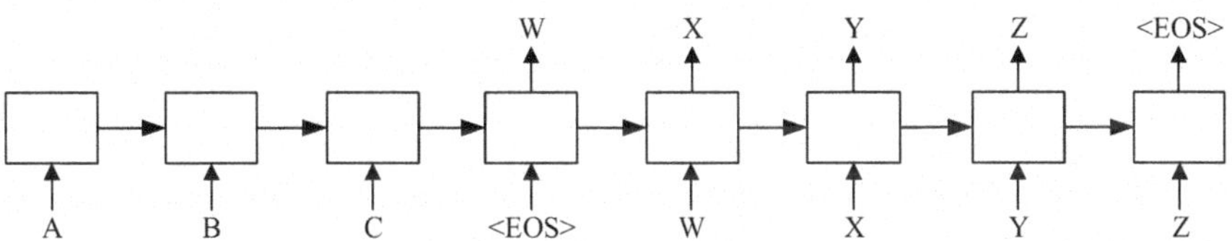

Fig. 5.9 Input-output diagram

the probability of $(y_1, \ldots, y_{T'})$ is calculated by the decoder, where the initial hidden layer is set to the vector v:

$$p(y_1, \ldots, y_{T'}|x_1, \ldots, x_T) = \Pi_{t=1}^{T'} p(y_t|v, y_1, \ldots, y_{t-1}) \tag{5.15}$$

In this equation, each $p(y_t|v, \; y_1, \ldots, y_{t-1})$ is a softmax function.

1. Use two LSTM models, one for the encoder and the other for the decoder.
2. Since the deep model performs better than the shallow model, a 4-layer LSTM structure is used. The input sequence is reversed, resulting in the original input *ABC* becoming *CBA*.
3. Assuming that the target language is *XYZ*, the LSTM model maps *CBA* to *XYZ*. This is due to the fact that *A* is in close proximity to *X*, while *B* and *C* are in close proximity to *Y* and *Z*, respectively. In fact, the model employs short-term dependencies, which are readily optimizable.

5.4.6 Automatic Question and Answer

A. **Task definition**

Question Answering (QA) is the process of using computer algorithms to automatically respond to user queries, with the aim of addressing their specific knowledge gaps. In contrast to existing search engines, question answering systems represent an advanced form of information service. The system provides users with accurate natural language answers, rather than a list of documents sorted based on keyword matching. In recent years, with the rapid development of artificial intelligence, automatic question answering has become a research direction that has attracted much attention and has broad development prospects.

To answer user queries correctly, the automatic question answering system must first understand the user's natural language question, then extract the key semantic information, and finally obtain the answer through retrieval, matching, and reasoning in the existing corpus, knowledge base, or question and answer library. The answer is then returned to the user. The above process involves a multitude of crucial techniques, including lexical analysis, syntactic analysis, semantic analysis, information retrieval, logical reasoning, knowledge engineering, language generation, and numerous other key technologies. Traditionally, automatic question answering has predominantly focused on restricted areas and responses to narrowly defined queries. However, with the rapid development of the Internet and big data, research is increasingly focusing on expanding these boundaries [1].

B. **Method overview**

The research lineage of automatic question answering is rooted in the very inception of artificial intelligence. In 1950, Turing Alan M, the foundational figure in artificial intelligence, published a paper titled "Computing Machinery and Intelligence" in the journal "Mind." His paper commenced with the proposition of an imitation game for machines, aiming to ascertain if a machine could indeed "think," followed by the introduction of the iconic Turing Test to gauge machine intelligence. Echoing this in the NLP sphere, question answering systems stand as one of four pivotal tasks designed to affirm a machine's grasp of natural language comprehension—the other trio being machine translation, paraphrase, and text summarization. The study of automatic question answering not only fosters advancements across AI-related fields but also carries profound scholarly importance.

The existing automatic question answering technologies can be roughly divided into the following three categories based on the different target data sources: retrieval question and answer, community question and answer, and knowledgebase question and answer.

(1) Retrieval questions and answers

The retrieval of questions and answers for research purposes is a field that is constantly evolving in conjunction with the advancement of search engine technology. The launch of the Text Retrieval Conference (TREC) QA mission in 1999 marked the advent of significant research progress in the field of search-based question answering. The task of TREC QA is to identify the answers to the questions posed, using a specific WEB dataset. This methodology employs a question answering system with retrieval and answer extraction as the fundamental processes. The specific process includes three distinct phases: question analysis, text retrieval, and answer extraction. The existing retrieval question answering methods can be classified according to the extraction method employed, which falls into two main categories: question answering methods based on pattern matching and question answering methods based on statistical text information extraction.

(2) Community Q&A

The advent of Web 2.0 has witnessed a surge in the popularity of Internet services based on User-Generated Content (UGC). This has led to the emergence of community question and answer (Q&A) systems, exemplified by platforms such as Yahoo! Answers, and Baidu Knows. The emergence of Q&A communities has opened up new avenues for the advancement of Q&A technology.

The fundamental objective of community Q&A is to identify historical queries similar in semantics to the user's query and retrieve the corresponding answer. This is achieved by leveraging the vast repository of historical query and answer data. Consider a scenario in which a user submits a query question. The question and answer pair data used by q_o to retrieve the relevant information is represented by $S_{Q,A} = \{(q_1, a_1), (q_2, a_2), \ldots (q_n, a_n)\}$. In a similar question and answer pair retrieval process, the objective is to identify a question and answer pair (q_i, a_i) that can effectively address the user's query q_o from $S_{Q,A}$.

In response to this problem, traditional information retrieval models, such as vector space and language models, can be applied. However, in comparison to traditional document retrieval, the characteristics of community Q&A can be described as follows: user questions and existing questions are relatively short, and there is a "lexical gap" between the two. Based on keyword matching, it is challenging for the retrieval model to achieve satisfactory accuracy in Q&A accuracy. At present, various works have proposed the introduction of a single-language translation probability model to address this issue within the existing retrieval framework. The probability of semantic conversion between two different words in the same language is obtained through the massive monolingual question and answer corpus, utilizing the IBM translation model to address the lexical-semantic gap to a certain extent. For example, related words with high probability corresponding to "weight loss" include "slimming," "running," "diet," "health," "distant movement," etc.

Additionally, there are numerous studies examining the significance of words in question retrieval and question matching based on syntactic structure.

(3) Knowledge Base Q&A

Although search-style question and answer and community question and answer are applied in some specific fields or business fields, their fundamental basis remains the matching of keywords and the utilization of shallow semantic analysis technology. It is challenging to achieve the deep logical reasoning of knowledge and the advanced objectives of artificial intelligence. Consequently, over recent years, researchers in academia or industrial contexts have increasingly focused their attention on the knowledge graphs and knowledge bases. The objective is to organize Internet text content into a graph structure, wherein entities serve as the fundamental semantic unit (node). The edges of the graph represent the semantic relationships between entities. The existing large-scale knowledge bases on the Internet include DBpedia, Freebase, YAGO, etc. These knowledge bases are predominantly graph-structured, comprising "entity-relation-entity" triples as the fundamental unit. Based on such structured knowledge, the objective of the question answering system is to identify and reason out matching answers on the knowledge base, in accordance with the semantics of the user's question. This process is referred to as a knowledge base-oriented question answering system or a knowledge base question and answer.

The most effective methodology for executing operations such as queries, matching, and inference on structured data involves the use of structured query statements, including SQL, SPARQL, etc. However, these sentences are typically authored by experts, which presents a challenge for ordinary users in terms of comprehension and accurate implementation. For the average user, natural language remains the most natural way of interaction. Consequently, the conversion of the user's natural language inquiries into structured query sentences represents the fundamental aspect of the knowledge base question and answer. The key is to understand natural language questions. The prevailing approach currently uses semantic analysis to transform users' natural language queries into structured semantic representations, such as the λ paradigm and the Dependency-based Compositional Semantics Tree (DCS-Tree). Semantic analysis grammars or methods that are analogous to this approach include Category Compositional Grammar (CCG) and Dependency-based Compositional Semantics (DCS).

While numerous semantic parsing methods have the potential to yield favorable outcomes within defined domains, a considerable number of essential elements, including the vocabulary and rule sets in CCG, are typically crafted manually within these frameworks. The aforementioned methods will encounter difficulties when facing large-scale knowledge bases, such as those pertaining to unfamiliar vocabulary. It is not feasible to obtain the requisite vocabulary in advance or manually when faced with an unfamiliar knowledge base. At present, numerous works have endeavored to address the aforementioned issues. For instance, the dictionary in CCG has been expanded through the utilization of the data back-labeling method. Additionally, the correspondence between the fact base and the knowledge base at the instance level has been mined to determine the semantic form of words.

However, the processing paradigm of the aforementioned method remains anchored in symbolic logic, thereby limiting its flexibility. In the analysis of the semantics of question sentences, there is a vulnerability to the semantic gap between symbols. Concurrently, the conversion of natural language question sentences into structured semantic expressions necessitates the execution of multiple steps. Furthermore, the occurrence of errors at multiple stages of the process has a considerable impact on the accuracy of the resulting questions and answers.

In recent years, there has been a notable advancement in the field of deep learning technology and related research, which has developed rapidly, with significant breakthroughs occurring in numerous domains, including image, video, and voice recognition. These techniques have gradually begun to be applied in the field of NLP. The method's principal benefit is its capacity to capture the semantic information embedded in text, including words, phrases, sentences, paragraphs, and chapters. This information is then projected into a low-dimensional semantic space. This approach addresses the issue of the semantic gap in traditional natural language processing by employing numerical calculation between vectors in a low-dimensional space. Consequently, an increasing number of researchers are investigating the potential of deep learning technology in addressing NLP problem challenges, including sentiment analysis, machine translation, syntactic analysis, etc. This phenomenon is not limited to other domains, such as knowledge base question answering systems. In comparison to the conventional symbol-based approach to knowledge base question answering method, the representation learning-based method demonstrates enhanced robustness. The efficacy of this approach has been gradually exceeding that of traditional methods. The fundamental premise of these methods is the assumption that knowledge base queries and responses can be conceptualized as a semantic matching process. Through representation learning, it is possible to transform user's natural language problem into a numerical vector in a low-dimensional space, thereby creating a distributed semantic representation. Furthermore, the entities, concepts, categories, and relationships present within the knowledge base can also be represented as numerical values within the same semantic space vector. Consequently, the conventional knowledge base question answering process may be conceptualized as a similarity calculation process between the semantic vectors of the question sentence and the semantic vectors of the entities and edges within the knowledge base.

The question answering system based on deep learning endeavors to establish a joint learning model through the establishment of a high-quality question answer corpus. Concurrently, the semantic representation of the corpus, knowledge base, and question sentence as well as their semantic mapping relationship, are learned. The objective is to model the complex question and answer process through numerical operations between vectors. The advantage of this approach is that it transforms the complex steps involved in traditional question semantic analysis, text retrieval, answer extraction, and generation into a process that can be learned. While the method has yielded some promising outcomes, there are still significant challenges to be addressed. For example, the following issues require further attention: (1) Resource issues. The method of deep learning relies on a large amount of training corpus. The acquisition of high-quality question answer pairs remains a significant challenge. (2) Additionally, the majority of existing question answering methods based on deep learning are mostly designed for straightforward queries, such as those involving a single relationship. Their capacity to address complex questions remains limited. The question of how to use deep learning methods to solve complex problems remains a topic of ongoing interest.

C. **Automatic question answering method based on a memory network**

Memory networks represent a neural network framework proposed by Facebook researchers Weston J et al. They address the challenge of long-term memory limitations in neural networks by incorporating long-term memory components. Based on this framework, various models of memory networks have been developed. The long-term memory module is used for prediction, which can be either read or written. The memory network is frequently employed in a question answering system, wherein the long-term memory module serves as the role of the knowledge base, and the memory network output is a text reply.

The memory network is constituted by a memory array, designated as m, which may be either a vector array or a string array. Additionally, it comprises four components: input I, generalization G, output O, and answer R. Among the aforementioned components, component I is responsible for transforming the model input into a feature representation within the model's internal feature space. Upon receipt of new input, component G updates the array m, thereby facilitating the storage of information within the memory network. The output of component O corresponds to the model's internal feature space, as defined by the model input and the memory array m output. The feature representation can be understood as a form of reading memory. Component R then converts the representation of the internal feature space output by component O into a specific format, such as text. This process can be understood as a transformation of abstract memory into a concrete representation. The modular structure of the framework allows for the modification and adaptation of individual modules to suit different application scenarios. The model, based on the aforementioned framework, is capable of accommodating both long-term (large) memories and memories that are easily readable and writable. Based on this framework, various models of memory networks have been developed subsequently.

In 2015, Sukhbaatar S et al. proposed end-to-end memory networks as a solution to the problem of end-to-end training, which had been identified in the previous article. The proposed memory network is capable of end-to-end training and outperforms

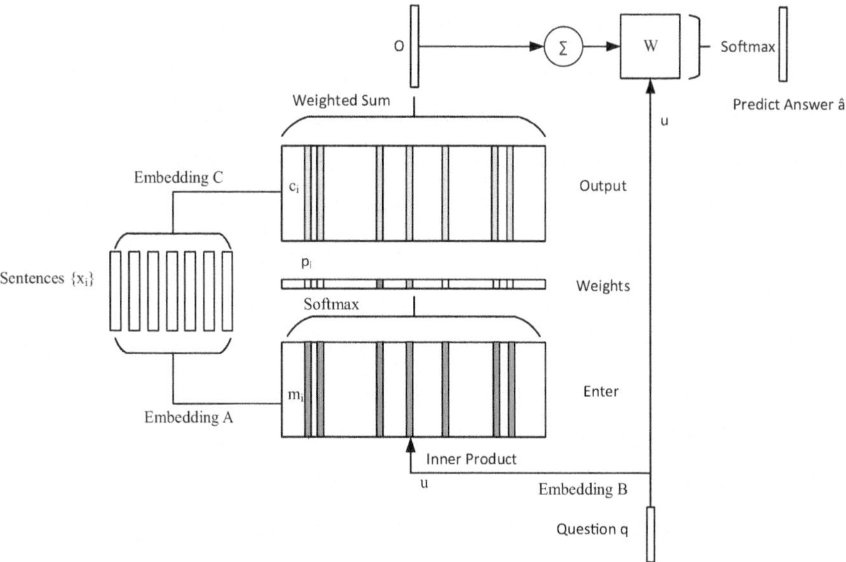

Fig. 5.10 Single-layer architecture of end-to-end memory network model

the original in the training phase. Additionally, the memory network requires less supervision information. The following section introduces a question and answer method based on the end-to-end memory network, which is designed to solve simple problems.

The end-to-end memory network model is divided into a single-layer and a multi-layer architecture. As displayed in Fig. 5.10, the input of the single-layer architecture model has two parts. The input set $\mathrm{S} = (x_1, x_2, \ldots, x_i, \ldots, x_n)$ is used to represent contextual knowledge, and the input vector q represents the question, and the output vector $\hat{a}$ is used to represent the predicted answer.

Initially, the context set S obtains the encoding vector m_i of the memory vector through the hidden layer matrix Embedding A. Similarly, the problem vector q obtains the coding vector u of the problem vector through the hidden layer matrix Embedding B, and then calculates the inner product of the two. p_i represents the degree of correlation between each memory coding vector m_i and the problem.

$$p_i = Softmax\left(u^T m_i\right) \tag{5.16}$$

Furthermore, each input x_i has an output vector c_i. The way c_i and m_i are obtained is similar, and c_i is obtained through the hidden layer matrix Embedding C. Perform a weighted summation on each output memory vector c_i according to p_i to obtain the code vector o of the answer vector.

$$\mathrm{o} = \sum_i p_i c_i \tag{5.17}$$

Consequently, the final answer is generated by decoding the encoding vector o of the answer vector. The predicted answer $\hat{a}$ is obtained through matrix W to be trained, and the cross-entropy loss function is used as the objective function for training.

As shown in Fig. 5.11, the multi-layer model structure can be conceptualized as a stack of multiple single-layer models. The input of the upper layer is the sum of the vectors of o_k and o_k in the lower layer. This can be expressed as the addition of u_k and o_k, which represent the problem code of the upper layer.

$$u_{k+1} = u_k + o_k \tag{5.18}$$

Each layer incorporates embedded matrices A^k, C^k, which facilitate the mapping of x to both memory units and coding units. However, this approach results in a considerable increase in the number of parameters with each additional layer of the model. In order to reduce the number of parameters and facilitate training, two methods are employed to reduce the number of parameters: the same adjacent embedding matrix and sharing parameters between layers.

The memory network structure exhibits robust scalability, with each of its modules offering considerable scope for improvement. It is a deep learning framework that is highly suitable for knowledge base question answering (KB-QA). The use of more complex memory networks represents a promising way to solve KB-QA in the future. Furthermore, the memory network framework allows for significant potential for improvement through the introduction of additional capabilities, the incorporation

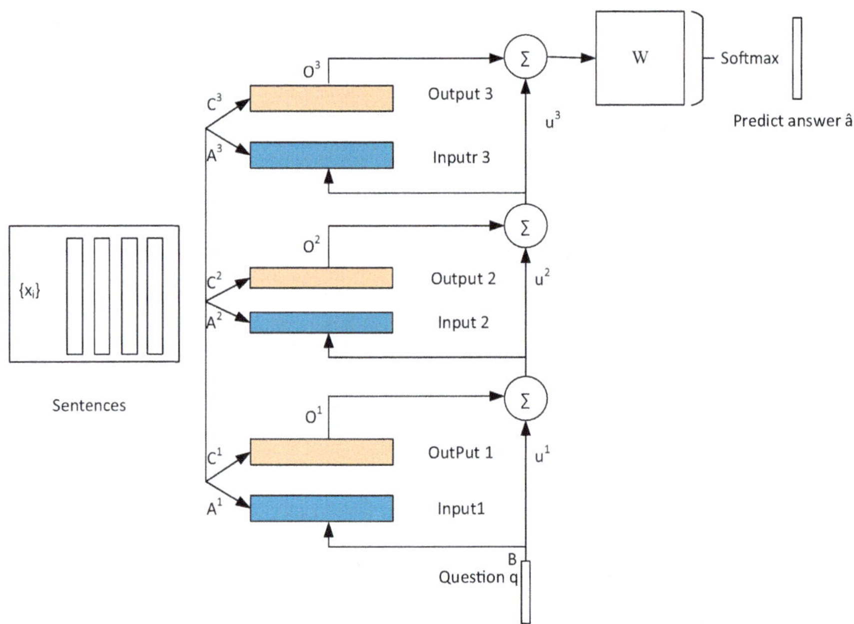

Fig. 5.11 Multi-layer architecture of end-to-end memory network model

of more reasonable models as components of the memory network, the integration of reasoning and attention mechanisms, the incorporation of forgetting mechanisms in memory selection, and the integration of multi-source knowledge inventory into memory.

In essence, automatic question answering has been a subject of investigation within the field of artificial intelligence for over six decades and has emerged as a robust and effective evaluation method for AI technology. In general, the evolution of automatic question answering technology has been characterized by a progression from a limited domain to an open domain, from a single round of question and answer to multiple rounds of dialogue, from a single data source to multiple data sources, and from shallow semantic analysis to deep logical reasoning. It seems reasonable to suggest that, given the rapid development of related technologies such as NLP, deep learning, knowledge engineering, and knowledge reasoning, automatic question answering may achieve a considerable degree of breakthrough in the future. The implementation and evolution of practical applications such as IBM Watson and Apple Siri give rise to the expectation that this technology will be more widely used in the near future.

5.5 Application Cases

5.5.1 IBM Watson Large Question Answering System

A. **System introduction**

The International Business Machines Corporation (IBM) Watson system, named after Thomas Watson Sr, the founder of IBM, gained prominence in 2011. The system defeated two human champions on the American TV quiz show, "Jeopardy!," and has been famous ever since.

The "Jeopardy!" competition is conducted in a distinctive question and answer format that encompasses a vast array of subject matter, including history, literature, art, popular culture, technology, sports, geography, word games, etc. According to various clues provided in the form of answers, contestants must give short and correct answers in the form of questions. In contrast with the general question and answer program, "Jeopardy!" employs a unique structure wherein questions are posed as answers and answers are presented in question form. To excel in this program, participants must possess a comprehensive understanding of history, literature, politics, science, and popular culture. Additionally, they must demonstrate the ability to analyze subtle meanings, ironies, and riddles. Computers are not equipped with the capacity to perform such intricate cognitive tasks.

It is noteworthy that Watson is inaccessible during the course of the game. In contrast to AlphaGo's battle with Lee Sedol, Watson is constrained to using the integrated fundamental package and expansion package stored on the hard disk as its repository of knowledge, in a manner analogous to that of human competitors. In this case, Watson and his opponent proceeded to the second round with a tie score. In the final episode, Watson was successful in overcoming two of the highest-ranked participants, namely Brad Rutter, who had previously won the top prize, and Ken Jennings, the individual who had amassed the most victories. This ultimately secured the first-place title for Watson.

Watson is, in essence, a computer-based question answering (Q&A) system developed by IBM. At its introduction, IBM described Watson as "an application of open question answering technologies such as advanced NLP, information retrieval, knowledge representation, automatic reasoning, and machine learning, and based on the DeepQA technology developed for hypothetical cognition and large-scale evidence collection, analysis, and evaluation." Although some deep learning technologies, such as transfer learning, are used to solve problems, they differ from AlphaGo in that they do not fully leverage deep learning technologies in the same way. The underlying concept is not necessarily that of deep learning; rather, it is more aligned with the principles of the Society of Mind.

B. **Overall architecture**

The first step in the question and answer process is the analysis of the problem. During the process of problem analysis, the system endeavors to understand the nature of the problem at hand and performs an initial analysis to determine the manner in which the rest of the system will address the problem.

Hypothesis generation involves searching the source of the system and extracting answer-size fragments from the search results, thus obtaining the results of the question analysis and generating candidate answers. Each candidate's answer is regarded as a hypothesis, and the system is tasked with substantiating its veracity to a specified degree of confidence.

The principal objective of effectively managing the trade-offs between resource allocation and precision is to apply a lightweight scoring algorithm, which is less resource-intensive, to a larger initial candidate set to trim the denser scoring components to smaller ones before they see their final candidate set. For example, a lightweight scorer may compute the probability that the candidate's answer is a LAT instance. We call this step soft filtering.

Candidate answers that pass the soft filtering threshold undergo a rigorous evaluation process, including collecting additional supporting evidence for each candidate answer or hypothesis, and applying various in-depth scoring analyses to evaluate supporting evidence. Multiple candidate answers to a specific question may be nearly identical, and the merging and ranking step helps take advantage of the relative differences between candidates.

The overall architecture of the IBM Watson Q&A system is shown in Fig. 5.12.

C. **Features**

At the IJCAI2016 conference, Professor James Hendler from Rensselaer Polytechnic Institute was invited to give a lecture to introduce the working principle of IBM Watson in detail.

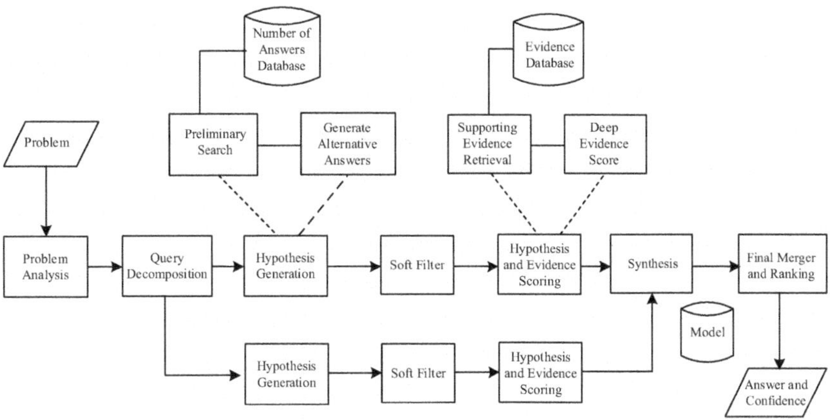

Fig. 5.12 Overall architecture diagram of the IBM Watson question answering system

Imagine Watson as a circular office corridor. Each office is staffed with a team dedicated to specific tasks. Let's streamline the entire process from the very beginning.

1. Question In
2. Question Analysis

   (1) What does this question mean?
   (2) What are we looking for?
   (3) Are there any additional pieces of valid information?
   (4) Do certain words within the question reference other words?

   In this approach, DeepQA aims to understand the issue, discern the nature of the inquiry, and conduct preliminary assessments to determine the most appropriate method for addressing the issue.

3. Primary Search

   (1) Can you find files in the database that may be related to this issue?
   (2) How many files were found?
   (3) Where did they come from?

4. Search Result Processing and Candidate Answer Generation

   (1) Is there a possible answer to this question in these documents?
   (2) How many alternative answers are there?

   When asking a question, a document is opened. As the document moves around the office, more information is added.

5. Context-Independent Answer Scoring

   (1) Is it possible that this choice is the correct answer?
   (2) Is this choice the correct answer form?

6. Soft Filtering
   (1) Which options are obviously wrong?
   (2) If so, do they take too much time to process?
   Each possible answer is given a score, indicating how well the answer supports the candidate.
7. Supporting Evidence Retrieval
   (1) Can you find any information in the database that can prove that a particular answer choice is correct?
   (2) How much information is available for each choice?
8. Search Result Processing and Context Dependent Scoring
   (1) How about choosing a job as an answer to a question?
   (2) Now there is more information. How many points can be given to each choice?
9. Final Merging and Ranking
   (1) Is there any additional information that can change the score?
   (2) What is the total score for each choice?
   (3) Which option has the highest score?
   (4) What is the option with the second-highest score?
   DeepQA has also observed this phenomenon: different surface forms are usually supported by different evidence and get completely different but potentially complementary scores. This leads to a method: the answer scores are combined before ranking and confidence calculations.
10. Answer Out
   The most useful answer is returned, and Watson tries to learn from its judgment.

This is the basic principle of Watson's work. In the following speech, Professor James Hendler also mentioned that Watson is constructed based on "relevant knowledge," and the implementation is presented in Fig. 5.13.

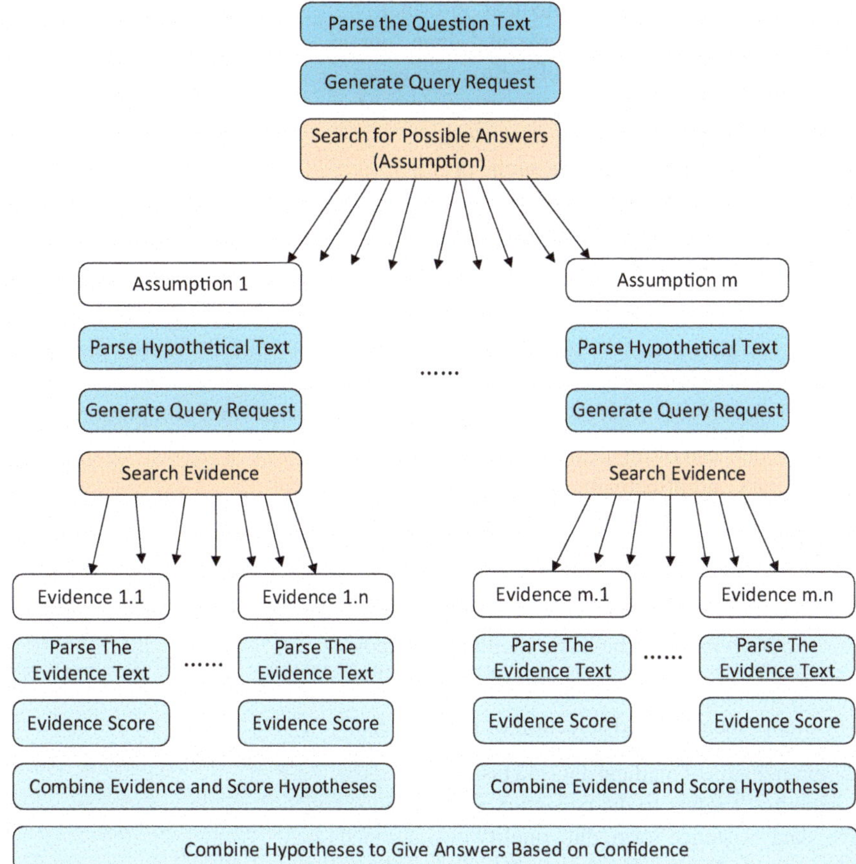

Fig. 5.13 Implementation process of Watson question answering system

By interpreting vaguely worded questions and searching for answers through its general knowledge database, Watson demonstrated the ability to understand natural language, which is one of the most difficult problems that computers need to overcome. This seems to indicate that computers will be able to truly "understand" complex information and talk to humans in the near future, and may even continue to evolve to surpass humans in most human-specific areas.

Watson integrates hundreds of algorithms to analyze the evidence for alternative hypotheses across multiple dimensions, such as type, time, space, popularity, paragraph support, source reliability, and semantic relevance. Each analysis produces some characteristics or scores, which reflect the degree of support of the evidence for the alternative answers in the corresponding dimensions. If a single scorer is removed in the final system, it will not have a significant impact on the test set of thousands of questions. In fact, no scorer can have an impact of more than 1%.

But in combination, Watson achieved an average accuracy of 92% when answering 40%-70% of the questions.

(1) Features

- **Hardware**

Watson is a system optimized for complex analysis. Integrating the massively parallel processor POWER7 and IBM DeepQA software makes it possible to answer "Dangerous Edge" questions within 3 seconds. Watson is a cluster server composed of 90 IBM Power 750 servers (including additional input and output ports, network and cluster controller nodes in 10 cabinets), with a total of 2880 POWER7 processor cores and 16TB of memory. Each Power 750 server uses a 3.5GHz, 8-core POWER7 processor with four threads per core. Only the powerful parallel computing power of the POWER7 processor can barely run the IBM DeepQA software installed by Watson.

John Rennie said that Watson can process 500GB of data per second, which is equivalent to reading 1 million books in 1 second. IBM R&D director and senior consultant Tony Pearson estimated that Watson's hardware cost nearly $3 million, and its 80 TeraFLOPs processing power ranked 94th out of the world's top 500 supercomputers and 49th out of the world's top 50 supercomputers. Rennie also said that the game's data is stored in Watson's memory because the hard disk's access speed is too slow.

- **Software**

Watson's software is written in a variety of languages, including Java, C++, and Prolog. It uses the Apache Hadoop distributed computing framework, the Apache UIMA (Unstructured Information Management Architecture) framework, IBM DeepQA software, and SUSE Linux Enterprise Server 11 operating system. It uses more than 100 different techniques for natural language analysis, source identification, finding and generating hypotheses, mining evidence, and merging and overturning hypotheses.

D. **Miscellaneous**

Watson's sources of information include encyclopedias, dictionaries, news, and literary works. Watson also uses databases, taxonomies, and ontologies, especially DBPedia, WordNet, and Yago.[15]

The IBM team provided Watson with millions of documents, including dictionaries, encyclopedias, and other reference materials that could create a knowledge base. Although Watson did not connect to the Internet during the game, it still has 200 million pages of structured and unstructured information on its 4TB disk for its use, including the full text of Wikipedia.

---

[15] http://www.aaai.org/Magazine/Watson/watson.php.

5.5.2 *Baidu Machine Translation System*

A. **System introduction**

Machine translation is the use of computers to automatically translate one language into another. As early as 1946, when the first modern computer was born, American scientist Warren Weaver proposed the idea of machine translation. Machine translation involves disciplines such as computer science, cognitive science, linguistics, and information theory. It is one of the ultimate goals of artificial intelligence. The study of machine translation technology has important academic significance.

In the context of the Internet and globalization, the online game among major countries has become fierce, and network information security faces unprecedented challenges. The development of a machine translation system with independent intellectual property rights to obtain real-time and accurate multilingual political, economic, cultural, military, and other information is an important basic guarantee for China's information security, which is of great significance to the protection of national security, the development of the national economy, and the implementation of international strategy.

Big data on the Internet brings new opportunities and challenges to machine translation research, making it possible to automatically acquire and update massive translation knowledge in real time. Traditional translation models and methods are in urgent need of innovation. Baidu Translation has made major breakthroughs in mass translation, knowledge acquisition, translation models, and multilingual translation technology. It has solved the problems of high research and development costs, long cycle time, low quality of traditional methods, and responds accurately and in real time to massive and complex translation requests on the Internet. As a result, our country has mastered the core technology of Internet machine translation and occupied the technological high ground in this field.

B. **Overall architecture**

Figure 5.14 shows the core technology of Internet machine translation based on big data.

(1) **Features**

The Baidu translation project represents a significant contribution to the field of machine translation.

1. A translation model based on Internet big data was proposed. Guided by this model, an adaptive training and multi-strategy decoding algorithm has been proposed, which has the potential to overcome the limitations of translation in multiple fields and text styles. Furthermore, it has the capacity to optimize and integrate the translation cloud platform and algorithms, and to respond to the millions of complexities from various translation requests that are received on a daily basis.

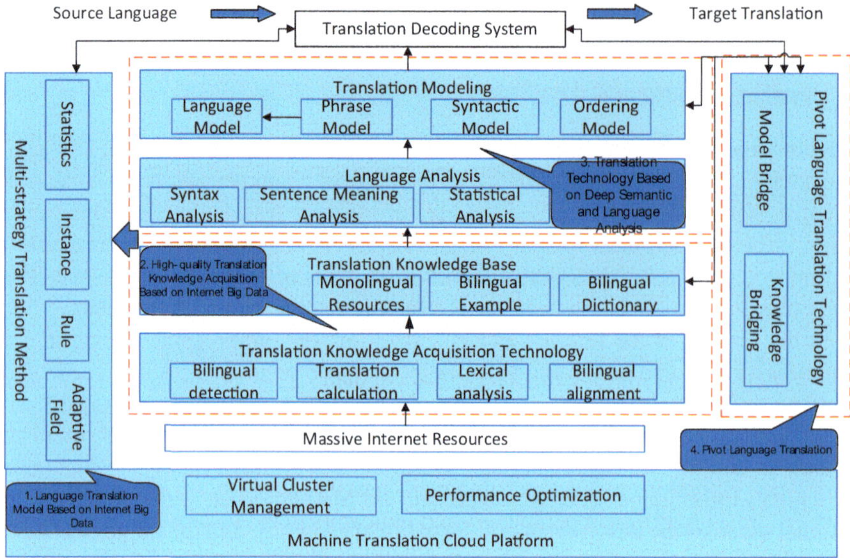

Fig. 5.14 Baidu machine translation technology architecture

2. The project has developed high-quality translation knowledge acquisition technology based on Internet big data. This has broken through the bottleneck of the small scale and high cost of traditional translation knowledge acquisition and established international standards in the field of language content processing. The scale of high-quality translation knowledge accumulated by the project is 100 times the scale of data released by the authoritative international organization NIST.
3. Language analysis and translation technology are proposed based on deep semantics. It resolved the recognized challenges of disambiguation and order adjustment in machine translation. The model proposed a tree-to-string-based syntactic statistical translation model for the first time in the world, which effectively uses source language syntax information to address phrase generalization and long-distance translation ordering issues.
4. The concept of the translation knowledge bridge and model bridge technology has been proposed based on the pivot language. This breakthrough surpasses the traditional constraints imposed by the limited linguistic scope of machine translation, enabling the translation of less-common languages that have sparse resources. It facilitates the swift implementation of multilingual translation capabilities, with the agility to integrate a new language within a mere 11 days. To date, the system supports 28 languages and 756 translation directions.

The above-mentioned technology has been implemented in numerous pivotal sectors of the nation's economy, with over 7,000 enterprises and third-party applications, including prominent names such as Baidu, Huawei, and Kingsoft, adopting it. In addition, Baidu's machine translation project was awarded the second prize of the National Science and Technology Progress Award in 2015. The application is as follows:

1. Based on the above outcomes, a "multilingual information collection, processing, and analysis system" has been implemented, it supports the automatic collection, translation, and analysis of more than 10 languages, including foreign languages such as English, Japanese, German, French, Korean, Arabic and Turkish, as well as Chinese minority languages such as Tibetan, Uyghur and Mongolian. The results of this system are applied to important national departments such as the 55th Research Institute of the General Staff of the People's Liberation Army, effectively supporting the development of core businesses of relevant units, and promoting the development of multilingual intelligence translation and analysis to maintain national security and social stability. It played a significant role.
2. The findings of this study have been implemented into the "Baidu Translator," which now supports 28 languages, including but not limited to Chinese, English, Japanese, Korean, and Thai, with 756 translation directions, thus conferring a competitive advantage in the market. The platform is capable of providing multilingual high-quality translation, with a user base of over 500 million worldwide. It is estimated that over 100 million translation requests are responded to on a daily basis. Third-party reviews indicate that, of the 32 translation directions launched at that time, 28 of the translation directions of Baidu Translation surpassed Google translation. In the technical appraisal of scientific and technological achievements organized by the Chinese Institute of Electronics, academicians and experts reached a unanimous consensus: "It has reached the leading international level in terms of translation quality, translation language direction, and response time."
3. The translation technology is integrated into the Baidu search engine to facilitate the retrieval of multilingual information and enhance the user experience. In July 2014, General Secretary Xi Jinping and Baidu CEO Li Yanhong jointly launched a Portuguese search engine to provide users with cross-language search capabilities. This technology is also applied to Baidu's Arabic and Thai search engines.
4. The "Baidu Translation" platform provides free services for more than 7,000 third-party applications, promotes the development of the national economy, and helps the internationalization of Chinese companies. It provides translation platform services for a large number of small and medium-sized enterprises, lowers the threshold for entrepreneurship and innovation, and promotes the prosperity and development of related industries. Huawei integrated the translation service into the camera translation application of its Ascend Mate mobile phone, enhancing the market competitiveness of the mobile phone. Mobile phones with translation functions are sold to more than 30 countries and regions, including France, Germany, Russia, Spain, and the United Kingdom. "Kingsoft PowerWord" uses the above translation technology to realize the leap from word query to sentence automatic translation and enhances its products' value. Its installed

capacity is 46.6 million sets, and it has a total output value of 45.61 million yuan. The B2B cross-border e-commerce platform "Dunhuang.com" uses "Baidu Translator" for cross-border trade, serves more than 1 million domestic suppliers, and facilitates the sale of their products to 224 countries and regions worldwide. This has contributed to the advancement of China's foreign trade.

5.5.3 *Microsoft Robot Xiaoice*

A. **System Introduction**

Microsoft Xiaoice is an emotional computing framework proposed by Microsoft (Asia) Internet Engineering Institute in 2014. By employing a range of algorithms, cloud computing, and big data, it adopts intergenerational upgrades to gradually form a complete artificial intelligence system that develops in EQ's direction. As of July 26, 2018, the sixth generation of Xiaoice has been developed. It is one of Microsoft's three global artificial intelligence product lines. In which it covers over 40 platforms in China, Japan, the United States, India, and Indonesia, with more than 660 million users. Its product forms involve conversational artificial intelligence robots, intelligent voice assistants, artificial intelligence content providers, and a series of vertical field solutions.

Microsoft divides the evolution of artificial intelligence interactive technology products into three stages. The first stage is basic artificial intelligence interaction: there is one or more interaction methods, such as text, voice, image, and video. However, different interaction methods are separated. The second stage is the primary senses, which, in the artificial intelligence system, a core engine (such as Xiaoice's EQ core dialogue engine) is used to unify the various interactions mentioned above, so that different senses can be mixed and used. Microsoft Xiaoice has entered this stage since its third-generation release in 2015. Microsoft released the fifth generation of Microsoft Xiaoice, entering the third stage (advanced senses). The advanced senses of Microsoft Xiaoice are formed by the organic fusion of multiple primary senses, so they have strong interaction capabilities and greatly increase the requirements for comprehensive technical reserves and data.

B. **Overall architecture**

The overall architecture of Xiaoice is presented in Fig. 5.15. It consists of three layers: the user experience layer, the dialogue engine layer, and the data layer.

The user experience layer: serves to establish a connection between Xiaoice and a number of popular chat platforms (such as WeChat and QQ). It is capable of communicating with users in two modes: full-duplex mode and alternate dialogue mode. This layer also includes a series of components for processing user input and Xiaoice's response, including speech recognition and synthesis, image understanding, and text normalization.

The dialogue engine layer consists of a dialogue manager, an empathy calculation module, a core chat, and dialogue skills.

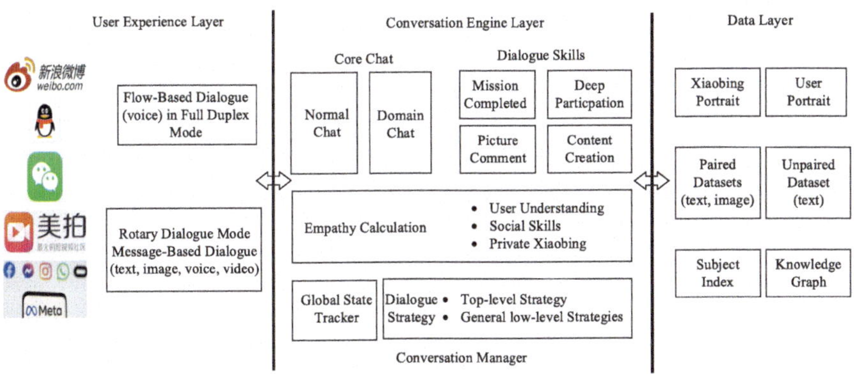

Fig. 5.15 The overall architecture of microsoft Xiaoice

The data layer consists of a set of databases that store collected human conversation data (text pairs or text-image pairs), non-conversational data, and knowledge graphs for core conversations and skills, as well as personal files for Xiaoice and all registered users.

(1) In-depth analysis of the dialogue engine layer

The dialogue engine layer mainly includes four major components: dialogue manager, empathetic computing, Core Chat, and skills.

– **Conversation manager**

The dialog manager serves as the central controller of the dialog system. It consists of a global state tracker and a dialogue policy. This operation can be used as a top-level strategy activation technique or as Core Chat.

The global state tracker uses working memory to track the conversation state. The working memory is empty at the beginning of each conversation. Subsequently, the dialogue between the user and Xiaoice, along with the entities and empathy tags detected from the text according to the empathy calculation module, is stored with text strings.

Xiaoice uses a layering strategy as follows:

(1) The top-level strategy manages the entire conversation by selecting Core Chat or skills activated based on the conversation state in each conversation round.
(2) A set of low-level strategies, each corresponding to a specific skill, is used to manage the conversation segment.

The dialogue strategy aims to optimize long-term user engagement through an iterative, trial-and-error process based on feedback from Xiaoice users.

The topic manager simulates the behavior of humans in changing topics during a conversation. It consists of a classifier and a topic retrieval engine. The classifier determines whether to transition to a new topic in each dialogue round.

If Xiaoice lacks sufficient knowledge about the topic to engage in a meaningful conversation, or if the user becomes disinterested, it will prompt a topic shift.

– **Empathy calculation**

Empathy calculation reflects Xiaoice's emotional intelligence. Given a user input query Q, the empathy calculation considers the context C, rewrites Q into the context version Qc, uses the query empathy vector eQ to encode the user's feelings and states in the dialogue, and is specified by the response empathy vector eR Respond to R.

The output of the empathy calculation module is expressed as a dialogue state vector $s = (Qc, C, eQ, eR)$. The dialogue strategy is used to select the skill and the input for the generated activation skill (for example, Core Chat).

The empathy calculation module consists of three parts: contextual query understanding, user understanding, and interpersonal response generation.

– **Core chat and conversation skills**

Core Chat is a very important part of Xiaoice IQ and EQ. Together with the empathy calculation module, it provides basic communication capabilities that take text input and generate interpersonal responses as output.

Core Chat is composed of two parts, namely Normal Chat and Domain Chat. Normal Chat is responsible for participating in open conversations covering a wide range of topics. Domain Chat is responsible for deeper conversations in specific areas (such as music, movies, and celebrities).

Xiaoice has 230 dialogue skills, which, in conjunction with Core Chat, constitute Xiaoice's IQ component.

C. **Features**

(1) The first generation of Xiaoice

In addition to intelligent dialogue, Microsoft Xiaoice also has practical skills such as group reminders, encyclopedias, weather, constellations, jokes, traffic guides, and dining reviews. A Microsoft team analysis revealed that the incorporation of "Microsoft Xiaoice" into a WeChat group resulted in a fourfold increase in group activity. Moreover, as the number of interactions with the group members increases, additional functions will be gradually revealed.

(2) The second generation of Xiaoice

The second generation of Xiaoice is completely exclusive to users. The cross-platform mobile Internet application helps users complete more and more tasks and constantly improves and upgrades themselves.

Users can adopt Xiaoice easily and conveniently, specify Xiaoice's new name and profile picture, and then complete the adoption.

After adoption, users can use Xiaoice on more and more third-party platforms. According to the time steps of technical docking, in the short term, users can use it on platforms such as TouchPal Number Assistant, Sina Weibo, JD Wireless, Xiaomi Michao, NetEase Yixin, and Tencent WeChat.

On August 13, 2014, Microsoft China and Weibo jointly announced that Microsoft Xiaoice on the Weibo platform was officially upgraded to the second-generation exclusive Xiaoice. The second generation of Xiaoice on Weibo will be more exclusive and private. Just post @ Xiaoice on Weibo to adopt it. Additionally, the second generation of Xiaoice will undergo an upgrade in the following areas: artificial intelligence level, private chat corpus, new skills, and the points system.

(3) The third generation of Xiaoice

Microsoft has indicated that the third-generation Xiaoice integrates Microsoft's multiple global artificial intelligence image and speech recognition technologies. In addition to the original long-range emotional dialogue capabilities, it also has a new artificial intelligence sensor that can see, hear, and speak. Specifically, the third-generation Xiaoice supports the image recognition function, which can "see" the pictures and even video content sent by users and conduct corresponding dialogues based on the pictures' content. This is mainly due to Microsoft's breakthroughs in image recognition technology. As previously reported by Microsoft, the company's image recognition technology is approaching human-level performance.

Besides, the third-generation Xiaoice is capable of verbal communication in addition to textual responses. As reported, the voice of Xiaoice is set by a 17-year-old cute girl. The third generation of Xiaoice will continue to support the "Sakura Change," whereby the user must utter the phrase "Sakura Change" to Xiaoice, who will then transform the user into a Japanese high school girl. Once this transformation is complete, Xiaoice will be able to communicate with the user in highly fluent Japanese. However, this is contingent upon Xiaoice's ability to comprehend the Japanese language being spoken.

In a recent development, Xiaoice, which has resumed its presence on the WeChat platform, will now be accessible in the form of a public account and open a commercial version. Third-party service account and subscription account managers can upgrade their official accounts to artificial intelligence official accounts through commercial versions. Additionally, Microsoft has indicated that the service is provided at no cost.

(4) The fourth generation of Xiaoice

Microsoft has revealed that the fourth generation of Xiaoice will include a real-time emotional decision-making dialogue engine, multiple new senses, three languages (Chinese, Japanese, and English), and a functional plug-in platform corresponding to different fields. Additionally, at the press conference, Microsoft disclosed the most recent developments pertaining to the internationalization of Xiaoice and pertinent information regarding its strategic partners.

(5) The fifth generation of Xiaoice

After nearly four years of growth, Xiaoice began to have its "future direction." On August 22, 2017, the fifth generation of Microsoft Xiaoice was officially launched. Microsoft announced that Xiaoice would fully enter the Internet of Things (IoT) domain, collaborating with numerous IoT vendors to use Xiaoice. "Xiaoice is a chat robot, but not just a chat robot," stated Shen Xiangyang, Microsoft's executive vice president. "Chatting is just a user experience, but the real core of our product design philosophy is to create an emotional computing framework with many living spaces, auxiliary equipment, and related equipment. Xiaoice can communicate with humans in any place and scene."

Microsoft released the fifth generation of Xiaoice, marking its entry into the third stage—advanced senses. These advanced senses are the result of an organic integration of various primary senses, significantly enhancing Xiaoice's interactive capabilities. Moreover, there is a considerable increase in the demand for comprehensive technical expertise and data. The full-duplex speech, an advanced sense, necessitates proficiency in both text and speech processing, including Speech Recognition and Text-To-Speech. Both components must adhere to higher quality standards to deliver a superior interactive experience, approximating natural human interaction. For instance, if the full-duplex voice experience is likened to a phone call, the previous intelligent assistant voice interactions would be akin to using a walkie-talkie. Furthermore, advanced senses greatly expand the potential application scenarios for the artificial intelligence system, allowing Microsoft Xiaoice to actively maintain engagement with human users. The high-level senses introduced in the fifth generation of Xiaoice have completed their initial deployment. The full-duplex voice feature has been integrated into a partnership with Xiaomi's IoT open platform, enabling control over a range of smart devices within the platform. Additionally, the real-time streaming media service has been implemented in public spaces in several major cities across China and Japan.

The fifth generation of Microsoft Xiaoice officially uses the Generative Model, which is the industry's first landing product and has fully entered the IoT domain. Some cooperative products have already been launched, while others have commenced sales.

In May 2017, Microsoft announced that Xiaoice had used numerous aliases to publish poems in various poetry forums and publications, and published the first collection of poems created by artificial intelligence. After that, Microsoft officially launched a joint creation product for poetry, and anyone can use Microsoft Xiaoice to complete their poetry creation. At this press conference, Microsoft disclosed that Microsoft Xiaoice had entered a number of creative fields, including poetry and audio children's books, songs, news, etc. Among them, the quality of audio children's books exceeds 98% of human creators. The time is only 1/500 of that of a human of the same level, and the cost is only 1/80000 of that of a human of the same level. Microsoft Xiaoice tested the water on various audiobook platforms through a small number of accounts and achieved the expected results. Microsoft mass-produced audiobooks and put them on the market. Microsoft's Xiaoice training achieved a 48kHz sampling rate in terms of songs while greatly expanding the range. The song of the same name, "I am Xiaoice," has been released on the QQ Xiaoice channel. Furthermore, Microsoft Xiaoice conducted an auditory analysis of the melody, integrating this with an investigation of landmarks in various urban centers, to generate visual content aligned with the city and the song's prevailing ambience. The "Skyline" clothing launched by Microsoft and SELECTED has entered the SELECTED store for sale through this technology.

Microsoft Xiaoice accelerates the integration with the Bing search engine and launches a new smart media business platform solution 3.0. As a columnist for the "Qianjiang Evening News," Microsoft Xiaoice wrote column articles through big data and successfully predicted several globally important results of the event. This time, Microsoft disclosed that Xiaoice also wrote some of the news content on Baidu Baijia and Today's Toutiao. At the press conference, Microsoft announced version 3.0 of the upgraded smart media business platform solution.

The accelerated integration of Microsoft Xiaoice and the Bing search engine, the upgraded Smart Media Business Platform Solution 3.0, includes six new modules, including media production, media knowledge graph, smart interaction, and full platform interaction. With the global big data capabilities of the Bing search engine, Microsoft Xiaoice's new solution can build a media knowledge graph based on global news data sources for the media, accurately mine the knowledge, meaning behind each piece of content in the global news, and build the relationship between each other, to help the media to sort out the stories behind the time and content more comprehensively and quickly.

(6) The sixth generation of Xiaoice

The sixth generation of Xiaoice continues to upgrade the emotional computing framework. The first major technological breakthrough is the introduction of the empathy model in the dialog. In dialog with users, Xiaoice can not only create responses, but also better control the dialog process through the skills of confirmation and verification. Microsoft announced that the empathy model test has been completed, and it has been launched in five countries covered by Microsoft Xiaoice.

Furthermore, Microsoft Xiaoice also integrates text, sound, and vision, and this year developed real-time vision based on full-duplex speech. The real-time visual capabilities of Xiaoice were also demonstrated at the press conference. On this display, Xiaoice can direct users to complete face detection through real-time interaction of vision and voice, and can conduct open-domain conversations in the process.

Microsoft Global Executive Vice President Shen Xiangyang served as the platform for this press conference. According to him, Xiaoice has become one of the world's largest conversational artificial intelligence systems, with 660 million human users, 120 million monthly active users, and a CPS (per conversation round) reach of 23 rounds, accessing 57 direct user scenarios.

5.5.4 BFS Public Opinion Analysis System

A. **System introduction**

BFS Public Opinion Analysis System is a user-oriented, highly integrated public opinion analysis system developed by the Information Security and Countermeasure Technology Laboratory of the School of Information and Electronics, Beijing Institute of Technology, for government agencies, enterprises, and major universities. The system includes twelve functions, such as hot topic extraction, entity-relationship extraction, sentiment classification, and the complete process of public opinion analysis. The system can be accessed and used for free in the Information System and Security and Countermeasures Experimental Center.

B. **Overall architecture**

The BFS public opinion analysis system comprises the following three subsystems: data management subsystem, semantic analysis subsystem, and user interaction subsystem. The overall system architecture is shown in Fig. 5.16.

C. **Features**

The public opinion system is divided into 11 functional modules, and the functional structure diagram is depicted in Fig. 5.17.

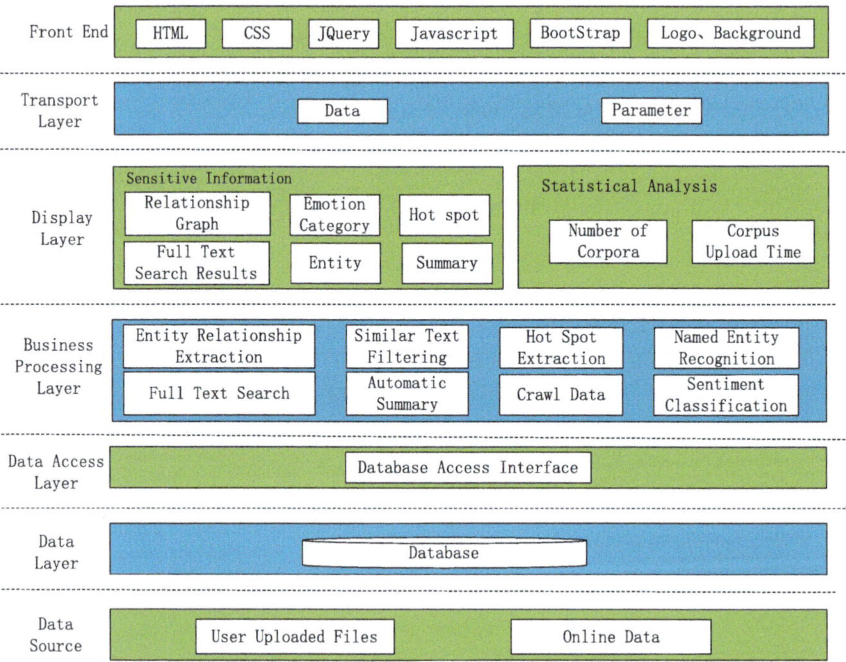

Fig. 5.16 The overall architecture diagram of the public opinion system

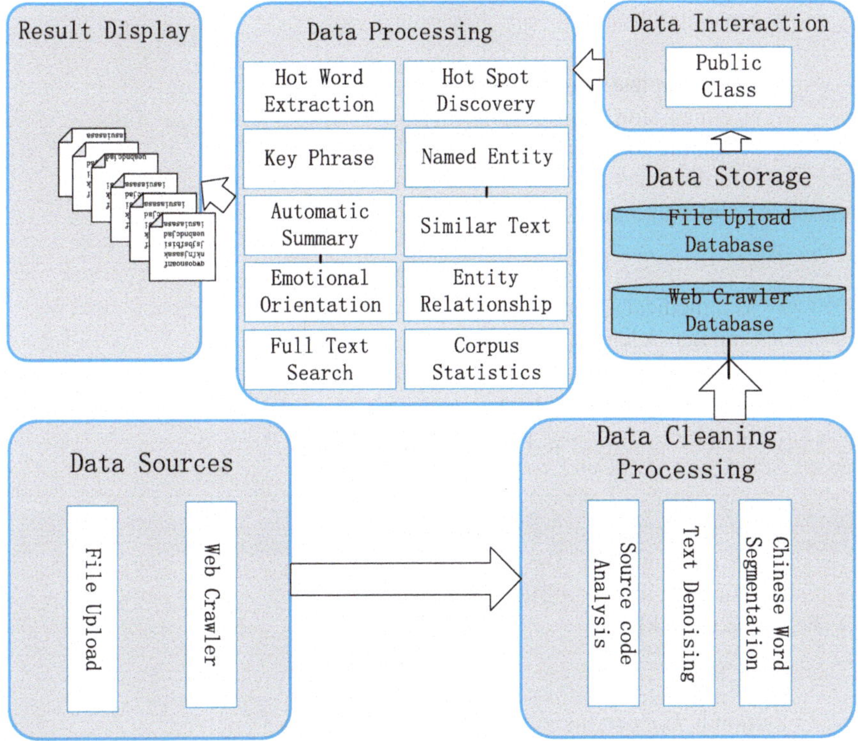

Fig. 5.17 Public opinion system functional structure diagram

(1) Data addition module

Build a file adding module based on B/S architecture. This module provides a multi-file (only supports txt, html, htm files) upload function. This process will perform preprocessing such as Chinese word segmentation and stop word removal. If you upload the source code of the webpage, the source code will be analyzed. This program will be resident in memory, rather than a subroutine, and will perform sequential processing.

This module provides the functions of file upload, text preprocessing, and database storage. The graphical interface of this module is shown in Fig. 5.18.

(2) Network data crawler module

This module provides the following functions: single file upload, Baidu keyword-related data crawling, URL-related data crawling, web source code analysis, data preprocessing, and database storage. The graphical interface of this module is shown in Fig. 5.19.

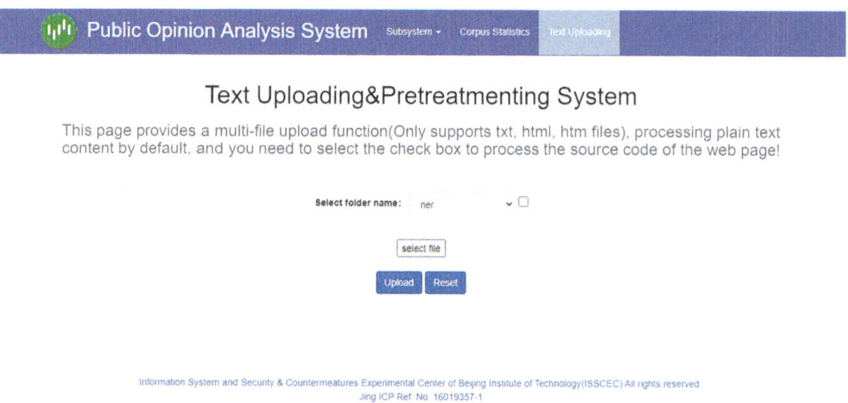

Fig. 5.18 Results of the data add module displayed on the page

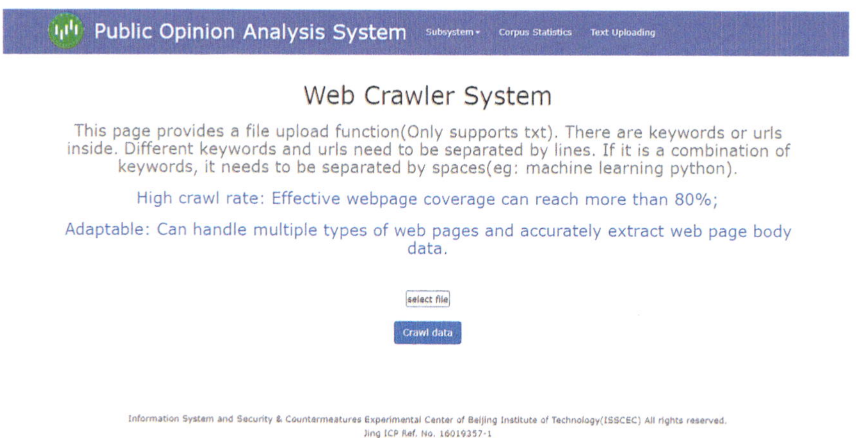

Fig. 5.19 Results of the web crawler module displayed on the page

(3) Hot word extraction module

This module offers the functionality of selecting the text sets within the database, setting the number of hot words, and extracting hot word phrases from the text set. The graphical interface is shown in Fig. 5.20.

(4) Hot topic extraction module

This module provides the functionality of selecting the text set in the database and setting the number of hot topics to dig out hot topics in the text set and extract hot topic phrases. The page is displayed as shown in Fig. 5.21.

The topic phrases are obtained through the processing of the hot topic system, in accordance with the user-selected text set and the pre-defined number of topics.

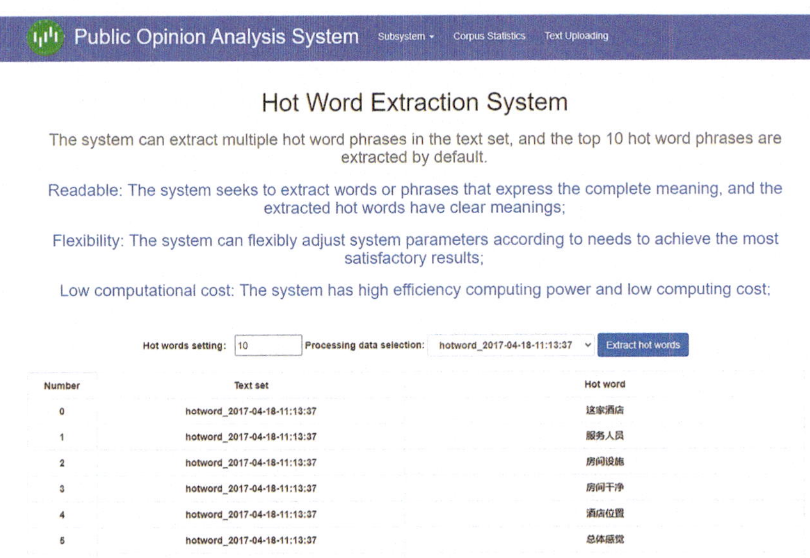

Number	Text set	Hot word
0	hotword_2017-04-18-11:13:37	这家酒店
1	hotword_2017-04-18-11:13:37	服务人员
2	hotword_2017-04-18-11:13:37	房间设施
3	hotword_2017-04-18-11:13:37	房间干净
4	hotword_2017-04-18-11:13:37	酒店位置
5	hotword_2017-04-18-11:13:37	总体感觉
6	hotword_2017-04-18-11:13:37	服务不错
7	hotword_2017-04-18-11:13:37	环境不错

Fig. 5.20 Results of the hot word extraction module displayed on the page

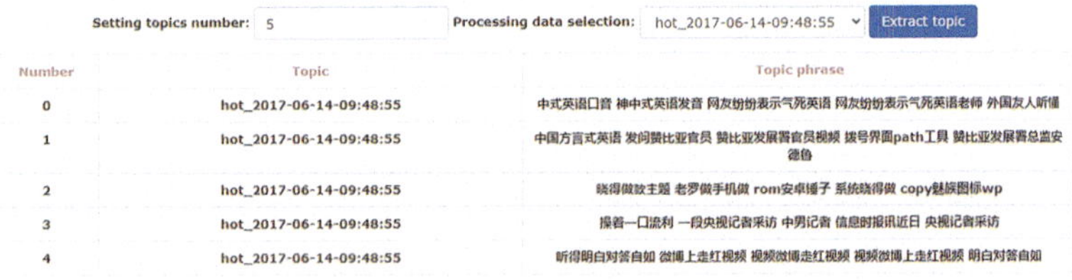

Number	Topic	Topic phrase
0	hot_2017-06-14-09:48:55	中式英语口音 神中式英语发音 网友纷纷表示气死英语 网友纷纷表示气死英语老师 外国友人听懂
1	hot_2017-06-14-09:48:55	中国方言式英语 发问赞比亚官员 赞比亚发展署官员视频 拨号界面path工具 赞比亚发展署总监安德鲁
2	hot_2017-06-14-09:48:55	晓得做款主题 老罗做手机做 rom安卓锤子 系统晓得做 copy魅族图标wp
3	hot_2017-06-14-09:48:55	操着一口流利 一段央视记者采访 中男记者 信息时报讯近日 央视记者采访
4	hot_2017-06-14-09:48:55	听得明白对答自如 微博上走红视频 视频微博走红视频 视频微博上走红视频 明白对答自如

Fig. 5.21 Results of the hot topic extraction module displayed on the page

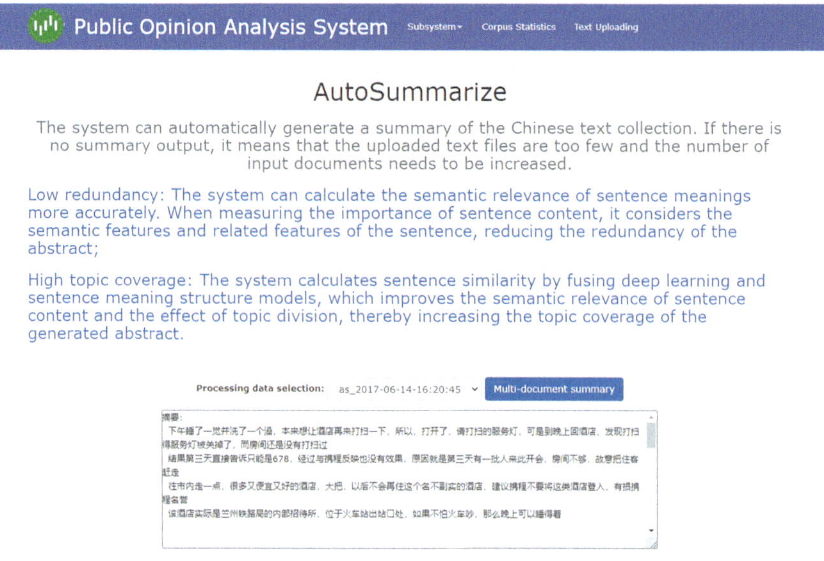

Fig. 5.22 Results of the automatic summary module displayed on the page

(5) Automatic summary module

This module provides the text set's selection function in the database and realizes the automatic generation of a summary for the selected text set. The page is displayed as shown in Fig. 5.22.

(6) Key phrase extraction module

This module is a key phrase extraction system that can extract the top *N* phrases from a single text. The page is displayed as shown in Fig. 5.23.

(7) Named entity recognition module

The user selects the text set to be processed from the "processing data selection" and clicks "extract entities" to start extracting named entities in a single text. The types include names of persons, places, organizations, companies, and product names. The page is displayed as shown in Fig. 5.24.

(8) Similar text filtering module

This module can calculate the degree of similarity between the selected text and other texts in the text set. It can filter out texts with a higher degree of similarity by a threshold. The page is displayed as shown in Fig. 5.25.

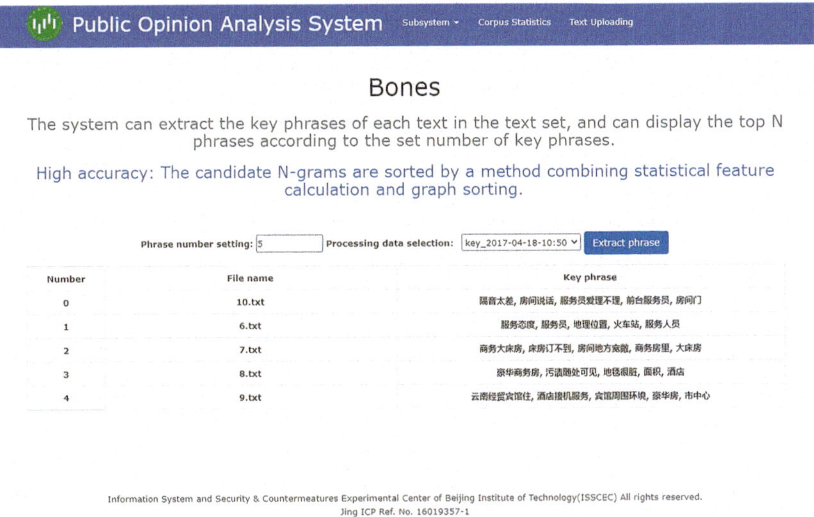

Fig. 5.23 Results of the key phrase extraction module displayed on the page

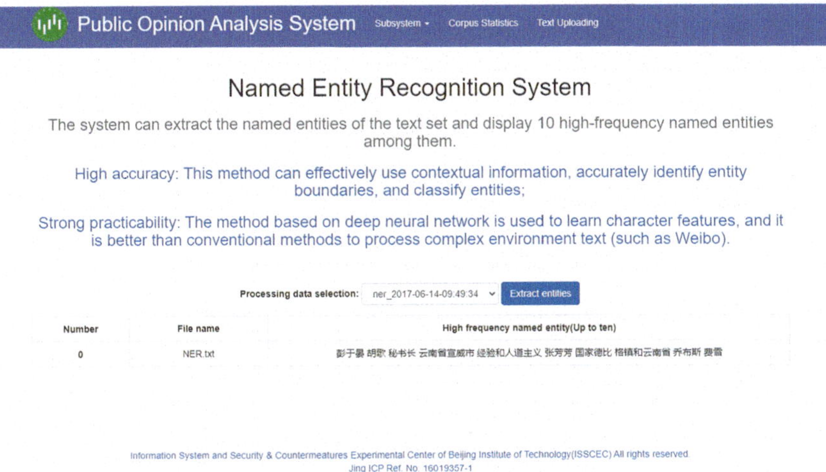

Fig. 5.24 Results of the named entity recognition module displayed on the page

(9) Emotional orientation analysis module

This module provides the Chinese text set's selection function in the database and determines each text's emotional tendency in the text set. The page is displayed as shown in Fig. 5.26.

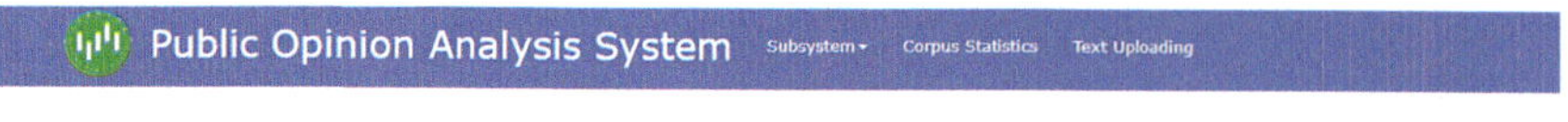

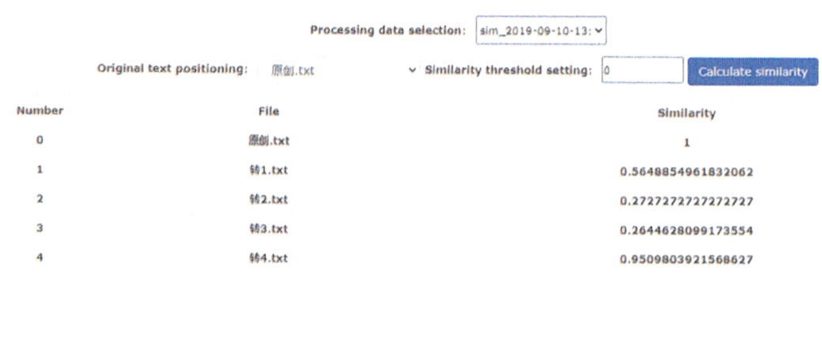

Number	File	Similarity
0	原创.txt	1
1	转1.txt	0.5648854961832062
2	转2.txt	0.2727272727272727
3	转3.txt	0.2644628099173554
4	转4.txt	0.9509803921568627

Fig. 5.25 Results of the similar text filtering module displayed on the page

Number	File name	Emotinal orientation
0	1.txt	-1
1	2.txt	-1
2	3.txt	-1
3	4.txt	-1
4	5.txt	-1
5	6.txt	-1
6	7.txt	1
7	8.txt	1
8	9.txt	1
9	10.txt	-1

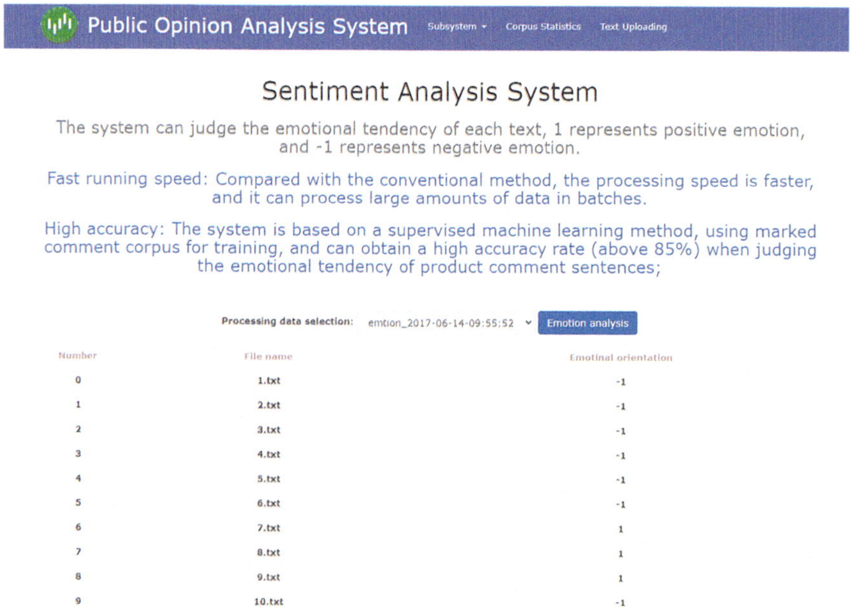

Fig. 5.26 Results of the emotion analysis module displayed on the page

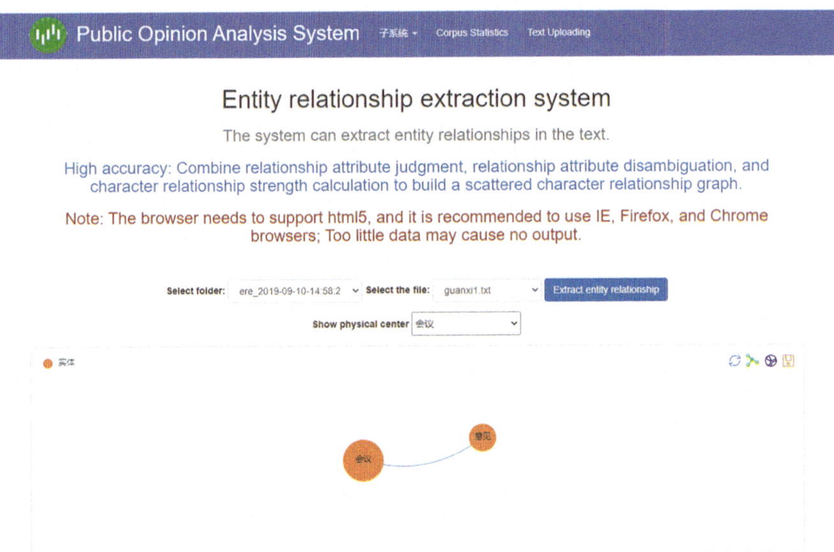

Fig. 5.27 Results of the entity-relationship extraction module displayed on the page

(10) Entity relationship extraction module

This module provides the function of selecting text sets and corresponding files in the database. It can extract the entity of a specific text and the corresponding entity-relationship, and display it in the form of a relationship graph. The page is displayed as shown in Fig. 5.27.

(11) Full text search module

This module provides a multi-file upload function, a directory selection function, indexing of the documents in the selected directory, a search function, and input keywords to query the documents containing the keywords. The page is depicted as shown in Fig. 5.28.

(12) Corpus quantity statistics module

This module provides the function of displaying the number of corpora used in each part. The page is depicted as shown in Fig. 5.29.

5.6 Summary

In the era of big data, text data emerges as a voluminous, intricate, and information-rich resource. It encompasses the vast repositories found in social media platforms like Sina Weibo and WeChat Moments, as well as basic resource data such as those represented by HowNet and Wikipedia. The extraction of valuable insights from

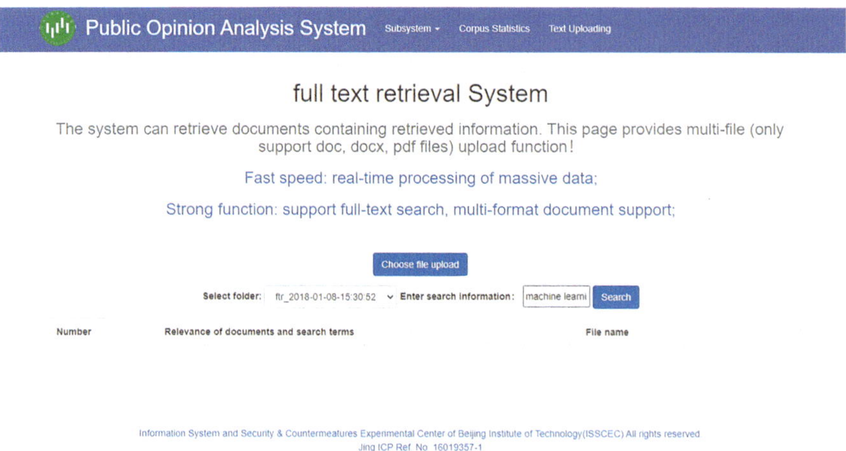

Fig. 5.28 Results of the full-text search module displayed on the page

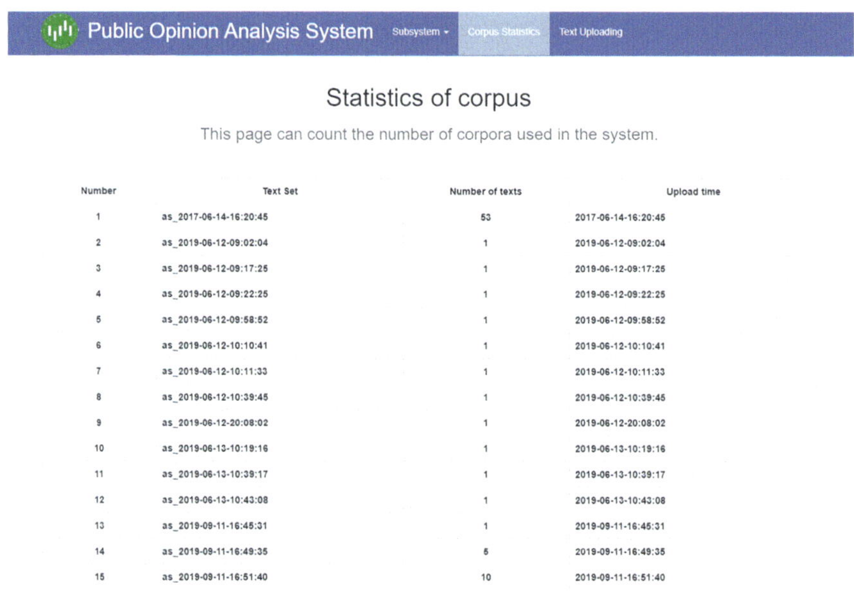

Number	Text Set	Number of texts	Upload time
1	as_2017-06-14-16:20:45	53	2017-06-14-16:20:45
2	as_2019-06-12-09:02:04	1	2019-06-12-09:02:04
3	as_2019-06-12-09:17:25	1	2019-06-12-09:17:25
4	as_2019-06-12-09:22:25	1	2019-06-12-09:22:25
5	as_2019-06-12-09:58:52	1	2019-06-12-09:58:52
6	as_2019-06-12-10:10:41	1	2019-06-12-10:10:41
7	as_2019-06-12-10:11:33	1	2019-06-12-10:11:33
8	as_2019-06-12-10:39:45	1	2019-06-12-10:39:45
9	as_2019-06-12-20:08:02	1	2019-06-12-20:08:02
10	as_2019-06-13-10:19:16	1	2019-06-13-10:19:16
11	as_2019-06-13-10:39:17	1	2019-06-13-10:39:17
12	as_2019-06-13-10:43:08	1	2019-06-13-10:43:08
13	as_2019-09-11-16:45:31	1	2019-09-11-16:45:31
14	as_2019-09-11-16:49:35	5	2019-09-11-16:49:35
15	as_2019-09-11-16:51:40	10	2019-09-11-16:51:40

Fig. 5.29 Results of the corpus quantity statistics module displayed on the page

such text data is inherently dependent on Natural Language Processing (NLP) technology, a field dedicated to understanding and addressing the linguistic components of human-computer interaction. NLP encompasses a spectrum of methodologies, from statistical empirical methods to rule-based rationalist approaches, and now, increasingly, to the incorporation of deep learning techniques.

From the perspective of basic resources, the era of big data has brought massive resources to NLP. This includes corpus resources like the LDC Chinese Treebank, Taiwan Academia Sinica Corpus, BFS-CTC Corpus, WordNet, FrameNet, and CNKI, which stand as representatives of the language knowledge base resources. Additionally, there are the knowledge graph resources like DBpedia, YAGO, and Freebase, which further enrich the field.

Current mainstream NLP technologies include a range of applications, from named entity recognition and entity-relationship extraction to sentiment classification, text summarization, machine translation, and automatic question answering. These technologies have introduced deep learning methods, formed deep learning methods, and blended empirical and rationalist approaches with the latest methods. For instance, in named entity recognition tasks, cutting-edge techniques often integrate deep neural networks, such as multi-layer LSTM and CNN, with CRF models. Similarly, in entity-relationship extraction, innovative methods utilize LSTM to assimilate syntactic tree knowledge, combining word embedding representations to generate context-aware vectors.

NLP technologies that are attuned to big data also form the bedrock of today's large-scale intelligent application systems. A prime example is IBM Watson, a sophisticated question answering system that has ingested a vast array of knowledge across domains like history, literature, art, popular culture, science, sports, and geography, even outperforming human champions in televised quiz shows. At its core, Watson represents the pinnacle of a computerized question answering system crafted by IBM. The convergence of big data resources and NLP technology is a catalyst for advancement, promising to unlock unprecedented value.

5.7 Exercises

1. Please briefly describe the opportunities and challenges that big data brings to natural language processing.
2. Please briefly describe the concept or definition of natural language processing.
3. Please briefly describe the basic concepts and relationships of the corpus, language knowledge base, and associated open data.
4. What is the data magnitude of the basic resources of natural language processing big data?
5. What are the mainstream natural language processing and big data analysis technologies?
6. What are the mainstream machine learning algorithms related to named entity recognition technology?
7. In the evaluation task 8 of SemEval-2010, which types of entity relationships are expanded to?
8. Briefly describe the remote supervision method used in entity-relationship extraction.
9. What are the application scenarios of emotion classification technology?

10. Please elaborate on the main process of text summarization based on deep learning.
11. If you are provided with the corpus of knowledge question and answer pairs (similar to questions and multiple answers that Baidu knows or Zhihu), how will you complete an automatic question and answer system, and what type of question answer system you have completed?
12. What are the key technologies involved in "Artificial Intelligence Robot Xiaobing" and briefly describe the basic process?
13. What valuable information can you get with an online public opinion analysis system?
14. What are the natural language processing technologies used in smart homes (such as multiple home devices that work with Google Home)?

References

1. Devlin J, Chang MW, Lee K, et al. Bert: pre-training of deep bidirectional transformers for language understanding. 2018. arXiv:1810.04805.
2. Zong C. Statistical natural language processing. Tsinghua University Press;2008.
3. Wang R, Luo S, Wu Z, Pan L. The application and development trend of deep learning in Chinese semantic analysis. Comput Technol Dev. 2019 29(09):110–116.
4. Li Q. FrameNet-a lexicon project based on frame semantics. China Sci Technol Inf. 2005;(16):39–38.
5. Long B, Guo W. Automatic construction of Chinese semantic frame network based on FrameNet. Modern Comput. (Professional Ed.);2010 (02):4–7.
6. Fujii A. Corpus2Based word sense disambiguation. Tokyo Institute of Technology;1998.
7. Dong Z, Dong Q. HowNet and Chinese studies. In: Contemporary linguistics, no 1;2001.
8. Qiao L, Yang L, Hong D, et al. Overview of knowledge mapping technology. J Comput Res Dev. 2016; 53(03):582–600.
9. Xu Z, Sheng Y, He L, Wang Y. Overview of knowledge graph technology. J Univ Electron Sci Technol China 2016; 45(04):589–606.
10. Dong X, Gabrilovich E, Heitz G, et al. Knowledge vault: a web-scale approach to probabilistic knowledge fusion. In: Proceedings of the 20th ACM SIGKDD international conference on Knowledge discovery and data mining. 2014. p. 601–610.
11. Zheng S, Wang F, Bao H, Hao Y, Zhou P, Xu B. Joint extraction of entities and relations based on a novel tagging scheme. In: Proceedings of the 55th annual meeting of the association for computational linguistics (Volume 1: Long Papers). 2017. p. 1227–1236.
12. Zhou Z. Machine learning. Beijing: Tsinghua University Press;2016.
13. Nenkova A, Vanderwende L. The impact of frequency on summarization. Microsoft Research, Redmond, Washington, Tech. Rep. MSR-TR-2005. 2005. p. 101.
14. Vanderwende L, Suzuki H, Brockett C, Nenkova A. Beyond SumBasic: Task-focused summarization with sentence simplification and lexical expansion. Information Processing & Management. 2007; 43(6):1606–1618.
15. Daumé III H, Marcu D. Bayesian query-focused summarization. In: Proceedings of the 21st international conference on computational linguistics and 44th annual meeting of the association for computational linguistics. 2006. p. 305–312.
16. Haghighi A, Vanderwende L. Exploring content models for multi-document summarization. In: Proceedings of human language technologies: The 2009 annual conference of the North American chapter of the association for computational linguistics. 2009. p. 362–370.

17. Celikyilmaz A, Hakkani-Tür D. A hybrid hierarchical model for multi-document summarization. In: Proceedings of the 48th annual meeting of the association for computational linguistics. 2010. p. 815–824.
18. Salton G, Buckley C. Term-weighting approaches in automatic text retrieval. Information Processing & Management. 1988; 24(5):513–523.
19. Erkan G, Radev DR. LexRank: Graph-based lexical centrality as salience in text summarization. J Artif Intell Res. 2004; 22:457–479.
20. Gong Y, Liu X. Generic text summarization using relevance measure and latent semantic analysis. In: Proceedings of the 24th annual international ACM SIGIR conference on research and development in information retrieval. 2001. p. 19–25.
21. Wan X, Yang J, Xiao J. Towards an iterative reinforcement approach for simultaneous document summarization and keyword extraction. In: Proceedings of the 45th annual meeting of the association for computational linguistics. 2007. p. 552–559.
22. Wan X, Yang J. Multi-document summarization using cluster-based link analysis. In: Proceedings of the 31st annual international ACM SIGIR conference on research and development in information retrieval. 2008. p. 299–306.
23. Gillick D, Riedhammer K, Favre B, Hakkani-Tür D. A global optimization framework for meeting summarization. In: 2008 IEEE international conference on acoustics, speech and signal processing. 2008. p. 4989–4992.
24. Li P, Wang Y, Gao W, Jiang J. Generating aspect-oriented multi-document summarization with event-aspect model. In: Proceedings of the 2013 conference on empirical methods in natural language processing. 2013. p. 1137–1146.
25. Nallapati R, Zhai F, Zhou B. SummaRunner: A recurrent neural network based sequence model for extractive summarization of documents. In: Proceedings of the AAAI conference on artificial intelligence. 2017; 31(1).
26. Zhou Q, Yang N, Wei F, Huang S, Zhou M, Zhao T. Neural document summarization by jointly learning to score and select sentences. In: Proceedings of the 56th annual meeting of the association for computational linguistics (Volume 1: Long Papers). 2018. p. 654–663.
27. Ouyang Y, Li W, Li S, Lu Q. Applying regression models to query-focused multi-document summarization. Information Processing & Management. 2011; 47(2):227–237.
28. Hong K, Nenkova A. Improving the estimation of word importance for news multi-document summarization. In: Proceedings of the 14th conference of the European chapter of the association for computational linguistics. 2014. p. 712–721.
29. Shen C, Li T. Learning to rank for query-focused multi-document summarization. In: 2011 IEEE 11th international conference on data mining. 2011. p. 626–634.
30. Wang L, Zhu S, Li T, Ding W. Query-focused multi-document summarization via combining query-sensitive features and graph-based features. Knowledge and Information Systems. 2013; 37(2):291–313.
31. Conroy JM, O'leary DP. Text summarization via hidden Markov models. In: Proceedings of the 24th annual international ACM SIGIR conference on research and development in information retrieval. 2001. p. 406–407.
32. Shen D, Sun JT, Li H, Yang Q, Chen Z. Document summarization using conditional random fields. In: Proceedings of the 20th international joint conference on artificial intelligence. 2007. p. 2862–2867.
33. Sipos R, Shivaswamy P, Joachims T. Large-margin learning of subsequence features for cross-document summarization. In: Proceedings of the 18th ACM SIGKDD international conference on knowledge discovery and data mining. 2012. p. 843–851.
34. Rush AM, Chopra S, Weston J. A neural attention model for abstractive sentence summarization. In: Proceedings of the 2015 conference on empirical methods in natural language processing. 2015. p. 379–389.
35. Gu J, Lu Z, Li H, Li VO. Incorporating copying mechanism in sequence-to-sequence learning. In: Proceedings of the 54th annual meeting of the association for computational linguistics (Volume 1: Long Papers). 2016. p. 1631–1640.

Chapter 6
Big Data and Medical Information Processing

6.1 Introduction

With the rapid development of medical information system (hospital information system) since the 1980s, the field of medical information processing has aggregated huge and complex health data, such as electronic medical records, personal health data, genomics data, which are widely used in all aspects of clinical research and treatment. Researchers and medical workers can use these data to better understand the effects of clinical treatment and the characteristics of research drugs, and provide more effective support for human health. With the continuous development and progress of various types of detection devices and new sensors in modern medicine, the number, format, and types of data collected and stored in medical informatics have increased tremendously. With the deepening and development of my country's new medical reform, medical informatics has ushered in the widespread application of cloud computing and electronic medical records. The combination of big data technology and medical information processing has created huge social and economic benefits.

The digitization of medical information is a prerequisite for mining of value from big data. Data sources include genetic data resources, medical image resources, electronic health records, medical voice records, etc. However, the complexity of medical data, characterized by polymorphism, incompleteness, and redundancy, has rendered traditional medical statistical methods inadequate for addressing the intricacies of modern healthcare challenges. The availability of vast quantities of medical data provides a robust foundation for medical information processing technology, while also presenting new challenges in knowledge discovery. By focusing on the needs of different entities such as medical institutions, government departments, pharmaceutical companies, insurance companies, and medical information, big data application can permeate every facet of disease prevention and health management, including telemedicine, precision medicine, intelligent diagnosis, and comprehensive health

S. Luo and L. Pan, *Big Data Analysis*, Big Data Management,
https://doi.org/10.1007/978-981-95-5725-7_6

decision-making. Therefore, the exploration of medical information processing technologies, when synergized with big data analytics, can effectively promote the interdisciplinary development of computer science, biomedicine, and bioinformatics. This integration holds profound significance for theoretical research and possesses substantial practical application value in the healthcare sector.

This chapter discusses four aspects of medical information processing: basic cognition, basic resources, analysis technology, and application cases. The basic cognition section covers the basic concepts, research history and status, the basic methods, and the difficulties of medical information processing. The basic resources section explains resources such as genetic data resources, medical image resources, electronic health records, medical voice records. The analysis technology section covers the basic tasks of medical information processing and its implementation methods. The application case section lists typical cases of combining big data and medical information processing.

6.2 Foundations of Medical Data Processing

6.2.1 Basic Concepts

Medical information, also known as health information, is a type of scientific and technological information that refers to all medical concepts, knowledge, skills, technologies, and behaviors. The content exchanged and transmitted between the sender and receiver of medical information can be called medical information. Medical information is a variety of health information, such as medical information, health information, drug information, nursing information, medical management information, and patient psychological information.

As a part of the information, medical information has essential and special characteristics due to the specificity of its field and type of user, which are mainly manifested in the following aspects:

(1) **Knowledge**: From the nature of the discipline, medical information belongs to both scientific and social communication. Whether in the medical system and scientific research, or various fields of society, medical treatment, production, management, or the publicity and popularization of medical and health knowledge, its knowledge will not change.
(2) **Diversity**: Taking drug information as an example, the content of drug information includes all production, research, management, and use related to drugs. The carrier forms of drug information are also diverse, such as pharmaceutical journals, drug books, drug retrieval tools, pharmacopeia drug collection, etc. The utilization of drug information also includes clinical, drug research and development, drug management, and other fields. Therefore, drug information is characterized by the diversity of information sources, information carriers, information content, and information use.

(3) **Practicality**: The origin and development of practical medicine come from social practice. Therefore, medical information is the product of human social practice. This information is derived from a multitude of sources, including animal experiments, medical researchers' personal experiences, the results of thousands of clinical trials and screening processes conducted by medical scientists and technicians. Consequently, medical information is a reflection of and a tool for social practice.
(4) **Timeliness**: The value of medical information is not contingent on its long-term, regular, and continuous provision. However, as time progresses, the value of the information in question diminishes gradually until it is no longer of any consequence. Given the ephemeral nature of medical information, failure to utilize and develop knowledge in a timely manner may result in the inefficient allocation of medical resources.
(5) **Publicity**: With the improvement of living standards, people pay more and more attention to their health problems, and medical information has also received more and more attention. Therefore, the basic purpose of medical information service is to organize the flow of medical information in medical education, clinical practice, academic research, public health, government services, and public health.

Medical information processing refers to the use of mathematical statistics, expert models, data mining, image processing, and other comprehensive methods with the help of a manual or computer to complete the whole process of medical information analysis and processing. There are many forms of medical information, including image, sound, video, text, and the most simple data. These medical information resources play an essential role in hospital decision management, medical treatment, and scientific research. Extracting valuable information from this complex information has become one of the problems that should be solved. Medical data has redundancy, incompleteness, fuzziness, and noise. These characteristics determine the difference and peculiarity between medical information processing and other data processing. Medical information processing forms an input dataset through the preprocessing of medical information, uses various medical information processing methods to obtain experimental results of medical information processing, and performs analysis to draw conclusions.

Two important research directions in medical information processing are clinical information processing and biological information processing. For clinical diagnosis datasets, the main task of information processing is to use feature selection and machine learning algorithm to explore the disease risk factors that affect the occurrence and development of a certain disease. Then it builds a disease classification and prediction model, and provides decision support for doctors in clinical diagnosis and reasonable treatment plans. For bioinformatics datasets, the task of information processing is to find differentially expressed genes or genetic sites for various types of cancer, diabetes and its complications, leukemia, Alzheimer's syndrome, and other complex human diseases, to study the pathogenesis of diseases and find effective treatment methods.

6.2.2 A Brief Research History

The first introduction of mathematical models to medical information processing dates back to 1959. Ledley et al. in the United States set a precedent for computer-aided diagnosis. They established computer diagnostic models based on Boolean algebra and Bayes' theorem for diagnosing lung cancer cases. In 1966, the concept of "computer-aided diagnosis" was first proposed in conjunction with the above research.

In the 1960s, the development of data and database technology has greatly promoted the progress of medical information processing. The new database management software, decision Support System (DSS), enables managers to make more effective use of data information in the decision-making process. In 1970, the first online analytical processing tool with a data mining analysis function, Express, was created. The development of medical data brings forward the practical demand for medical information processing. At the same time, the development of database technology provides technical support for medical information processing.

In 1976, Stanford University in the United States successfully developed an artificial intelligence medical expert consultation system (MYCIN) to diagnose and treat bacterial infections, diagnose blood-related infections, and prescribe antibiotics. In the 1980s, foreign medical expert systems achieved rapid development, such as the Pathfinder system for diagnosing lymph node diseases and the MUNIN diagnostic system for neuromuscular diseases. Subsequently, medical expert systems for glaucoma, AIDS, skin cancer, breast cancer, lung cancer, liver disease, and other diseases appeared.

Since the 1970s, research on clinical information processing has been carried out in China. Most of them use the method of probability and statistics, through disease and clinical data, namely symptoms, signs, physical examination information, etc. The statistical analysis calculates the frequency of common occurrence and the probability between the disease and the final diagnosis. In the 1980s, with the development of technology, Artificial Intelligence (AI) was applied to construct the medical information processing model combined with an expert's reasoning process and knowledge. The two methods need certain mathematical models to achieve. The commonly used mathematical models are weighted sum, Bayesian, and fuzzy mathematics. The application of information processing in the field of preventive medicine has also been significantly increased. In 1983, the Institute of Health, Academy of Medical Sciences, used a microcomputer to process the national nutrition survey data and analyzed the residents' food composition and nutritional status of the residents of China. Similar work also includes national hypertension sampling and epidemiological surveys in areas with high incidence areas of gastric cancer.

In the 1980s, medical image processing methods were also widely used and developed. A digital image system was used to determine the blood flow velocity in coronary vessels and to calculate the stenosis of cardiac annular vessels. In the 1990s, medical image processing was mainly used in coronary artery analysis, breast cancer analysis, cerebral hemorrhage, brain tumor, stroke, and other medical analysis and visualization tasks. The Image Research Center of San Francisco Medical College in California applies the image processing technology of military satellite ground reconnaissance to the diagnosis of cancer cells, and realizes the recognition of breast cancer sites from X-ray images with high resolution and real-time performance. With the development of visualization technology, the medical image is more inseparable from the medical image information processing, which plays an important role in clinical diagnosis, teaching, and scientific research.

In the 1990s, the first electronic health records were initially created. Seven hospitals in a community in the United States established a community hospital information network to share patients' medical and health information resources. Thus, the duplication and omission of patient information management can be avoided, and patients can receive the most centralized and effective treatment. Researchers can also conveniently obtain all the information on population health care. Japan established the Electronic Medical Record Development Committee in 1995 to implement the structure of medical data. Its MERIT-9 project studied the information exchange mode of text and image in the electronic medical record. Following the trend of hospital information construction in developed countries, China began to explore the construction mode of domestic electronic medical records.

Since 1986, the academic activities of machine learning have been very active, making its application scope expand continuously, and the intersection with many subjects, including medicine, is also deepening. For example, in post-genomic biology, the clustering method has been successfully applied to analyze large-scale gene expression data. In clinical medicine, a Bayesian neural network is used to obtain the relationship between taking antipsychotic use and cardiomyopathy. In medical image processing, the neural network is used to classify breast cancer.

With the rise of AI, deep learning, big data, data mining, cloud computing, image recognition, knowledge base, and knowledge map in the twenty-first century, more and more medical information technology research uses machine learning, data mining, and other technologies to build a knowledge base, and then perform information processing. In 2008, Google launched the flu prediction service, which can effectively track the signs of flu outbreaks by detecting users' search content on Google. Watson, an AI system developed by IBM, passed the medical licensing exam through self-study. Watson can read 248,000 papers and 106,000 clinical reports within a mere 17 s. It is an outstanding representative of the cognitive computing system.

In medical diagnosis, medical information processing is mainly focused on the analysis and processing of diagnostic data. The diagnosis of the dataset is formed through the extraction and preprocessing of diagnostic data. The method of classification and prediction, tailored to the unique attributes of the data, facilitates real-time disease forecasting. This serves as a valuable auxiliary diagnostic tool for medical students. For example, in 2015, Otoom et al. [1] proposed to obtain user data based on wearable sensors and use SVM, BayesNet, and tree model to monitor, diagnose and predict coronary heart disease patients in real-time, and give an emergency alarm. In 2016, Khushboo et al. [2] proposed to apply data mining classification technology

(Bayesian, k-nearest neighbor, and support vector machine) to thyroid disease analysis. In 2017, the AI system constructed by Bejnordi et al. [3] can quickly read pathological photos and diagnose whether breast cancer has lymph node metastasis, greatly improving the diagnosis speed and reducing the burden of pathologists.

In the biomedical field, medical information processing mainly focuses on analyzing gene sequence patterns and gene functions, especially DNA data analysis. By analyzing the causes of pathogenic genes, new disease diagnosis methods, prevention and treatment, and new drugs will be found. In 2017, China launched the "China 100,000 people genome project",[1] which is the first large-scale national program in human genome research. In 2018, DeepMind launched the AlphaFold study, which predicts the 3D shape of proteins based on gene sequences. In 2019, Ogden's team [4] published the research results of using AI to obtain superior gene therapy vectors in science.

6.2.3 Basic Method

The main purpose of medical information processing is to extract information from medical data and effectively discover potential knowledge. In the face of complex medical information, the traditional statistical methods can only be used to analyze the data with significant mathematical characteristics and statistical laws. Data mining has become the basic medical information processing method by processing massive and complex data, extracting valuable laws and knowledge, and constructing the prediction model. The basic methods of medical information processing are as follows:

(1) Comprehension and selection of medical data

Comprehension of medical data is a task that needs to be done before medical data preprocessing. It refers to the necessary understanding of medical data, including data type, data existence form, data quality evaluation standard, etc. Most medical institutions use an Online Transaction Processing (OLTP) to collect data. These systems contain a large amount of medical data, such as medical records and medical decision support information. Consequently, the selection of data for medical information processing should be meticulously curated from these systems to ensure relevance and quality for analysis.

(2) Privacy protection of medical data

Medical data is inherently confidential, both hospitals and researchers have the responsibility and obligation to safeguard the privacy of patient information. To achieve this, it is essential to anonymize the patient records and remove any identifiers that could link the records back to the patients. Anonymization is the process of

[1] http://news.cctv.com/2017/12/28/ARTIjKmD33l9ejmCrSExIxyQ171228.shtml.

removing or replacing the patient's identification from a record, rendering it unidentifiable. Following anonymization, researchers are unable to ascertain any information about patients through observation records. There are some subtle differences between identity transformation and anonymization. In identity transformation, the transformed logo may still contain patients' real information, but only authorized researchers can obtain the hidden, real information.

(3) Multi-source medical data fusion

Medical data is usually composed of many heterogeneous data, such as examination images, diagnostic data, electronic medical records, and laboratory results. The characteristics of multi-source medical information, the coexistence of temporal and non-sequential, digital and non-digital, increase the difficulty of data fusion. Different techniques and measures should be adopted to deal with these medical data with different physical characteristics to converge or be consistent in attributes. Then the results of processing should be used comprehensively.

(4) Dimension and noise reduction of medical data

Medical data often exhibit redundancy, with many records containing repetitive, irrelevant, or contradictory information concerning disease outcome. For example, symptoms, laboratory results, and treatment measures may be the same for some diseases. Besides, different parameters and indicators in the medical dataset may be highly correlated. For example, there is a definite numerical relationship between height, weight, and body mass index BMI. Therefore, the pre-processing of sample selection and feature selection is needed.

The noise in medical data is inevitable, such as incomplete medical records, errors caused by data measurement, inconsistency between the obtained medical data and the actual theory, and the influence of objective factors. When processing medical data, noise reduction is needed to minimize noise's impact on the research object.

(5) Medical statistics

Medical statistics is a comprehensive probability theory and mathematical statistics of the principles and methods of medical research related to data collection, collation, and applied disciplines. Statistical analysis of medical data through statistical software SAS, SPSS, and stat can effectively solve some medical research problems, provide a theoretical basis, and provide the basis for medical data mining. In selecting medical statistical methods, different statistical methods should be used for different medical data according to the research purpose, and the statistical experimental results should be analyzed, compared, and evaluated.

(6) Medical data mining

The medical database is a huge information base, which has a large amount of data and diversity. It will take more time to extract valuable knowledge from such a database. We should consider using different data mining algorithms to improve the efficiency of data mining for different types, structures, and dynamic data changes. At the same time, the mining algorithm should be fault-tolerant and robust.

In medical data mining, a variety of data mining methods can be selected according to the specific medical data and different mining objectives. For example, Decision Tree algorithm is used to manage decision protocols, create classification patterns of metabolic disorders, acquire relevant knowledge of ear neuropathy, data mining of diabetes mellitus and differentiation of dementia severity; SVM is used for bone age estimation, fall monitoring, adjunctive diagnosis of dementia and depression based on human brain image; Apriori algorithm is used to analyze clinical diseases, the relationship between symptoms and prescriptions, the relationship between different diseases; and Deep Learning algorithm was used to find out the relationship between taking antipsychotics and the onset of cardiomyopathy and myocarditis, classify life-threatening arrhythmias, dynamically detect the depth of anesthesia of patients and controll the dosage of narcotics.

(7) Medical data analysis and conclusion

Medical information processing must provide theoretical support for medical activities and management decision-making through massive data integration and analysis. By analyzing the medical data of the experimental results, we can compare the latest medical research conclusions and verify the effectiveness of the method from a clinical point of view.

6.2.4 Facing Difficulties

With the maturity of information processing technology and the continuous expansion of its application in the medical field, the advent of big data in medical information processing has brought forth a series of challenges at various levels:

(1) Data level

The medical information processing database needs to be strengthened, mainly including: First, the cost of manual annotation of basic data resources of medical information processing is high, and it is difficult to accurately label. Accurate annotation is the prerequisite of the high-quality medical model. The deviation or incompleteness of the dataset will affect model training. Second, medical data records are incomplete, fuzzy, and redundant. Due to the limitation of cases and medical records, the medical database cannot fully reflect all patients' disease information. At the same time, the record and description of disease information are uncertain and fuzzy. Also, there are many duplicate records in the medical database. Third, medical information has different patterns. Integrating a large number of heterogeneous data, such as medical records, images, laboratory reports, and laboratory results, to dig out valuable information is a problem to be solved in medical information processing. Fourth, data acquisition and privacy protection are also difficulties in medical information processing. There is a lack of reasonable data sharing and data circulation mechanisms in the field of medical information processing. The fuzzy ownership of medical big data restricts data sharing. Data privacy protection and data security should also be considered.

(2) Algorithm level

Although the research of medical information processing algorithm has made more achievements at home and abroad, there are still some problems in the research work, mainly including: First, the research work of medical information processing algorithm is the deep integration of medical and information science fields, which requires both medical professional knowledge and in-depth understanding of information processing algorithm. The existing medical information processing often directly applies the classical information processing algorithm to the medical clinical data or biological information data processing task, which does not fully consider the characteristics of medical data and the applicability of the algorithm, resulting in the performance of the algorithm is not ideal. Second, the information processing methods need to consider the uncertainty of medical data at the algorithm level to solve the characteristics of medical data correlation, class imbalance, and noise are difficult problems in medical information processing. Third, the results of medical information processing algorithm are usually for a specific disease. It is difficult to make a comprehensive analysis of the patient's situation. In fact, different patients with the same disease may have different symptoms, biochemical test results, and treatment measures, which reduces the applicability of medical information processing decision results. Fourth, with the rapid development of medical information technology, the characteristics, the number of samples of clinical diagnosis data, and bioinformatics data have increased by leaps and bounds. Some databases (such as gene databases) have huge data volume, which increases the time and performance challenges of medical information processing algorithms.

(3) Application level

The application problems in the implementation stage of medical information processing still need to be solved, mainly including: First, medical information processing needs deep integration of medical and information science fields, there are problems such as talent shortage, lack of practical experience, solid foundation, talent training, and introduction mechanism needs to be further established. Second, the main responsibility is not clear. There are no laws and regulations for it in China. Define the legal status, responsibility-sharing mechanism, and regulatory object of AI application in medical information processing. The ethical boundary of medical information processing is complex. Excessive management and control will hinder its innovation and development, and the lack of management will bring the risk of unclear subject responsibility in the application. Third, the standards and laws and regulations are relatively backward, the medical big data used in medical information processing lacks sound legal norms, and the ownership, use right, privacy standard, security, the security of medical big data are relatively backward. There is no clear legal guidance on responsibility norms and whether the law can tolerate innovation. Fourth, the lack of application evaluation standards, the lack of quality standards, access system, evaluation system, and security system in medical information processing, which cannot effectively verify and evaluate the clinical data, algorithm, calculation, safety and effect, and the lack of corresponding scheme system and standards has caused certain obstacles for the market entry of medical information processing products.

6.3 Medical Big Data Resources

The medical domain is inherently complex, with a strong emphasis on practicality, experiment, and statistics, establishing it as a confirmatory science. This complexity imbues medical data resources with a distinctive nature. The fundamental resources necessary for medical information processing encompass a comprehensive spectrum of data from biomedical research, clinical diagnosis and treatment process, and medical and health care information. This includes quantitative data (such as physical sign parameters, test results), imaging data (such as Single-Photon Emission Computed Tomography (SPECT), ultrasound (B-mode), and computed tomography (CT) images), signal data (such as electromyographic signals, and Electroencephalogram (EEG) signals), textual data (such as patient identification records, symptom descriptions, detection, diagnosis results, etc.) and so on. These foundational resources exhibit characteristics such as polymorphism, incompleteness, vital timeliness, and redundancy. Additionally, they are marked by low mathematical characteristics, non-standardized formats, and asymmetric communication between healthcare providers and patients. Ethical and legal considerations further add to the complexity of these medical big data resources. Despite these challenges, medical big data resources are the basis of medical information processing tasks. Harnessing the full use of medical information processing technology to uncover the valuable insights hidden in the data is crucial. It can assist doctors in their diagnostic and therapeutic decision-making processes and carry out exploratory research on diseases, which is of great significance to the development of the medical information field.

According to different data types, medical information processing big data basic resources are mainly divided into genetic data resources, medical image resources, electronic health records, and medical voice records. The framework for these fundamental resources in medical information processing is shown in Fig. 6.1, in which gene data resources include the protein structure dataset, such as Protein Data Bank (PDB), the Saccharomyces cerevisiae genome dataset, such as Saccharomyces Genome Database (SGD), and the international thousand human genome project database, among others; Medical image resources include the multiple disease image dataset, such as The National Library of Medicine (MedPix), autism brain imaging dataset, such as ABIDE (Autism Brain Imaging Data Exchange), and colon cancer CT scan image dataset; Electronic health records include intensive care medical dataset (MIMIC), unstructured clinical record dataset (STRIDE (Standardizing Trials of Records and Data for Informatics and Electronic Health)), and the Centres for Disease Control and Prevention (CDC) dataset, among others; Medical voice records include the medical voice dataset, such as Tumor and Radiotherapy Group Oral (TORGO),

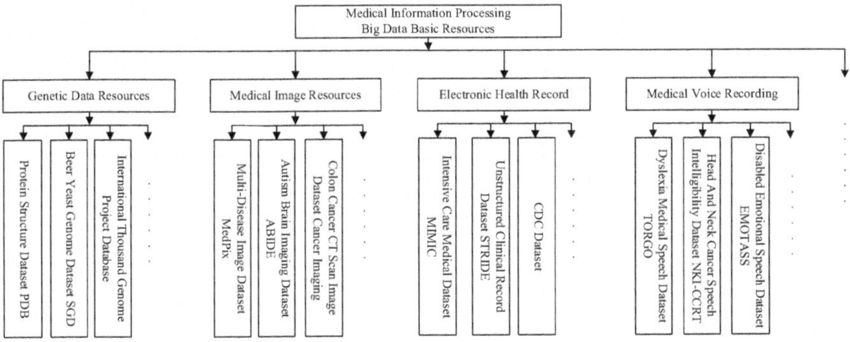

Fig. 6.1 Framework diagram of basic medical information processing big data resources

the head and neck cancer speech clarity dataset, such as The Netherlands Cancer Institute- Concurrent Chemoradiotherapy (NKI-CCRT), and the emotional voice dataset, such as Emotional Sensitivity Assistance System for People with Disabilities (EmotAsS), etc.

6.3.1 Genetic Data Resources

Genetic data resources constitute a pivotal segment of medical big data, characterized by their extensive content and diverse formats. They are distributed in information centers and sequencing centers around the world, as well as research institutions and universities with an affiliation to the fields of medicine, biology, and agriculture. The most important of these is the human genome database constructed by human genome research and sequencing centers worldwide. Genetic maps have been constructed from dozens of animal and plant genetic data resources with the genome project's general implementation. Besides, genetic data resources include chromosomes, genetic mutation, genetic disease, taxonomy, genetic regulation and expression, radiation hybridization, and other genetic datasets.

A. **Protein Data Bank**

Protein Data Bank[2] (PDB) is the complete protein structure dataset in the world. It consists mainly composed of three-dimensional structures of biomacromolecules measured by X-ray crystal diffraction and nuclear magnetic resonance. It is hosted by the Research Collaboration for Structural Bioinformatics (RCSB). PDB can be used for protein structure prediction and structural homology comparison, and is the underlying data basis for biological molecular structure research.

RCSB PDB (Research Collaboratory for Structural Bioinformatics PDB) operates the U.S. data center for the global PDB archive, and makes PDB data available free

[2] http://www.wwpdb.org.

of charge to all data consumers without limitations on use (policies). The vision of the RCSB PDB is to provide open access to the accumulating knowledge of the 3D structure, function, and evolution of biological macromolecules, thereby expanding the frontiers of fundamental biology, biomedicine, and biotechnology. Recognized experts in fields, including but not limited to structural biology, cell and molecular biology, computational biology, information technology, and education serve as advisors to the RCSB PDB.

B. **Saccharomyces Genome Database**

SGD[3] (Saccharomyces Genome Database) is a genomic dataset resource for the complete genome sequencing of Saccharomyces cerevisiae, including a lot of information on Saccharomyces cerevisiae's molecular biology and genetics. The whole genome information resources of the dataset can be accessed via the Internet, including genes and their products, phenotypes of some mutants, and various annotation information. The yeast genome was the first eukaryotic genome to be completed in 1998. The molecular structure of the SGD gene can be analyzed by searching for various types of gene sequences, which can be used for gene cloning, gene structure analysis, etc.

The SGD resource provides the gold standard for functional description of budding yeast, and a platform from which to study related genes and pathways in higher organisms. SGD also serves the yeast research community by maintaining the reference genomic chromosome sequence, overseeing S. cerevisiae genetic nomenclature, functioning as a central hub for researchers to share contact information, and supplying a list of yeast labs. All data in the SGD are freely available to researchers and educators worldwide via web pages designed for optimal usability.

C. **International Database of 1000 Human Genome Project**

The international thousand human genome project database is the largest human genome database for studying genetic variation. Since its launch in 2008, more than 1000 human genome data have been obtained in just four years. The database will eventually contain the entire genome information of 2500 individuals from 27 ethnic groups worldwide, with a data volume of 50 TB (50,000 GB) and 8 trillion DNA base pairs. Relevant data resources are open public resources, serving as a critical resource for association analysis of various diseases, thereby facilitating the elucidation of major human disease mechanisms, and underpinning personalized approaches to prediction, prevention, and treatment. Besides, it enriches our comprehension of human population genetics and promote the study of human evolutionary history.

The 1000 Genomes Project took advantage of developments in sequencing technology, which sharply reduced the cost of sequencing. It was the first project to sequence the genomes of a large number of people, to provide a comprehensive resource on human genetic variation. Data from the 1000 Genomes Project was quickly made available to the worldwide scientific community through freely accessible public databases.

[3] https://www.yeastgenome.org.

6.3.2 *Clinical Medicine Image Data Resource Collection*

With the continuous improvement of imaging technology, medical images have been widely used in clinical medicine. Physicians are now able to gain a direct visual insight into the human body, allowing them to comprehend both the normal anatomical structures and the alterations that occur in disease states. The various imaging modes, principles, and equipment can be divided into anatomical imaging modes to describe physiological morphology and functional imaging modes to describe human function or metabolism. From the research object, the research object of medical image analysis is extensive, not only limited to the disease with prominent diagnostic characteristics, but also includes various images of different organs, anatomical morphology, and functional process.

A. **National Library of Medicine Announces MedPix®**

MedPix[4](MedPix: The National Library of Medicine)is an open online dataset covering various disease images, including more than 12,000 patient cases, 9000 themes, and nearly 59,000 images, and integrates image and text metadata. The medical staff and other medical students are interested in the data. The dataset's content is classified by disease location (organ system), pathological category, patient file, image classification, and image title. It can be retrieved by patient symptoms, signs, diagnosis, organ system, image morphology, image description, keywords, contributing authors, and many other search options.

B. **Autism Brain Imaging Data Exchange**

Autism Brain Imaging Data Exchange[5] (ABIDE) is an autism brain imaging dataset consisting of 539 patients with Autism Spectrum Disorder (ASD) and 573 non-affected individuals. More than 1% of children worldwide have been diagnosed with ASD. However, due to the complexity and heterogeneity of ASD, the best diagnosis and treatment remain elusive. Large-scale data samples are essential to address these challenges. However, a single laboratory cannot obtain a large enough dataset to reveal the brain mechanisms behind ASD. The ABIDE program gathers functional and structural brain imaging data collected from laboratories worldwide to accelerate research and understanding of autism. The ABIDE project consists of two large datasets, ABIDE I and ABIDE II, each of which brings together data from more than 24 international brain imaging laboratories and is available free of charge to researchers worldwide.

C. **Cancer Imaging Archive**

The Cancer Imaging Archive[6] is a Computed Tomography (CT) scan image dataset for colon cancer diagnosis, including data of patients with no polyps, 6–9 mm polyps,

---

[4] https://medpix.nlm.nih.gov.

[5] http://preprocessed-connectomes-project.org/abide/.

[6] https://www.cancerimagingarchive.net/.

and more than 10 mm polyps, a total of 825 cases were included. The description of the polyp and its location in the colon segment was provided for each case.

The Cancer Imaging Archive was released in 2008 after the American College of Radiology Imaging Network (ACRIN) submitted a proposal to the National Cancer Institute (NCI) Executive Committee for conducting a national CT colonography trial. A case study showed that publicly accessible shared image data could provide valuable research assets for a wide range of image processing research communities. NCI Executive Committee strongly supported the proposal. Since its release, the dataset has received widespread attention and has been applied in various medical researches.

D. **Digital Retinal Images for Vessel Extraction**

Digital Retinal Images for Vessel Extraction[7] (DRIVE) is a color image dataset of the retina, containing 40 images, including seven pathological images, such as exudate, hemorrhage, pigment epithelial cell changes, etc., showing signs of mild early diabetic retinopathy. The image is taken with a Canon CR5 nonmydriatic 3CCD camera from a 45-degree field of view. The image size is 768 * 584 pixels. Each color channel is 8 bits and has a field of view of 540 pixels in diameter. It is stored in the compressed JEPG format. This dataset is usually used for a comparative study of vessel segmentation in retinal images.

E. **Open Access Series of Imaging Studies**

The Open Access Series of Imaging Studies[8] (OASIS) is a brain Magnetic Resonance Imaging (MRI) dataset for Alzheimer's disease research, including two cross-sectional datasets and longitudinal datasets.

The cross-sectional dataset included data from young, middle-aged, non-demented, and demented older elderly subjects, including 416 subjects aged 18 to 96 years, with 3 or 4 T1-weighted MRI images for each subject. The subjects were all right-handed, male and female. Where the subjects over 60 also included 100 patients diagnosed with mild or moderate Alzheimer's disease. The longitudinal dataset consisted of data of non-dementia and dementia older people, including 150 subjects aged 60 to 90 years. Each subject received 3 or 4 T1-weighted MRI images at two or more visits, and the interval between visits was at least one year. The subjects were all right-handed, male and female. In this dataset, 72 subjects showed no dementia during the whole data collection process. 64 subjects showed dementia during the entire data collection process, of which 51 subjects had Alzheimer's disease. Another 14 subjects showed no signs of dementia at the first visit, but developed dementia at the subsequent visit.

F. **Lung Image Database Consortium**

Lung Image Database Consortium[9] (LIDC) is a dataset of lung images, including chest CT scan and pathological marker annotation. The web-accessible dataset

[7] https://www.kaggle.com/datasets/andrewmvd/drive-digital-retinal-images-for-vessel-extraction.

[8] http://www.oasis-brains.org/.

[9] https://wiki.cancerimagingarchive.net/display/Public/LIDC-IDRI#.

contains 1018 samples and was jointly created by seven academic centers and eight medical imaging companies for the detection and aiding lung cancer diagnosis. The data of each subject included clinical chest CT images and relevant XML files. The XML files were the relevant results after two-stage image annotation by four experienced chest radiologists. LIDC dataset can be used by all researchers through the Internet, and has a wide range of practical applications in research, teaching, and training.

G. **Belarus Tuberculosis Portal**

Belarus Tuberculosis Portal[10] (TB) is a database of X-ray images of tuberculosis patients. Tuberculosis is a significant problem in public health in Belarus. Still, Belarusian tuberculosis experts' ability to track such patients can be significantly improved by using radiation images and clinical data of patients to enhance the feasibility of treatment plans and better record the treatment results. Over the past few years, Belarus has established several computer-aided diagnoses and treatments of lung diseases, especially a fully digital telemedicine system, which can screen lung diseases by X-ray. As a result of these projects, this extensive X-ray image database was created, containing about 1 million scanned images and corresponding results. It also includes computer-based image retrieval, such as grid-based image retrieval, computer-based image analysis for lung cancer diagnosis, and project-based image retrieval.

H. **Digital Database for Screening Mammography**

Digital Database for Screening Mammography[11] (DDSM) is a breast X-ray image dataset, which can be used to screen for breast cancer. The project is funded by the U.S. Army Medical Research and Materiel Command's Breast Cancer Research Program. Massachusetts General Hospital, the University of South Florida, and the Sandia National Laboratory are collaborating on the project. The DDSM dataset contains about 2500 studies, each including two images of each breast, patient-related information (such as age, American College of Rheumatology (ACR) breast density rating, abnormal ACR keyword description, etc.), image information (such as scanner, spatial resolution, etc.), and the location and type of suspicious area. The project also provides software for accessing breast X-ray images and running automatic image analysis algorithms.

6.3.3 Electronic Health Records

The health record is an objective and actual record and archive of personal health, health care, and treatment information within a particular time, process, phenomenon, and actual event range. Health records' written contents usually include chief

---

[10] https://grantome.com/grant/NIH/AAI12021001-1-0-5/.

[11] http://www.eng.usf.edu/cvprg/mammography/database.html.

complaint, test results, diagnosis, treatment plan, and clinical findings. Test results can consist of laboratory results reports and many other examination results, such as X-ray, pathology, ultrasound, pulmonary function, endoscopy, etc.

Electronic Health Record (EHR) is with the development of computer science; people's demand for well-structured and easy to retrieve data is increasing, which promotes the electronic health record. EHRs have clear handwriting, easy retrieval, and optimized structure. They have the potential for further improvement, but at the same time, it puts forward higher requirements for data collection.

A. **Medical Information Mart for Intensive Care**

Medical Information Mart for Intensive Care[12] (MIMIC) is a medical intensive care dataset published by the Laboratory of Computational Physiology, Massachusetts Institute of Technology, BEDMC Israel, and Philips Medical, contains data from more than 60,000 Intensive Care Unit (ICU) patients from 2001 to 2012, and is organized into 26 data tables, which are categorized into four categories according to their contents: patient basic information table, outpatient related information table, inpatient ICU clinical information table, and auxiliary dictionary term information table. The data includes demographic data, vital signs measurement data, laboratory measurement data, surgical data, medication data, nursing staff notes, impact reports, etc. It can provide doctors with accurate, timely, and complete patient information, help doctors diagnose patients more accurately, reduce medical errors, and provide safer medical interventions.

B. **Stanford Translational Research Integrated Database Environment**

The Stanford Translational Research Integrated Database Environment (STRIDE) is an unstructured clinical record dataset, containing 20 million records collected by Stanford Medical Center over 19 years. These records, derived from 1.2 million patients, included a wide array of medical reports, including pathology, radiology, and transcriptional reports. Each patient's record contains at least ten notes and spans a time window of at least one year. The dataset environment integrates clinical data, biological specimen data, and research data management platforms, connecting multiple logical databases that have been in operation since 1994. This dataset can be used to evaluate a series of clinical and epidemiological complications quantitatively.

C. **Centers for Disease Control and Prevention Dataset**

Centers for Disease Control and Prevention[13] (CDC) serves as a pivotal institution dedicated to the health and safety of the population, which contains data from multiple fields, such as biological monitoring, child vaccination, influenza vaccine, health statistics, pregnancy and vaccination, venereal diseases, smoking and tobacco use, treatment for adolescents, traumatic brain injury, web metrics, Morbidity and Mortality Weekly Report (MMWR), violent injury, etc.

---

[12] http://physionet.org/physiobank/database/mimic3cdb/.

[13] https://en.wikipedia.org/wiki/centers_for_disease_control_and_prevention.

D. **DocGraph**

DocGraph[14] is an innovative concept that functions as a "doctor recommendation social graph," essentially serving as a sophisticated network that maps the relationships and recommendations between doctors. This data-driven approach has a multitude of applications, which can be particularly beneficial in various sectors of the healthcare industry; for example: for hospitals, doctor recommendation data can cover many essential patient metrics, such as readmission rates. To understand which referral mode is related to poor coordination of care between outpatient facilities, doctors can use the dataset to determine where to set up a clinic; for government officials, doctors can use the dataset to decide where to open a clinic. The dataset can be used to map the actual care of healthcare organizations in the community and help break the monopoly on data monitoring and validation.

E. **University of California Dataset**

The University of California has many datasets related to disease monitoring and diagnosis, including:

(1) Liver disease dataset: Created by BUPA Medical Research Co., Ltd., it contains 345 samples, and each sample has 7-dimensional attributes, which are average red blood cell volume, alkaline phosphatase, alanine aminotransferase, aspartate aminotransferase, glutamyl transpeptidase, the amount of alcohol consumed per day, fields created by BUPA researchers to divide training and testing.
(2) Thyroid disease dataset: 10 independent datasets created by Garavan Institute, containing 7200 samples, each with 21-dimensional attributes.
(3) Breast cancer dataset: Contains 286 samples. Each sample has 9-dimensional attributes, including age, whether menopause, tumor size, diseased breast, diseased area, etc.
(4) Heart disease dataset: Created by the Hungarian Institute of Cardiology, Zurich University Hospital in Switzerland, University Hospital Basel in Switzerland, and VA Medical Center. It contains 303 samples, each with 75-dimensional attributes, including ID, name, age, types of chest pain, smoking, fasting blood sugar, diabetes history, family history of coronary artery disease, whether the electrocardiogram is normal, etc.
(5) Lymphography dataset: Created by the Yugoslav Ljubljana Tumor Research Institute, it contains 148 samples. Each sample has 18-dimensional attributes, including whether the lymphatic vessels are normal, changes in nodes, changes in structure, and whether they have special forms, etc.
(6) Parkinson's dataset: Created by the University of Oxford and the National Speech Center in Denver, Colorado. It contains 197 samples, each with 23-dimensional attributes, including average fundamental frequency, maximum fundamental frequency, fundamental frequency change, examiner's health, etc.

[14] https://www.docgraph.com/.

6.3.4 Medical Phonetics

Medical voice recording can assist in pathological voice recognition in an objective and non-invasive way. Pathological speech includes: speech defects, the most common congenital disabilities are missing lip, cleft palate, and other phenomena. Vocal cord defects refer to vocal cords that are hoarse or discouraged due to trauma or dysplasia, incomplete closure, etc. Voice disorders, mostly in patients in adults, due to fright, concussion, or psychological stimulation, the nervous system of speech is damaged, leading to speech reversal or misunderstanding, making it difficult to speak. Speech disorders are near related to the field of language research, such as "aphasia," "stuttering," etc.

A. **TORGO**

TORGO[15] is a medical speech dataset for dysarthria. It was created by the Department of Computer Science and Language Pathology of the University of Toronto and the Toronto Children's Rehabilitation Hospital. The subjects were patients with dysarthria. The dataset collected acoustic and articulation presentations from dysarthria speakers, including patients with Cerebral Palsy (CP), Amyotrophic Lateral Sclerosis (ALS), and corresponding controls. CP and ALS were the two most common causes of dysarthria. The TORGO dataset aims to collect detailed physiological information of patients to learn "hidden" pronunciation parameters through pattern recognition. The dataset contains many video data captured by the digital camera used to export 3D pronunciation information. All acoustic data are downsampled to 16 kHz.

B. **NKI-CCRT**

The Netherlands Cancer Institute-Concurrent Chemoradiotherapy(NKI-CCRT),[16] the dataset collected data from 55 patients with advanced head and neck cancer who received chemotherapy and radiotherapy (CCRT). Speech intelligibility was recorded and evaluated before and after treatment (10 weeks and 12 months).

C. **EMOTASS**

Emotional Sensitivity Assistance System for People with Disabilities (EmotAsS)[17] is the first dataset of emotional voice recordings from people with disabilities, including a wide range of mental, neurological, and physical disabilities. The dataset, provided by the InterSpeech Challenge 2018, includes recordings from 15 adults with disabilities (age range 19–58 years, average age 31.6 years). The task is to classify five emotions of the atypical presentation. The recording is conducted in the daily work environment and includes about 11 K words and about 9 h of speech.

---

[15] http://www.cs.toronto.edu/~complingweb/data/torgo/torgo.html.

[16] https://www.aclweb.org/anthology.

[17] http://emotion-research.net/sigs/speech-sig/is18-compare.

6.4 Big Data Techniques in Healthcare

Among the various data analysis technologies, big data analysis technology is one of the most popular. It is possible to extract hidden, unknown, and potentially useful information from massive data and to establish calculation programs to identify rules or patterns by big data analysis technology. The application of big data analysis technologies has become pervasive across a range of domains, including telecommunications, insurance, environmental science, finance, and medicine, among others. Since its inception, big data analysis technology has been applied to the field of medical data research. The shift from clinical diagnosis data to bioinformatics data has led to a surge in the application of big data in medical informatics. Medical information processing has become the most active research of big data analysis, focusing on processing and analyzing medical data to improve patient outcomes and streamline healthcare operations. By employing analytical techniques to extract meaningful insights from vast medical datasets, medical professionals can make data-informed decisions, provide the best scheme for diagnosis and treatment, and significantly improve medical research's ability.

According to the different types of medical information data essential resources, the big data processing technology in the medical field can be mainly classified into four primary categories: gene sequence analysis technology, medical image processing technology, electronic medical record analysis technology, and medical voice processing technology. The framework of big data analysis technology for medical information processing is shown in Fig. 6.2. Among these, gene sequence analysis focuses on the study of gene sequence patterns and gene function, especially DNA data analysis; medical image processing mainly focuses on lesion detection, image segmentation, image registration, and image fusion; electronic medical record analysis is divided into two main areas: the mining of Electronic Health Record (EHR) phenotypes of patients and the research of deep learning algorithm; and medical voice processing technology has a number of useful applications, including disease diagnosis, dysarthria speech recognition, medical speech transcription, and so forth.

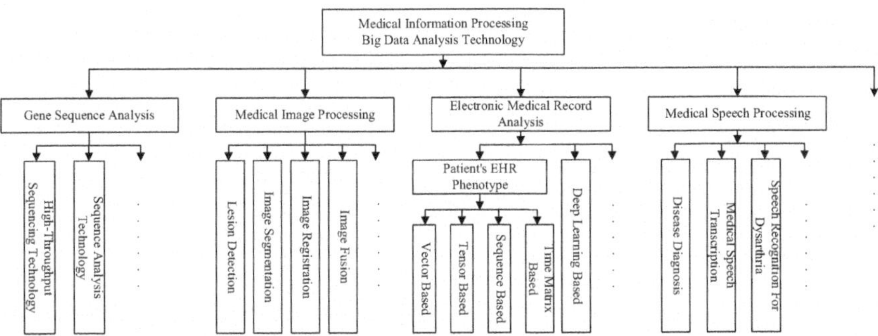

Fig. 6.2 The overall framework of medical information processing big data analysis technology

6.4.1 Gene Sequence Analysis

A. **Task Definition**

The rapid development of the biomedical field has given rise to widespread societal concern. The ongoing advancement of gene research and the continual evaluation of modern biotechnology have resulted in the accumulation of a substantial corpus of biomedical data in this field. The calculation and analysis of these data have become more and more critical.

Gene sequence analysis is an essential means for examining the intricate structure and function of genes, constructing gene maps, and identifying transgenes in gene engineering and molecular biology. It is also one of the essential technologies in this field. Gene analysis enables the revelation of primary structural changes in genes and genomes, the identification of gene and genomic variations, the analysis of artificially recombined genes, the confirmation of gene mutations, the examination of genetic diversity within species, and the delineation of new species.

One of the most significant research areas in the medical field is the integration of life science and computer science, with the objective of analysing human gene sequences by the application of big data analysis methods. The results of gene analysis can be used as basic materials for research institutions, clinicians, and other downstream medical service industries. The primary business of gene analysis companies is to collect human gene sequence information and apply it to infer the different diseases that different customers are prone to. However, they are usually unable to effectively process complex genetic data. Pharmaceutical companies, research universities, medical industry, and other downstream institutions use the data provided by gene analysis companies to further explore the connections between genetic variation and various diseases.

B. **Related Work**

(1) **High-Throughput Sequencing**

Genes play an important role in genetic research. With the breakthrough of molecular biotechnology and the advent of high-performance computers, genetic research has entered the quantitative level. Gene sequence analysis needs to obtain gene sequence first. Nowadays, the most widely used sequencing technology is high-throughput sequencing technology. High-throughput sequencing technology, also known as "Next-Generation Sequencing (NGS)" technology, can determine hundreds of thousands or even millions of sequences at a time. Compared to traditional Sanger sequencing technology, NGS has the advantages of high speed, high throughput, and low cost. High-throughput sequencing data are widely used in biology, medicine, genetics, and many other fields.

Many large-scale scientific research projects, such as the 1000 Genome Project, the Encyclopedia of DNA Elements Project, and the International Cancer Genome Project, are producing massive gene sequences on an unprecedented scale. As of February 2014, there were 100 billion base pairs of DNA sequence data have been registered in the US GenBank database alone. The size of short reading sequence data among all high-throughput sequencing reached thousands of PB. With the continuous improvement of sequencing technology and the constant reduction of sequencing costs, many gene sequences will be generated every day, making the amount of biological data exponential: doubling every 14 months on average.

With the development of high-throughput sequencing technology, various biological data show explosive growth, and this trend will be further enhanced with the development of biological sequencing technology. Faced with the big data analysis task in life science, data integration of different dimensions, data analysis of interdisciplinary and classical data mining algorithms face new challenges.

(2) **Sequence Analysis Technology**

After obtaining a gene sequence, bioinformatics analysis is needed to extract as much information as possible to guide further experimental research. The basic information of genes can be clarified by chromosome location analysis, intron/exon analysis, Open Reading Frame (ORF) analysis, and expression profile analysis. Through promoter prediction, CpG island analysis, and transcription factor analysis, we can identify cis-acting elements in the regulatory region, which can provide the basis for gene regulation research. Besides, through similarity search, functional site analysis, structural analysis, query gene expression profile clustering database, gene knockout database, genome upstream and downstream neighbors, etc., we can try our best to mine the network information database to infer the gene function.

Nucleic acid sequence analysis includes pairwise alignment and multiple sequence alignment. Double sequence alignment compares the similarity of two sequences and finds the corresponding positions of similar bases and amino acids. It is a powerful tool for sequence analysis by computer. It can be divided into global and local alignment, with the Needleman-Wunsch algorithm and the Smith-Waterman algorithm as the substitutes. Double sequence alignment tools include FASTA and BLAST, which are widely used similarity search tools. These two tools use the local alignment method, select the scoring matrix to score the sequence, and determine the significant local alignment through the size of the score and statistical significance analysis. Using FASTA and BLAST, we search the database and find the sequence which is similar to the query sequence.

In the study of biological problems, it is often necessary to compare more than two sequences simultaneously, which is called multiple sequence alignment. Multiple sequence alignment can be used to study a group of related genes or proteins, to infer the evolutionary relationship of genes, and also to find common patterns among a group of functional or structural related genes. The most widely used multi-sequence alignment tool is ClustalW.

Gene structure analysis is to analyze gene structure according to the mRNA sequence and genome sequence of genes. BLAST and BLAT are the primary analysis tools for gene structure analysis.

The routine analysis of gene sequence includes initial sequence quality control analysis and sequence splicing, OTU clustering and abundance analysis, OTU species

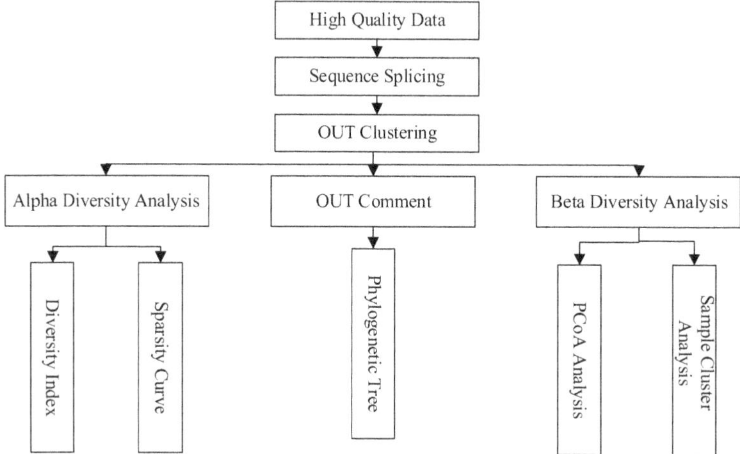

Fig. 6.3 The structure diagram of routine analysis of three gene sequences

annotation and abundance analysis, phylogenetic tree construction, Alpha diversity analysis (diversity index calculation and dilution curve), Beta diversity analysis (distance matrix analysis, PCoA analysis, and sample cluster analysis). The high-level analysis includes significant difference factor analysis, specific species analysis, multivariate statistical analysis based on sample species classification, and environmental impact factors. The conventional analysis structure of the gene sequence is shown in Fig. 6.3.

It is worth noting that when analyzing the sequence, we should first clarify the nature of the sequence, which is the mRNA sequence or the genome sequence? Is it obtained by computer splicing or Polymerase Chain Reaction (PCR) amplification and sequencing? Is it prokaryotic or eukaryotic? These determine the choice of analytical methods and the interpretation of analytical results.

C. **Human LncRNA Big Data Gene**

With the development of biological analysis technology and the rapid accumulation of biomedical data, it is possible to identify LncRNA from massive RNA sequences by machine learning. Machine learning can integrate sequence, structure, and expression data to predict new LncRNA. A support vector machine, random forest, Bayesian algorithm, decision tree, and other supervised learning methods have been successfully used to predict ncRNA. Chang et al. constructed a small bacterial ncRNA prediction model based on a support vector machine. They verified the effectiveness of the method by using small non-coding RNA data from Escherichia coli. Some also use unsupervised learning to analyze and predict ncRNA, and a context-sensitive hidden Markov chain model is used to predict the microRNA precursor. According to the known the microRNA precursor sequence, the model automatically integrates sequence features to predict unknown microRNA precursors.

Ensemble learning is a general machine learning method. It is usually based on integrating multiple classifiers to make the final decision, which effectively improves the classifiers' generalization ability. The commonly used ensemble learning methods are the AdaBoost algorithm and the Bagging algorithm.

In 2018, Yu Bin et al. [5] proposed a new LncRNA big data gene prediction method based on ensemble learning. Firstly, 86 features of sequence base frequency are extracted as the original feature set. Secondly, the optimal features are selected based on GA-SVM, and the accuracy of SVM 50-fold cross-validation is used as the fitness. Finally, a gene prediction model combining the AdaBoost algorithm and SVM (AdaBoost SVM) is built. The calculation process is presented in Fig. 6.4.

The steps of the AdaBoost SVM prediction method are described as follows:

1. Input the gene sequence data of human LncRNA (positive example) and mRNA (counterexample) into the training set, and corresponding class labels.
2. According to the base frequency characteristics, 86 features of LncRNA and mRNA gene sequences were extracted, and the base sequences were transformed into digital signals:

$$D = \{(x_i, y_i)|i = 1, 2, \ldots, 34000\}, x \in R^{86}, y \in \{-1, 1\}$$

3. The binary coded individuals are selected, and the test accuracy of SVM 50-fold cross-validation is used as the fitness. The feature subset is determined based

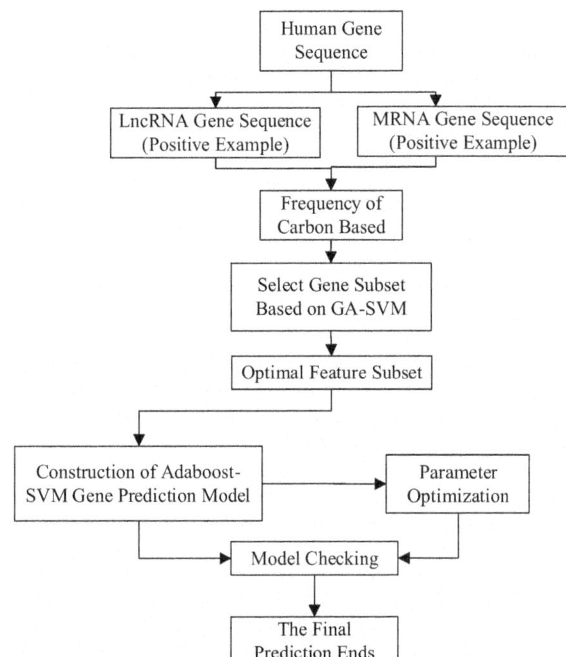

Fig. 6.4 Flow chart of LncRNA gene prediction based on AdaBoost-SVM

on the GA-SVM method. A new training set S_1, is selected by crossover and mutation, and the corresponding test set is S_2.

$$S_1 = \{(x_i, y_i)|i = 1, 2, \ldots, 34000\}, x \in R^{49}, y \in \{-1, 1\}$$
$$S_2 = \{(x_i, y_i)|i = 1, 2, \ldots, 14502\}, x \in R^{49}, y \in \{-1, 1\}$$

4. Cnstruct the AdaBoost-SVM prediction model. First, determine the initialization weight $\frac{1}{34000}$ of each training sample in the training set S_1, and input S_1 into SVM. Second, the training deviation is calculated as

$$\varepsilon_i = \sum_{i=1}^{n} D_i(i)\{h_i(x_i) \neq y_i\}$$

with

$$D_{t+1}(x) = \frac{D_t(x)}{Z_t} \times \begin{cases} e^{-\alpha_t}, h_t(x) = y, \\ e^{\alpha_t}, h_t(x) \neq y. \end{cases} = \frac{D_t(x)e^{(-\alpha_t, y_t h(x))}}{Z_t},$$

where Z_t is the normalization factor, $\alpha_t = \frac{1}{2}\ln(\frac{1-\varepsilon_t}{\varepsilon_t})$, and h_t is the weak classifier. When the weight is updated, the weight of the misclassified samples is appropriately increased, and the weight of the non-misclassified samples is appropriately reduced. Third, the training sample set is obtained by the resampling method, and the base classifier is trained by cyclic iteration for t times. Finally, according to the formula, the weight of the base classifier is increased

$$H(x) = \text{sgn}\left(\sum_{t=1}^{T} \alpha_t h_t(x_t)\right)$$

The category label of the prediction test set.

With the development of high-throughput sequencing technology, it is imperative to predict LncRNA by bioinformatics method based on machine learning.

6.4.2 *Medical Image Processing*

A. **Task Definition**

Digital medical images are stored, transmitted, and managed by medical Picture Archiving and Communication Systems (PACS) using Digital Imaging and Communications in Medicine (DICOM). Because of the investment and extensive use of the PACS system, medical image data is growing rapidly. How to effectively process and analyze medical image big data and extract or mine useful medical information and knowledge from large-scale image datasets has attracted people's attention.

From the perspective of large-scale image datasets, image information processing includes image acquisition, image storage, image compression, multimedia database, and other fields. From the perspective of mining valuable information and knowledge, image processing and analysis, pattern recognition, computer vision, image retrieval, machine learning, artificial intelligence, knowledge representation, and other fields are also involved. Information processing has achieved good research results in the retinal image, brain image, cell image, skin cancer image and other medical image analysis.

The key technologies of medical image information processing include data pre-processing, information fusion technology, etc. Data pre-processing refers to the medical image data containing many complex original data, with many fuzzy, incomplete, and redundant information. Therefore, before the medical information processing, it is necessary to clean and filter the information to ensure the data's consistency and certainty. For example, data pre-processing technology can be used to remove or reduce the influence of image noise, improve the quality of the target image, or extract the edge of the target tissue. Information fusion technology refers to the medical image information, including text, waveform signal, data, image, and other physical properties of different data, which should be processed by different technical methods to achieve the convergence or consistency of data attributes. Then the results are fused to analyze and mine its potential information and knowledge.

B. **Related Work**

The objects of medical image processing are various medical images with different imaging mechanisms. There are four types of medical imaging widely used in clinical practice, including X-ray imaging (X-CT), Nuclear Magnetic Resonance Imaging (MRI), Nuclear Medicine Imaging (NMI), and Ultrasonic Imaging (UI). The medical image diagnosis mainly involves observing a group of two-dimensional slice images to find the lesion, which often needs the doctor's experience to determine. Using computer image processing technology to analyze and process the two-dimensional slice image, realize the segmentation extraction, three-dimensional reconstruction, three-dimensional display of human organs, soft tissues, and lesions, which can assist doctors in qualitative and even quantitative analysis of lesions and other exciting areas, to significantly improve the accuracy and reliability of medical diagnosis. It can also play an important auxiliary role in various medical researches. Medical image processing mainly focuses on four aspects: lesion detection, image segmentation, image registration, and image fusion.

(1) **Pathological Examination**

Computer-Aided Detection (CAD) is a field that needs to be improved in medical image analysis. In the standard methods of CAD, the candidate lesion locations are usually detected by supervised methods or classical image processing techniques (such as filtering and mathematical morphology). The detection of the lesion location is incremental and usually consists of a large number of hand-made features. The direct way of deep learning is to train Convolutional Neural Network (CNN) to process a group of image data centered on the lesion site. Before the GPU era, the training time was described as "computing-intensive." In 1993, CNN was used to detect pulmonary nodules; in 1995, CNN was used to identify microcalcifications in mammography. In 1996, Sahiner et al. [6] applied CNN to medical image processing. ROI of tumor or healthy tissue were extracted from a mammogram. Setio et al. [7] detected pulmonary nodules in the 3D chest CT scan, extracted 2D patches centered on candidate images in nine different directions, and classified each candidate image using various combinations of CNN. According to the test results, it is slightly better than other classical CAD systems for the same task. The highlight is the use of nine views of volume objects to extract two-dimensional patches, a schematic diagram of the proposed system, and the fusion methods. Roth et al. used CNN to improve three CAD systems, which were used to detect colonic polyps, sclerosing spinal deformities, and lymphadenopathy in CT images. They also use detectors and 2D patches in three orthogonal directions, as well as up to 100 randomly rotated views. Random rotation of the "2.5D" view is the image obtained from the original 3D data decomposition. By using CNN to detect the 2.5D view image and then summarizing it, the detection accuracy can be improved. For the three CAD systems using CNN, the lesion detection accuracy is enhanced by 13–34%, which is almost impossible for CAD systems using non-deep learning classifiers (such as support vector machines).

(2) **Image Segmentation**

Medical image segmentation refers to dividing an image into several regions according to the similarity or difference between different areas. The processing objects are mainly images of various cells, tissues, and organs. Traditional image segmentation techniques include region-based segmentation and boundary-based segmentation. The former depends on the image's local spatial features, such as the uniformity of grey level, texture, and other pixel statistical characteristics. At the same time, the latter mainly uses gradient information to determine the boundary of the target. Combined with specific theoretical tools, image segmentation technology has been further developed. For example, the medical image segmentation method based on a 3D visualization system combined with the fast-marching algorithm and watershed transform can obtain fast and accurate segmentation results.

With the development of other emerging disciplines, some new image segmentation techniques have been produced. For example, based on statistics, fuzzy theory, neural network, wavelet analysis, snake model (dynamic contour model), combined optimization model, etc. Although new segmentation methods have been proposed, the results are not ideal. The hot spot of image segmentation research is the knowledge-based segmentation method. That is, some prior knowledge is introduced into the segmentation process by some means. To restrict the computer's segmentation process, the segmentation results can be controlled within our understanding scope and not too far off the mark. For example, when there is a big difference between the gray value of the liver mass and the healthy liver, the mass and the normal liver cannot be regarded as two independent tissues.

The medical image segmentation method research has the following obvious characteristics: any single image segmentation algorithm is difficult to obtain satisfactory results for general images, so it is necessary to pay more attention to the effective combination of various segmentation algorithms. Due to the complexity of human anatomical structure and functional system, although there are studies on automatic segmentation of medical images to distinguish the required organs and tissues or to find the method of the lesion area, the ready-made software package cannot complete the automatic segmentation, and it still needs the artificial intervention of anatomy. If the computer cannot complete the task of image segmentation, the human–computer interactive segmentation method has gradually become the research focus. The research of new segmentation methods mainly focuses on automatic, accurate, fast, adaptive, and robust. The comprehensive utilization (integration) of standard segmentation technology and modern segmentation technology is the future medical image segmentation technology in the direction of development.

Using the dataset of 2891 cardiac ultrasounds, Ghesu et al. combined with deep learning and edge space learning to detect and segment medical images. The combination of "effective exploration of large parameter space" and the implementation of sparsity in a deep network improves the computational efficiency. Compared to the reference method, the average segmentation error is reduced by 13.5%. Brosch et al. [8] studied the segmentation of multiple sclerosis lesions in MRI images. A 3D deep convolutional encoder network is proposed, which combines convolution and deconvolution. The convolution network learns high-level features, while the deconvolution network performs pixel-level segmentation. The system was applied to two public datasets and a clinical trial dataset and compared to five public methods. Pereira et al. studied the segmentation of brain tumors on MRI, using a deeper architecture, data normalization, and data enhancement techniques, using different CNN architectures for tumor images. This method separately performs image enhancement and core region segmentation for suspected tumors and achieves the highest results on the 2013 public challenge dataset.

In 2018, a German medical rehabilitation institution proposed a representative prostate image segmentation method based on total convolution. The end-to-end training with CNN on prostate MRI images can complete the whole segmentation process at once. At the same time, a new objective function is proposed simultaneously, which is optimized according to the dice coefficient during training. In this way, we can deal with the imbalance between foreground and background, and add the random application of nonlinear data transformation and histogram matching. Experimental evaluation shows that this method achieves excellent results on public datasets and greatly reduces the processing time.

(3) **Image Registration**

Image registration is the premise of image fusion, which is recognized as a difficult image processing technology and critical technology to determine the medical image fusion technology development. In the first mock exam, the first mock exam often does not provide enough information for doctors. It is often necessary to combine multiple patterns or multiple images of the same pattern by registration and fusion to achieve the complementarity of the region of interest. If the information from various imaging sources is expressed in a single image, doctors can make a more accurate diagnosis or develop more appropriate treatment methods. Medical image registration includes image localization and transformation. The corresponding points of two images can be completely consistent in spatial position and anatomical structure by looking for a spatial transformation. In 1993, Petra et al. [9] reviewed the two-dimensional image registration methods. According to the registration benchmark characteristics, the image registration methods were divided into two types: image registration based on external features (with framework) and image registration based on internal features (without framework). Because of its non-invasive and traceability, the latter has become the research center of the registration algorithm.

In 2019, Huazhong University of Science and Technology researched structural non-rigid multimodal medical image registration based on Principal Components Analysis Net (PCANet). It proposed a structure representation method based on PCANet for multimodal medical image registration. Compared with the manually designed feature extraction method, PCANet can automatically learn interior features from many medical images through multi-level linear and nonlinear transformation. The multi-level image features extracted from each layer of PCANet are used to provide a useful structure representation for multimodal images. A large number of experiments on the Atlas, BrainWeb, and RIRE datasets show that this method can offer lower Target Registration Error (TRE) values and more satisfactory results than the Modality-independent Neighborhood Descriptor (Mind), the Sum of Squared Differences (SSD) on entropy images, Weber Local Descriptor (WLD), and Normalized Mutual Information (NMI) methods.

The new development of medical image registration technology is that the theory and informatics method are applied in the registration method. For example, the maximum mutual information is used as the registration criterion for image registration. The elastic deformation model based on mutual information has gradually become a research hotspot. In terms of registration object, it has developed from a two-dimensional image to three-dimensional multi-mode medical image registration. Some new algorithms in the medical image are also expanding, such as wavelet transform, statistical parameter mapping, genetic algorithm, etc. The development direction of medical image registration technology in the future is to improve the algorithm to a fast and accurate path, use optimization strategy to improve image registration, and explore non-rigid image registration.

(4) **Image Fusion**

The main purpose of image fusion is to improve the readability of images by processing redundant data between multiple images and to improve the clarity of images by processing complementary information between multiple images. Multi-modal medical image fusion combines valuable physiological function information with accurate anatomical structure, by providing a more comprehensive and precise clinical application data. The fused image's creation is divided into two parts: image data fusion and fusion image display. Image data fusion mainly includes the pixel-based method and image feature-based method. The former is to process the image point by point. The gray values of the two images' corresponding pixels are weighted and summed, and the gray scale is made larger or smaller. The implementation of the algorithm is relatively simple, but the implementation effect and efficiency are relatively weak. After the fusion, the image will appear a certain degree of blur. The latter needs to process the image feature extraction and object segmentation. The principle of the algorithm is complex, but the effect is ideal. The common display methods of the fusion image are pseudocolor display, sectional display, and a three-dimensional display. The pseudocolour display is generally based on a specific image and displayed with grayscale. Another image is superimposed on the reference image and presented with the color scale. The tomographic display method is often used for some specific images. This can synchronously display the fused three-dimensional data as a cross-section, coronal plane, and sagittal plane, which is convenient for the observer to diagnose. The three-dimensional display method is to synchronously display the fused data. The three-dimensional image display makes the observer observe the lesion's spatial anatomical position more intuitively, which is of great significance in the surgical design and radiotherapy planning.

In the research of image fusion technology, there are many new methods. Among them, the nonlinear registration based on wavelet transform, finite element analysis, and artificial intelligence technology in image fusion is the hot research direction. With the development of 3D reconstruction and display technology, 3D image fusion technology research has been paid more and more attention. The fusion and information expression of the 3D image will also be the focus of image fusion research.

Computer-aided image processing is an inevitable trend to develop an image analysis system that comprehensively utilizes image processing methods, combines human body constants and image features of some diseases to help or simulate doctors' analysis and diagnosis. Some image analysis software uses fixed-point man–machine interaction and automatic measurement and analysis, which can complete some measurement and auxiliary diagnosis work in a fixed-point manner. However, it is far from the level of intelligent analysis and expert system. Automatic identification of marker points and measurement analysis, as well as the fusion of medical image information and text information, in the future development direction of computer-aided diagnosis technology.

C. Biomedical Image Segmentation Based on Convolution Network

A deep convolutional network performs well in visual recognition tasks, but it has strict requirements on the training set and network size. Moreover, the typical use of the convolutional networks is for the classification task, where the output of the image is a class label. However, in many visual tasks, especially in biomedical image processing, the desired output should assign a class label to each pixel. Also, in biomedical tasks, thousands of training images are often unattainable.

Olaf Ronneberger et al. [10] proposed a CNN u-net for two-dimensional image segmentation. The network architecture of the u-net is shown in Fig. 6.5. The U-net is a 2D full CNN with a symmetric U-shaped network structure. The left side is the contraction path, and the right side is the expansion path. The entire network has four max pool layers, four up conv layers, and 20 convolution layers. The U-net architecture is an iterative structure. Each iteration has two convolutional layers and one pooling layer. The convolution kernel size in the convolution layer is 3 × 3. The activation function uses ReLU. After the two convolution layers, there is a max-tilt layer with a size of 2 × 2 and a step size of 2. Each downsampling doubles the number of characteristic channels. In each step of the shrink path, up conv is used first. Each time, the number of characteristic channels is halved, and the size of the feature image is doubled. After deconvolution, the result is spliced with the feature map of corresponding steps in the contraction path. If the size of the feature map in the shrink path is slightly larger, it is trimmed and then spliced. The results are convoluted twice by 3 × 3. The size of the last convolution kernel is 1 × 1, and the 64-channel feature map is transformed into a specific depth result, in which the network has 23 layers. The gradient descent method was used for training. The last layer uses the cross-entropy function and softmax. The formula for cross-entropy function is as follows:

$$E = \sum_{x \in \Omega} w(x) \log(p_{l(x)}(x))$$

To increase the weight of some pixels, it is introduced into the formula $w(x)$. A weight map is prepared for each labeled image to compensate for the different frequencies of each pixel type in the training set. The network pays more attention to learning the segmentation boundary between the organizations in contact with each

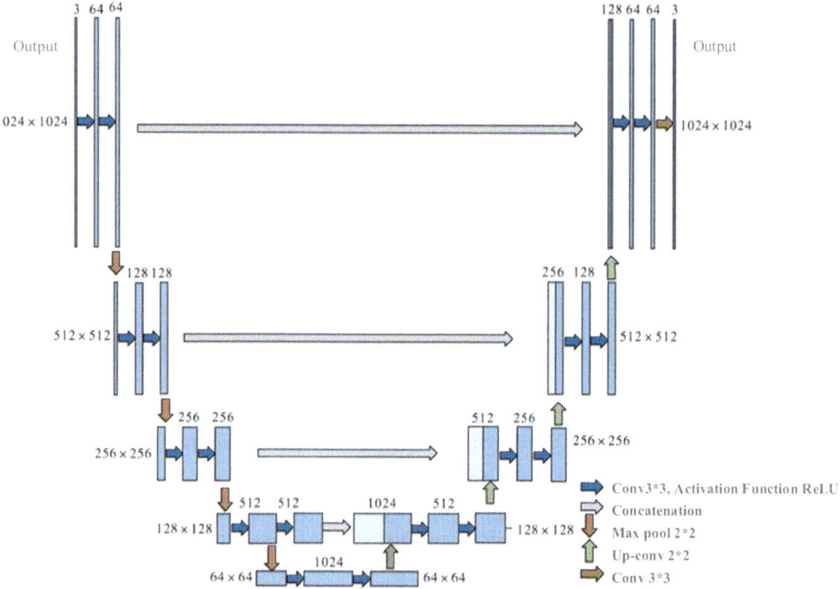

Fig. 6.5 U-net model architecture (32 × 32 pixels with the lowest resolution)

other. The formula for weight calculation is as follow:

$$w(x) = w_c(x) + w_0 \times e^{\left(-\frac{(d_1(x)+d_2(x))^2}{2\delta^2}\right)}$$

where w_c is a weight graph used to balance the class frequency, d_1 represents the distance to the nearest cell boundary, d_2 represents the distance to the second closest cell boundary, and sets $w_0 = 10$, $\delta \approx 5$ pixels based on experience.

The weight of the network is initialised by Gaussian distribution. The standard deviation of the distribution is $(N/2)^{0.5}$, and N is the number of input nodes of each neuron.

U-net is a classic network design method, which has a large number of applications in image segmentation tasks. There are also many new methods that improve on this basis and integrate new network design concepts, but few people have made a comprehensive comparison of these improved versions. Since the same network structure can show different performance on different datasets, it is still necessary to select the appropriate network combined with datasets in specific task scenarios.

6.4.3 Analysis of Electronic Medical Records

A. **Task Definition**

EHR is a systematic collection of patient health information stored in digital format, which can be shared among different healthcare equipment and institutions. Electronic Medical Record (EMR) data collected and managed include demographics, medical history, medications, allergies, immune statuses, laboratory test results, radiological images, vital signs, etc.

EHR is considered the key to high-quality nursing. Many medical institutions adjust the nursing management plan according to the data in EMR to improve the medical level and nursing quality. The analysis of various clinical data in EMR has helped clinicians identify and stratify patients with chronic diseases. EMR big data analysis can also be used to predict the risk of clinical disorders and improve the quality of medical care.

With the accumulation of EMRs, the analysis of such data enables researchers and healthcare institutions to promote personalized medical care. However, the original EMR data has its problems, such as high dimension, timeliness, sparsity, irregularity, and imbalance. These problems significantly increase the difficulty of directly applying traditional machine learning or statistical models to predict patients' potential disease. Consequently, it is essential to use a more robust model to solve the original challenges of EMR data in the risk prediction task.

Deep learning models have demonstrated the ability to extract essential features directly from the original EMR and have achieved excellent results in tasks such as computing phenotypes, diagnosis prediction, and risk prediction. For example, an attention-based Recurrent Neural Network (RNN) is used to predict heart failure disease. CNN is introduced to extract patients' time characteristics and predict disease risk, etc.

B. **Related Work**

In this section, we briefly review the work of improving the level of health care and the health status of residents based on EMR analysis technology. One part is about the EHR phenotype extraction of patients, and the other part is about the research and application of deep learning algorithms in healthcare.

Hripsak et al. [11] pointed out that electronic phenotype extraction refers to the problem of extracting effective phenotypes from patients with EHR. This is a key step before using EHR to perform any data-driven analysis task (such as comparative effectiveness studies, predictive modeling, etc.), and using the original EHR directly will face many challenges. The following is a summary of the research on electronic phenotype extraction according to the different representation methods of EHR.

The vector-based representation method constructs an information vector for each patient. Its dimension is equal to the number of different events in the EHR, and the values in each dimension are important statistical information of similar medical events in a specific period (such as sum, average, maximum, minimum, etc.). Using vector-based representation, each phenotype is usually assumed to be a linear combination of these original medical events, and some optimization programs calculate the coefficients of each medical event. The limitation of this representation is that it ignores the temporal relationship between events.

The tensor-based representation method constructs an EHR tensor for each patient. Each pattern of the tensor represents a particular type of medical entity (i.e., patient, medication, or diagnosis). Ho et al. [12] proposed a method based on non-negative tensor decomposition to extract phenotypes from these EHR tensors. This method explores the interaction between different medical entities. The limitation is that they did not consider the temporal relationship of events.

The sequence-based representation method constructs an EHR sequence for each patient according to the timestamp of each event. Then, frequent pattern mining can be applied to recognize temporal patterns as phenotypes. One problem is that the method usually returns many patterns (also known as the "mode explosion" phenomenon) due to the vast differences between patients' EHRs. Based on the time matrix representation method, the patient EHR is expressed as a time matrix; one dimension corresponds to time, the other dimension corresponds to medical events. Zhou et al. proposed a representation method for grouping medical events with similar temporal trends. However, they did not consider the temporal relationship between different events. Wang et al. proposed a convolution matrix decomposition method to detect translation-invariant patterns of the EHR matrix across patients. However, this method cannot determine the optimal pattern length, so it is necessary to enumerate all possible values. The method proposed by Cheng et al. is based on time matrix representation, uses CNN architecture to identify important phenotypes, and automatically weights them in the prediction phase.

Deep learning is a machine learning algorithm that uses the model architecture composed of multiple nonlinear transformations to abstractly model data. The deep learning model has achieved remarkable results in computer vision and speech recognition applications and has shown excellent ability to learn intricate patterns from high-dimensional, noisy sequential EHR data. The deep learning method uses a Multi-Layer Perceptron (MLP) to determine the phenotype and the mathematical representation of medical events in EHR. However, the MLP based model does not consider the temporal nature of EHR data. To model time-series EHR data, RNN is used to predict patient health status and classification. CNN is used to capture the local time dependence of EHR data and predict a variety of diseases and other related tasks. In healthcare, risk prediction is an essential and challenging task. Choi et al. used an attention-based RNN to predict the risk of heart failure. While Cheng et al. used the CNN model to analyze discrete patient EHR data and used the pre-training embedding of medical features in the CNN model to improve the prediction performance. They established a semi-supervised deep learning model with a generated confrontation network for the clinical risk prediction task of patients.

C. **Risk Prediction of EMR Based on Deep Learning Method**

Cheng et al. constructed a risk prediction model based on the time matrix representation of patient's EHR. Explicitly, the EHR record is modeled as an event matrix associated with time. The horizontal dimension corresponds to the timestamp, and the vertical dimension corresponds to the medical event. If the ith event is observed at the jth timestamp of the corresponding patient, the (i, j) element of the EHR matrix is 1. However, unlike images and videos, the EHR matrix of different patients has different sizes in the time dimension, so the standard CNN model cannot directly apply this event matrix representation.

In the following, the details of the methods to solve the above problems will be introduced. In this paper, the basic CNN model is presented first, and then the CNN architecture of time fusion is added.

(1) **Basic Model**

As shown in Fig. 6.6, the basic model architecture is slightly modified based on the CNN architecture. Each event matrix of length t is expressed by X and $X \in R^{d \times t}$. Let $x_i \in R^d$ be the d-dimensional event vector corresponding to the ith event item $x_{i:i+j}$. In general, let $x_i, x_{i+1}, \ldots, x_{i+j}$ be the concatenation of terms. One side convolution operation involves a filter $w \in R^{d \times h}$, which is applied to the window of H event features to generate new features. For example, the feature c_i is generated from the window of the event $x_{i:i+h-1}$, among which $c_i = f(w \cdot x_{i:i+h-1} + b)$, where $b \in R$ is a deviation term, and F is a nonlinear function, such as RELU, Tanh function, and so on. This filter is applied to each possible feature window in the event matrix $\{x_{1:h}, x_{2:h+1}, \ldots, x_{n-h+1:n}\}$ to generate a feature graph $c = [c_1, c_2, \ldots, c_{n-h+1}]$, where $c \in R^{n-h+1}$. Then the mean pooling is used on the feature map, and the average value $\hat{c} = \max\{c\}$ is taken to select the most important feature with the maximum value for each feature map. The last layer is the full connection layer with softmax classifier.

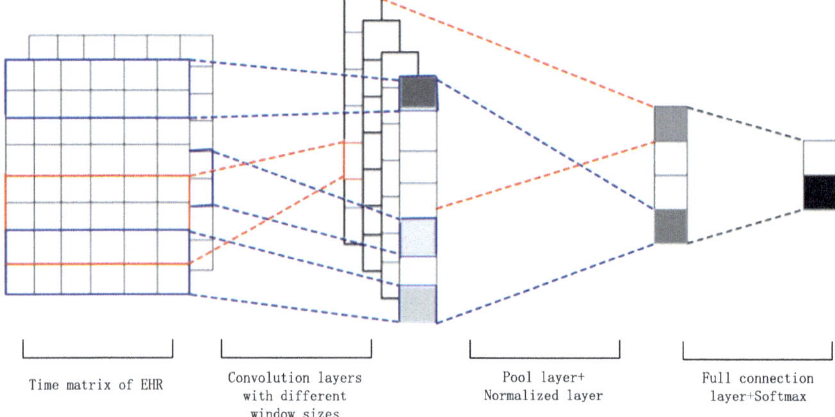

Fig. 6.6 Basic CNN model architecture

(2) **Time Fusion CNN**

Unlike static images and documents, EHR data vary significantly in the time domain, and the characteristics of the time dimension are also crucial for prediction. In this part, each data sample is treated as a series of fixed-size subframes. Since each subframe contains several consecutive time intervals, we can extend the model in the time dimension to learn the temporal characteristics. Three broad connection modes are described below.

Single frame: The single frame structure regards EHR records as a static matrix proposed in the basic model. The other part is to increase the normalization layer. Using a single frame structure can well understand the impact of static data on improving classification accuracy.

Early fusion: The early fusion model can merge the information in the whole time window at the input layer. This is achieved by modifying the filter on the first convolution layer in the single-frame model and extending it to the number of subframes K.

Late fusion: The late fusion model performs fusion at the last full connection layer. It first places several individual single frame networks (as shown in Fig. 6.7, where the yellow, red, green, and blue boxes represent the full connection layer, convolution layer, normalization layer, and pooling layer, respectively). These streams are then merged into the full connectivity layer. In this model, the local information in each subframe can be easily detected.

Slow fusion: The slow fusion model is a compromise between the two fusion methods, which can gradually fuse the time information of whole network. The higher level can gradually obtain more global information in the time dimension. This is achieved by extending the connection of all convolutional layers in the time dimension. Finally, the fully connected layer can extract the global pattern features by comparing the outputs of all layers.

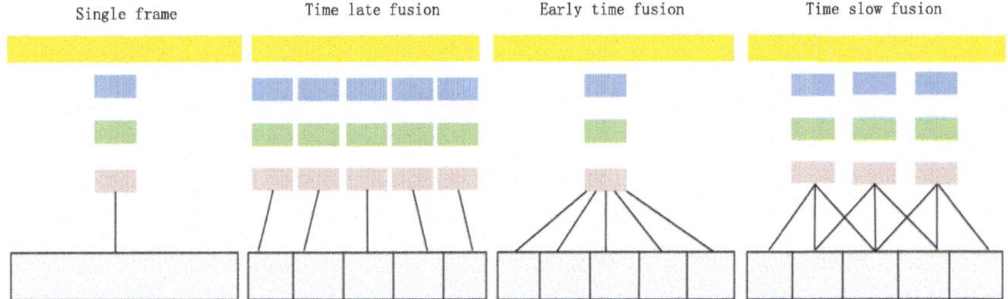

Fig. 6.7 The schematic diagram of the method of CNN network fusion information in the time dimension

To sum up, all three-time fusion models attempt to introduce temporal information into the CNN model. The later fusion model can obtain the local information of each subframe, while the early fusion model tries to obtain more global information from the data. Therefore, the eclectic temporal and global fusion can be achieved at the same time.

(3) **Regularization**

The *L2* norm constrained random deactivation (dropout) of the weight vector is used in the full connection layer for regularization. Random deactivation can set the weight of hidden cells to zero during forward and backward propagation. In other words, given the reciprocal layer $z = [\hat{c}_1, \hat{c}_2, ..., \hat{c}_{n-h+1}]$, for the output unit y in forward propagation, the random deactivation strategy $y = w \cdot (z \circ r)$ is the element by element multiplication operator, and $r \in R^m$ is the hidden vector composed of random variables of Bernoulli distribution. For non-hidden units, only gradient backward propagation is carried out.

6.4.4 Medical Speech Processing

A. **Task Definition**

Speech processing, also known as speech signal processing or voice processing, aims to identify speech signals, applied to mobile phones, computers, and even general life so that people and machine systems can communicate. The commonly used voice data in the medical field include patient recording data, medical dialog data, etc. It has useful applications in disease diagnosis, dysarthria speech recognition, medical speech transcription, etc.

B. **Related Work**

(1) **Disease Diagnosis**

Dysarthria, characterized as a neuromotor speech disorder, is a common symptom of various nervous system disorders, including stroke, Parkinson's disease, and brain injury. Clinically, dysarthria is evaluated by Speech-Language Pathologists (SLPs), who subjectively measure articulation and speech intelligibility. Due to the lack of professional speech-language pathologists, it is necessary to identify dysarthria objectively and accurately. Speech processing analysis is emerging as a useful and straightforward clinical decision support method, which provides a monitoring method for the automatic diagnosis of symptom severity. Carmichael et al. used a multi-layer perceptron and decision tree to classify different forms of dysarthria, using the Computerized French Dysarthria Assessment (CFDA) configuration file as the input. Essentially, they used acoustic signal processing technology to measure the vector of joint dysfunction values. Shirvan et al. used a genetic algorithm to analyze speech to extract and optimize Parkinson's disease characteristics, and used the k-nearest neighbor (KNN) classification method to classify the data. They applied KNN to a dataset of 192 speech signals recorded from 32 subjects (23 PD patients and 9 healthy subjects). The results show that Parkinson's disease (PD) can be classified with high accuracy. Wu et al. used an MLP artificial neural network to diagnose Parkinson's disease speech. They trained the MLP system using relatively large speech sample feature datasets collected from PD patients and healthy individuals to accurately classify the two speech types.

Early intervention in childhood can reduce the language barrier in adulthood. Traditionally, it is expensive and time-consuming to diagnose speech disorders based on speech pathologists' assessment of children's speech intelligibility. Rosdi et al. [13] proposed a classification method for automatic speech intelligibility detection of speech disorders using fuzzy Petri nets. This method can improve the ability of feature recognition and the ability of automatic speech intelligibility detector to distinguish children with speech disorders.

Du et al. proposed a deep learning method for feature extraction and representation, combined with the word bag method for classification, and automatically evaluated mental health from the acoustic aspect.

(2) **Speech Recognition for Dysarthria**

Most of the traditional speech recognition systems for dysarthria are based on structured methods. For example, the HMM-GMM method uses a Hidden Markov Model (HMM) to model the sequential structure of the speech signal. The Gaussian Mixture Model (GMM) is used to model the distribution of the spectrum representation of the waveform. However, the HMM-GMM based system needs to train a large amount of data, which is invalid for the low pitch of abnormal speech used in training. Therefore, these methods cannot be easily applied in the context of abnormal pronunciation.

Due to the development of Deep Neural Network (DNN) architecture, a significant breakthrough has been made in pathological speech processing as an effective alternative to HMM-GMM, which can better recognize dysarthric speech. For example, Kim et al. used RNN with long- or short-term memory to model the dysarthric speech in a speaker-independent way, and used CNN to extract effective local features.

Hu et al. proposed a speech recognition system for dysarthria based on Gated Neural Network (GNN), which allows robust integration of acoustic features with visual features and optional pitch based prosodic features. They designed a system for English and Cantonese, in which a speaker-independent GNN acoustic model was trained on the mixed training set using a series of 86-dimensional logarithmic filter banks and pitch features.

(3) **Medical Phonetic Transcription**

Medical dialog between patients (and possibly even nurses) and doctors has several different characteristics: (1) it involves multiple speakers (doctors, patients, and occasionally nurses), with overlapping conversations and different distances from the microphone, as well as different sound quality; (2) it covers various speech patterns, accents, background noise, and vocabularies from spoken to complex domain-specific languages. Automatic Speech Recognition (ASR) must deal with long-form content, which interweaves important clinical information with random chat sounds. Developing ASR systems for medical conversations is complicated by the lack of large amounts of clean, collected data on which to build systems.

Edwards et al. constructed a neural network based speech recognition system using relatively small medical speech data (270 h) and conducted a benchmark test on medical transcription experts. The speech recognition system was evaluated in clinical question answering tasks. The results show that domain adaptation using a language model can significantly improve the accuracy of interpreting clinical questions. Language model adaptation using crowdsourced input data has been shown to improve the accuracy of medical speech recognition systems. The latest development of the end-to-end speech model provides a powerful alternative. The Listening and Spelling (LAS) model is an end-to-end model that can learn the speech model as part of the ASR model itself. The LAS model is very robust to noise and does not need a speech model.

C. **Diagnosing Dysarthria Using Long- and Short-Term Memory Networks**

Mayle et al. proposed to use the RNN with Long Short-Term Memory (LSTM) units to determine whether a Mandarin speaker has dysarthria based on syllable pronunciation samples.

Given an audio segment X, which contains the pronunciation of Mandarin syllables, the model will produce a label Y, which is positive only if and only if the speaker has dysarthria. Figure 6.8 shows the path of a single training sample through the proposed model architecture.

Firstly, the original waveform X is pre-processed into the Mel Frequency Cepstrum Coefficient (MFCC) eigenvector $X\prime = \{x_1, ..., x_t, ..., x_T\}$, where the MFCC frame number T between the original input X can be different. The MFCC vector x_t is created with a sliding window of 25 ms (with a span of 10 ms). Each MFCC vector $x_t = \{\theta_1, ..., \theta_n, ..., \theta_N\}$ consists of N coefficients, where n = 13, unless otherwise specified. These values are normalized so that each coefficient θ_n has zero mean and unit variance across all training samples. The MFCC sequence $X\prime$ is taken as the input to the LSTM layer, and the last output h_T of the LSTM is provided to the logistic regression layer.

Several variants of the LSTM model are tested, including the addition of a layer and the use of bidirectional LSTM. For one-layer models, L2 regularization is used. The

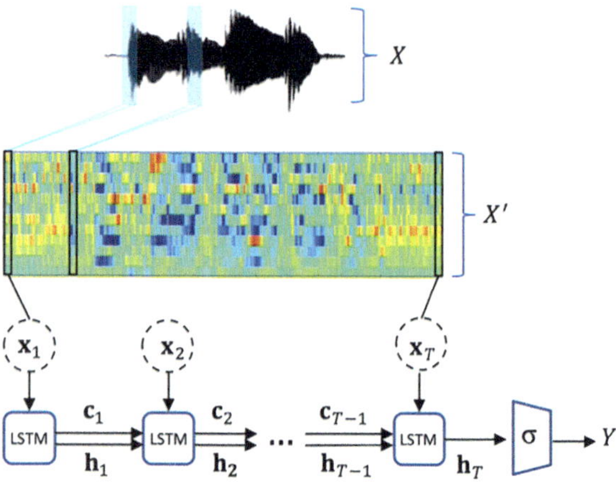

Fig. 6.8 LSTM network model architecture

two-layer model uses dropout between LSTM layers and between the last LSTM layer and logistic regression. The bidirectional LSTM network performs two simultaneous data transfers, from left to right, and from right to left. The output vectors generated by two passes are cascaded and fed into the logistic regression layer.

The training data consisted of a sample of 69 Mandarin speakers, including 31 native speakers (19 males and 12 females) with dysarthria, aged between 25 and 83 years (mean $\pm$ SD: 56.74 $\pm$ 16.40 years). In the first experiment, the entire set of syllables in the dataset was reorganized and divided into a training set, a test set, and a verification set with a ratio of 2:1:1. Because the dataset is divided by syllable level, different syllables of the same patient may belong to the training, verification, and test sets. Therefore, the syllables that appear during the test may come from patients observed during training. The performance of all LSTM architectures is better than that of a fully connected neural network architecture. When syllable level accuracy is evaluated, the bidirectional LSTM slightly lags slightly behind the unidirectional LSTM. The second experiment creates training and test sets by dividing speakers. This way, a person's syllables can't appear in both the training and test sets. To evaluate the syllable level and speaker level, three LSTM models were constructed. When the syllables with compound vowels are removed from the dataset, the Area Under the Curve (AUC) of all models exceeds 90%. In addition, it is found that the performance of the LSTM model can be improved by increasing the number of cepstrum coefficients. Although these methods may not be practical as independent medical tests, they show that the LSTM network can effectively realize the automatic diagnosis of articulation disorders.

6.5 Application Cases

The application of big data analysis in medical information processing can facilitate scientific decision-making for the diagnosis and treatment of diseases, summarize the curative effect of various treatment schemes, and enhance the hospital's capacity for evidence-based decision-making, medical treatment, scientific research, and teaching. The scope of its applications is expanding, including precision medicine, telemedicine, disease prediction, disease auxiliary diagnosis, disease prognosis evaluation, new drug research and development, health management, medical image recognition, etc. According to the actual application of medical information, several big data analysis technologies are often integrated to give full play to their technical advantages according to the specific medical basic resources and different application goals. Besides, with the improvement of people's living standards, the increase of health awareness, and the deepening of China's medical system reform, the potential market of medical information processing big data analysis and application is also comprehensive. Making better use of big data analysis and application to provide comprehensive and accurate diagnosis and health care measures has become a challenge to promote medical development.

The medical information processing application cases selected in this section under the background of big data are precision medicine, diabetes health promotion system, elderly comprehensive health assessment system, and telemedicine. Among them, precision medicine includes precise pre-hospital pre-treatment, in-hospital accurate disease diagnosis, treatment, and post-hospital precise rehabilitation management. Precision medical services based on big data include accurate diagnosis, precise treatment, and precise drugs. The diabetes health promotion system includes disease risk assessment, guidance scheme establishment, and intervention evaluation guidance. A comprehensive health assessment system for the elderly includes physical function and disease, health assessment of the state, mental state, and living environment. Telemedicine includes remote consultation, remote case diagnosis, telemedicine imaging diagnosis, remote monitoring, etc.

6.5.1 Precision Medicine

A. **Introduction to the Model**

With the completion of the "Human Genome Project" in 2003, personal genome, tumor genome, environmental genomics, and gene sequencing technology have been developed in an unprecedented way. Besides, biological and medical fields have a considerable amount of big data reserves, which have been greatly enriched and integrated into the close information exchange in the field of medical and health services. In this context, the medical diagnosis model has undergone a revolutionary change, and "precision medicine" has begun to be understood as a new medical model.

Precision medicine represents a novel medical concept and medical model based on individualized medicine, with the rapid progress of genome sequencing technology and the cross-application of bioinformatics and big data science. It is essentially a new medical concept and medical model through genomics, proteomics and other genomic technologies, medical frontier technology, big data, and information science technology. According to the type of disease, we can analyze, verify, and apply biomarkers to accurately find the cause of the disease and the target of treatment, accurately classify the different states of disease, and finally realize the personalized and precise treatment scheme for specific patients. In short, precision medicine is to formulate personalized and precise treatment schemes according to patients' individual characteristics and realize personalized medical service through molecular biological indicators.

With the implementation of precision medicine makes the medical model is changing from extensive to precise, promoting the further integration and intersection of medical science and technology, big data, and information science. In recent years, under the background of "precision medicine," domestic and foreign governments and enterprises have started the strategic and scientific research deployment. On January 20, 2015, former U.S. President Barack Obama released a proposal entitled "Precision Medicine Plan," which attempts overcome the problem of cancer drug resistance by collecting genomics and other molecular information. The Chinese government has established an expert group on China's precision medicine strategy to create precise treatment programs for malignant tumors, hypertension, diabetes, congenital disabilities, and rare diseases. Precision medicine in various diseases for molecular typing diagnosis, and based on the results of molecular biological detection of personalized treatment, has achieved significant results, effectively extending the patients' survival time and improving the quality of life.

As the next generation of diagnosis and treatment mode, precision medicine is of great significance to the biomedical industry. This diagnosis and treatment technology closely combines traditional medical methods with modern information technology. Based on population subdivision, it can scientifically interpret the big biological data closely related to human body function and the nature of disease, to realize "the right time, the right people, and the right drugs."

B. **Overall Structure**

Medical care's mission is to protect human health, and the significance of precision medicine is to achieve this goal more effectively. From the disease development process perspective, the medical system of precision medicine can be divided into pre-hospital precision pre-treatment, in-hospital accurate disease diagnosis, treatment, and post-hospital precise rehabilitation management. The overall structure is shown in Fig. 6.9.

C. **Functional Features**

In pre-hospital management and in-hospital diagnosis and treatment, the most complex and critical step in precision medicine is disease classification. In precision medicine, people tend to use genetic or biological methods to classify the disease

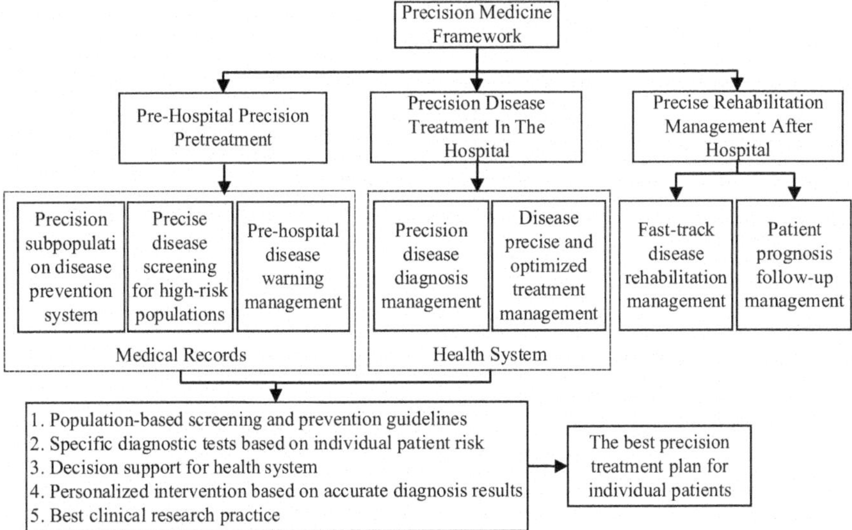

Fig. 6.9 Overall architecture of precision medicine

at the genetic or molecular level, and finally get the precise treatment plan. Therefore, the pre-hospital pre-treatment system contains a complete medical information database that stores the epidemiological data such as region, season, environment, and population characteristics, which is an essential part of pre-hospital management. With the help of information technology, information on family history, special habits and molecular genetic characteristics of a large sample population can be screened and processed in the medical information database, and a disease prevention framework for people or individuals can be established, thereby enhancing the effectiveness of prevention, reducing the incidence rate, and promoting people's health.

The system can be divided into two parts: accurate disease information diagnosis management system and accurate disease optimization treatment management system. These two systems are based on purely technical factors such as a large number of diagnostic instruments and the research methods of evidence-based medicine and personalized medicine. They can provide doctors with a large amount of disease data and help them diagnose the disease quickly and accurately. At the same time, they can also provide decision support for the selection of treatment methods.

The precise post-hospital rehabilitation management system contains the follow-up data of patients' prognosis. The purpose of the system is to accurately capture the patients' rehabilitation status and provide personalized rehabilitation guidance.

(1) **Technical Features**

In general, precision medical technology mainly includes genomics technology, information technology, and other fields. Genomics technology refers to the related technologies in the fields of genomics, proteomics, metabonomics, and transcriptomics that is the hot spot and foundation of biological research, and clinical precision medicine. Genomics technology mainly includes biochip technology, second-generation sequencing technology, panomics technology, nano string technology, etc. This book mainly introduces the information technology of precision medicine, the essence of which is to establish a biological big data information database. Precision medical services based on big data include accurate diagnosis, precise treatment, and precision medicine.

1) **Accurate Diagnosis**

Accurate diagnosis mainly refers to molecular diagnosis. The process includes: firstly, the complete collection of patients' clinical information is realized through an EMR system; at the same time, a biological sample library is used to collect patients' biological sample information. Secondly, a gene sequencing platform is used to collect molecular level information of patients. Finally, the bioinformatics analysis method based on big data is used to integrate and analyze all visual information displays, the final form of an accurate clinical diagnosis report. In the field of accurate diagnosis, better applications include tumor gene detection and precision surgery.

2) **Precision Therapy**

This link refers first to the collection of patient information and patient samples, then to the use of omics and big data analysis technology to complete the classification of large sample populations and specific disease types, and ultimately to the analysis, identification, validation and application of biomarkers. The objective is to accurately find the cause and treatment target and provide accurate support and basis for clinical decision-making. Precision therapy has become a new trend in cancer treatment, and it is gradually penetrating into the treatment of other clinical diseases.

3) **Precision Medicine**

Precision medicine is an essential embodiment of precision medicine. It refers to the research and development of targeted specific drugs according to the types of diseases. Individual differences in the genome are used to realize personalized drug use. Great progress has been made in the clinical efficacy of targeted drugs, yet challenges remain in the management of drug toxicity and treatment resistance.

The technical system of precision medicine mainly includes biological sample banks, bioinformatics, EMRs, and big data analysis technology. The first three are the prerequisites of precision medicine, and the last is the key to achieving precision medicine.

Biological sample bank, also known as a biological bank, provides important omics data and clinical medical information for precision medicine, and is an important part of precision medicine. The combination of statistics, molecular biology, and computer science methods with omics technology is employed for the purposes of disease research, the analysis of biological samples, and the identification and verification of biomarkers. Biological sample bank has various types, including tissue bank, organ bank, cell line bank, gene bank, disease database, etc.

Bioinformatics is a discipline that employs a combination of statistics, molecular biology, and computer science to facilitate the storage and analysis of biological data. The research areas of focus include genomics, proteomics, the simulation of spatial protein results, and drug design. The application of big data in bioinformatics primarily concerns the storage, computing framework, and efficient utilization of data, with the objective of collecting and analyzing biological information. The mainstream bioinformatics technologies include Hadoop and map-reduce. The combination of patient information and experimental results allows the use of bioinformatics methods to identify biomarkers, including proteins, genes, and metabolites, which can then be used to inform drug design scheme, diagnosis, and treatment scheme.

The biomarkers obtained must be combined with clinical data and patient sample data following the application of bioinformatics methods. It is therefore necessary to integrate EMRs with biological information data, clinical data, basic patient information, and other information functions in order to establish a foundation for molecular information analysis and other data analysis. The data stored in an EMR includes a range of information pertinent to the patient's health, including real-time health data, physical examination results, genetic testing data, historical disease data, and information regarding the patient's health consumption behavior. In precision medicine, EMRs can facilitate the aggregation of a vast array of data, enabling research groups to conduct more efficient and cost-effective studies.

(2) **Processing Model Features**

In this chapter, big data analysis techniques, such as data mining, are used to analyze the precise medical big data stored in digital platforms, such as medical clouds and server clusters, and visualize it for patients, doctors, biopharmaceutical companies, and other different users. The data mining framework of precision medicine is shown in Fig. 6.10.

DNA sequencing already has the analytical techniques that can accurately detect a single nucleotide, cell, or molecule. The routine application of high-throughput gene chip and protein chip indicates that people can get genotypes of different diseases. Mapping the molecular disease spectrum of specific patients has gradually become an

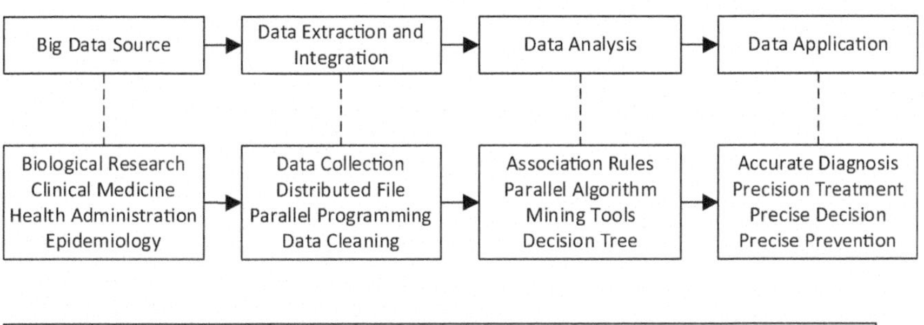

Fig. 6.10 Framework diagram of precision medical data mining

important part of clinical diagnosis and treatment. However, the above tests generate a large amount of data, and the management, storage, mining, and processing of these data need to be solved. With the development of big data technology, precision medicine has received strong technical support. For example, in clinical practice, by tracking, comparing Positron Emission Tomography (PET)/CT, Single-Photon Emission Computed Tomography (SPECT)/CT, MRI, and CT examination results, it is easy to monitor and evaluate the development of the disease and the effectiveness of drugs, and provide the basis for individualized treatment.

Gene big data analysis is an important aspect of precision medical big data analysis. The effective methods of gene big data analysis include path analysis and gene network reconstruction. Path analysis refers to the analysis of abnormal gene expression effects in a special gene set. The mainstream methods and tools are on express, go miner, clue go, etc. Gene network reconstruction refers to the analysis of signals measured by high-throughput sequencing technology, to reconstruct a regular network. The regular network will affect the biological process of most cells, which will affect the physiological state of the human body. The mainstream methods and tools for gene network reconstruction include Recon 2 and Boolean methods. With the continuous development of precision medicine, gene big data analysis needs to consider the combination of EMR data and biological sample database data.

Similar to traditional medical information big data processing, precision medical big data processing has the following three basic characteristics: (1) massive data; (2) rapidity of demand; (3) diversity of data types. The development of medical information has led to the continuous production and accumulation of medical data, as well as the increasing scale of medical data. New data types are also emerging, such as 3D / 4D image data. With the progress of big data processing technology, especially the gradual maturity of visualization and cloud computing technology, the acquisition, storage, analysis, and application of massive data, have become more convenient. The model of precision medical data processing is shown in Fig. 6.11.

Using various platform tools, people can realize real-time data collection. Precision medical data collection tools include EMRs, a drug management system, wearable physiological sensors, implantable physiological sensors, 3D medical images,

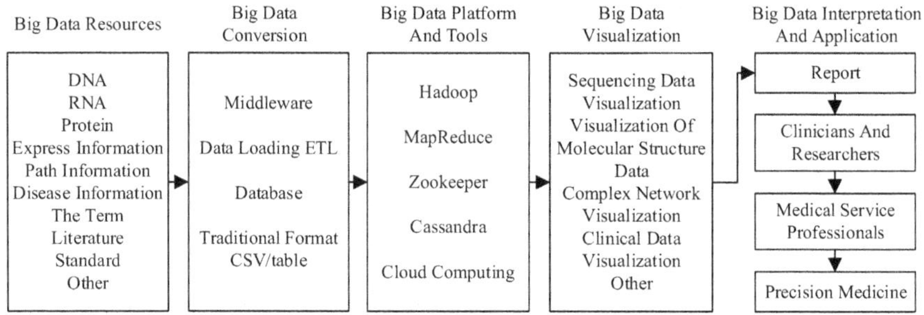

Fig. 6.11 Precision medical data processing model

gene sequencers, omics detection platforms, gene chips, and microbiological analysis. Cloud computing solutions and server cluster solutions are generally used for big data storage. When solving the problem of data fragmentation, we also need to consider how to ensure data security. Big data analysis mainly uses big data platforms and tools, such as Hadoop, MapReduce, Zookeeper, Cassandra, to perform data transformation protocol. The results of big data analysis need to be visualized. Data visualization of precision medicine mainly includes visualization of sequencing data, visualization of molecular structure data, visualization of the complex network, and visualization of clinical data. Finally, the biomedical report needs to be generated and transmitted to clinicians and researchers for professional interpretation and then transmitted to medical service professionals to develop personalized, precise medical schemes to meet patients' medical needs. For patients, doctors, and biopharmaceutical companies, the greatest benefit of applying big data analysis can improve the quality and efficiency of the treatment process, while saving time and energy costs.

(3) **Characteristics of Security System**

Precision medicine requires a high-performance security system, where information infrastructure generally refers to the software and hardware, and network facilities in the process of data generation, storage, analysis, and presentation.

1) **Data Generation**

In precision medicine, we need to use a large amount of population data for research and analyze patient parameters. It also requires individual data and drug data for reference matching. Generally, data generation facilities in precision medicine include intelligent health monitoring devices, EMR systems, gene sequencing platforms, electronic health record systems, etc.

2) **Data Storage**

In order to make precision medicine can be carried out smoothly, it is necessary to collect and store the massive data generated by the data generation facilities for further data analysis and use. To meet the precision treatment requirements, it is necessary to ensure the accuracy, comprehensiveness, and comparability of data, which places high demands on precision medicine data storage facilities. The main precision medical data storage facilities include medical cloud, server warehouse, etc.

3) **Data Analysis**

Data analysis facilities suitable for precision medicine mainly include genomics, bioinformatics, data analysis methods of big data, such as cancer genomics technology, gene chip technology, high-performance computing technology, data mining technology, etc.

4) **Data Presentation**

Genetic data and clinical data need to be processed and analyzed by relevant software to generate bioinformatics reports, which can be fully interpreted by to obtain valuable information, provide accurate and personalized services for patients. Commonly used data presentation tools include Excel forms, Google documents, integrated genomic data viewer, clinical data integration cloud platform, etc.

6.5.2 *Diabetes Health Promotion System*

A. **System Introduction**

With the development of the social economy and the improvement of people's living standards, lifestyle changes, and the aging of the population, the prevalence of diabetes is increasing worldwide and has become one of the most common chronic non-communicable diseases. According to the latest statistics from the International Diabetes Federation (IDF), approximately 8.8% (7.2–11.4%) of adults aged 20–79 years worldwide have diabetes, which means that there are at least 415 million diabetics worldwide. Diabetes and its complications are increasing worldwide. Diabetes has become a major global public health problem of the twenty-first century. In order to realize the early detection of high-risk individuals of type 2 diabetes, and improve the intervention effect through personalized intervention, and achieve the purpose of health promotion of high-risk population of type 2 diabetes. A diabetes health promotion system for type 2 diabetes mellitus has been established by the Information System and Security Confrontation Experimental Center of School of Information and Electronics, Beijing University of Technology.

B. **Overall Structure**

The overall architecture of the diabetes health promotion system for type two diabetes is illustrated in Fig. 6.12.

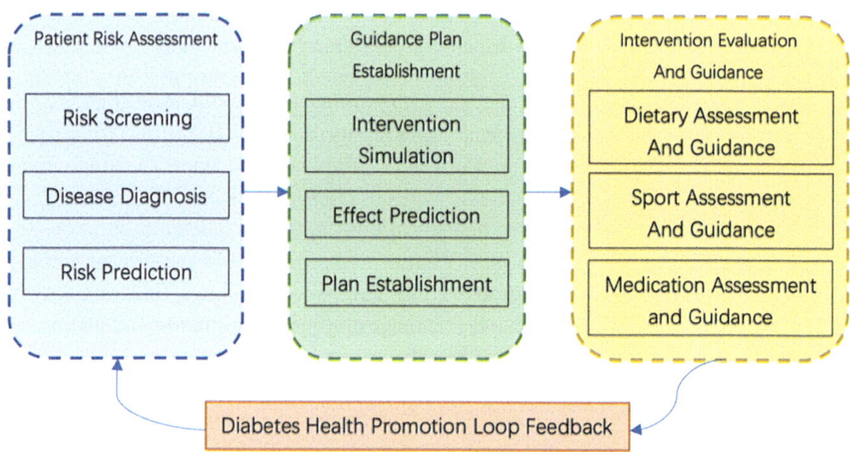

Fig. 6.12 Overall architecture diagram of the diabetes health promotion system

C. **Function Features**

The socio-economic development and living standards of residents around the world are constantly improving, lifestyle has changed, and population aging is becoming more and more serious. The prevalence of diabetes is increasing worldwide and has become one of the most common chronic non-communicable diseases. Recent statistics from the International Diabetes Federation (IDF) indicate that approximately 8.8% (roughly 7.2–11.4%) of adults aged 20–79 years worldwide have diabetes. Diabetes and its complications are on the rise worldwide, and diabetes has become a major global public health problem of the twenty-first century. In order to realize the early detection of individuals at high risk of type 2 diabetes, and improve the intervention effect through individualized intervention, at the same time to achieve the purpose of health promotion of high-risk population of type 2 diabetes. The Diabetes Health Promotion System for Type 2 Diabetes Mellitus was established in the Information System and Security Countermeasure Experimental Center of the School of Information and Electronics of Beijing University of Technology.

(1) **Risk Assessment**

Disease risk assessment is the analysis and inference to describe and estimate the risk status, influencing factors, and change trends of diabetes mellitus in the future. Disease risk assessment is a research hotspot of experts and scholars, but the research mainly focuses on the assessment of death risks. There are few and imperfect assessment models for disease occurrence risk. Only cardiovascular disease risk assessment and diabetes risk assessment are more mature. Diabetes risk assessment is used to describe and estimate the possibility of diabetes or death due to diabetes in an individual. It is often used to screen people over the age of 18. Through diabetes risk assessment, it can reduce the workload and work pressure of medical staff, make individuals have a comprehensive understanding of health risks, stimulate individuals to change bad lifestyles, formulate individual health intervention measures, and evaluate the effectiveness of intervention measures. The risk assessment method is divided into risk classification and risk prediction.

1) **Risk Classification**

The risk classification of diabetes refers to routine physical examination indicators to realize the discovery of undiagnosed potential diabetes population and the assessment of the current disease risk of the uninfected patients, aiming to understand the strategy of early detection and early treatment of diabetes. The sample size used in risk classification related research is concentrated in 500–30,000 people, most of which are in the range of 1000–5000, a total of 16 items. The number of variables used in related research is mostly 3–8. The most commonly used variables are demographic variables and physical variables. Demographic variables include age, gender, race, physical variables include waist circumference, BMI, and waist-hip ratio. Some studies also include lifestyle-related variables, including diet, smoking, and exercise. Based on the invasive and non-invasive indicators, a diabetes risk assessment model can be constructed for auxiliary medical diagnosis. For example, both the Oman

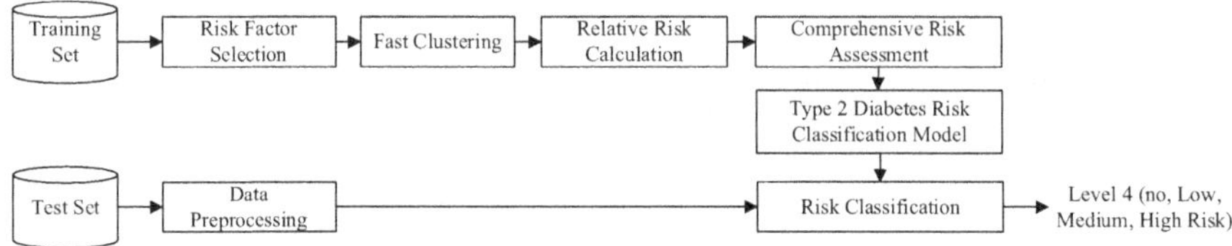

Fig. 6.13 Algorithm framework of type 2 diabetes risk classification model

diabetes risk assessment and the China diabetes risk classification models divide the disease risk into four risk levels (no risk, low risk, medium risk, and high risk). Finland, the diabetes classification model, divides the disease risk into five levels (very low, low, moderate, high, and very high) according to the individual risk score. This section will introduce the risk classification model of type two diabetes by taking the diabetes health promotion system for type two diabetes developed by the Information System and Security Confrontation Experimental Center of School of Information and Electronics, Beijing Institute of Technology.

The research and development of the model is supported by the National Science and Technology Support Plan of China's 12th Five-Year Plan and the National Natural Science Foundation of China. Based on 16,246 research subjects, a 4-level risk assessment model for type 2 diabetes is constructed. The algorithm principle is shown in Fig. 6.13.

The purpose of risk factor selection is to screen the characteristics related to the occurrence of diabetes, using a C4.5 decision tree and multiple logistic regression to complete, and comprehensively consider the difficulty of feature collection, to determine the characteristics used in the final assessment. The original feature dimension of the data is 96 dimensions. After manual attribute screening, 57 dimensions are retained for risk factor selection after removing useless features (such as ID numbers). The research individuals were divided into five groups: (1) total population; (2) all males; (3) all females; (4) under 50 years of age; (5) 50 years of age and older. C4.5 was used to construct a decision tree for the five groups of samples, and the classification label was whether the individual had diabetes. In the multiple decision trees constructed, the closer the feature's split point is to the root node, the higher its importance. Simultaneously, a multiple logistic regression model, features with a statistically significant correlation ($P < 0.05$) will be regarded as risk factors. Finally, age, diastolic blood pressure, high-density lipoprotein, waist circumference, gender, cholesterol, family history, body mass index, and triglycerides were selected by the two algorithms at the same time and determined as the input features of the grading model. In addition, it is worth noting that in the C4.5 decision tree, fasting blood glucose and its split value of 5.85 mmol/L appeared multiple times near the root node (6 times in the first 8 layers, 4 times in the first 2 layers), indicating that this value has a very high reference value for judging the risk of diabetes, i.e., individuals with fasting blood glucose greater than 5.85 mmol/L should be judged as high risk.

Rapid clustering can re-divide research individuals into multiple clusters based on risk factors, so that individuals of the same race are as close as possible in the risk factor feature space. Individuals of different races are as far away as possible, which helps achieve personalized risks for people with different characteristics. Scoring standards improve the accuracy of judgment. This step first needs to normalize the data and then use the rapid clustering method to calculate individuals' differences. Through multiple clustering experiments with different cluster numbers, the R^2 value is used to evaluate the clustering results to determine the best number of clusters (3 types) and avoid the impact of random initialization of the clustering algorithm.

The relative risk calculation is performed separately in the three groups of populations after rapid clustering, including the following two steps: (1) calculate the risk of diabetes for each individual; (2) calculate the Relative Risk (RR) within the group according to the age and gender distribution of the individual's group. Logistic regression is used to calculate risk of individual and population. The population is divided into 24 groups based on age (20 to > 75, one group every 5 years) and gender to calculate the individual relative risk RR value.

The comprehensive risk assessment considers the individual's fasting blood glucose level and its relative risk, combined with certain discriminatory rules in each group, and provides the individual's diabetes risk classification results. The process is as follows:

Algorithm 1: Comprehensive Risk Assessment Process

```
if (GLU > 5.85) {
   risk status = high risk;}
else {
   if (RR > 2.2) {
determine the degree of fitting the relative criteria {
   if (degree == 3) risk status = high risk;
   else if (degree == 2) risk status = medium risk;
   else if (degree == 1) risk status = low risk;}
}
else {
   risk status = non risk;}
   }
```

The above determination process is implemented by C + + language. Based on the risk assessment model, a B/S architecture (browser/server) system is constructed to provide users with convenient diabetes Risk assessment System for type 2 Diabetes (RSD).[18] The system interface is presented in Fig. 6.14.

2) **Risk Prediction**

Risk prediction is to predict the possibility of diabetes in a certain period in the future. It is a prognostic study based on cohort data. It aims to understand the future trend of individual disease risk, improve individual lifestyle and diet, and reduce or even prevent type 2 diabetes.

[18] http://www.isclab.org.cn/rsd/RSDAssess.php.

BFS-2型糖尿病风险评估系统

BFS - Diabetic Risk Assessment System

本系统用于评估20~85岁人群中个人2型糖尿病的当前患病风险程度和未来10年内2型糖尿病的患病几率。判定系统由国家科技支撑计划支持研发。系统可以提示每一个体当前不同患病风险等级，以及未来10年内患病的概率。

糖尿病是可以预防控制的。了解自己当前的糖尿病患病风险是预防糖尿病发生的第一步。每个人都可以通过加强运动、合理膳食，有效地控制糖尿病患病风险，降低糖尿病发生的几率。

本系统评估与预测结果是基于您目前的状态完成的，您的状态变化后（如加强运动、控制饮食或是体重增加），评估与预测结果会发生相应的改变。再次评估结果一方面可体现您为维护健康付出的努力，另一方面也会提示您注意可能增长的风险。

BFS-2型糖尿病风险评估系统愿为民众健康维护提供技术支持与服务。也诚挚地希望得到每位使用者的意见反馈。

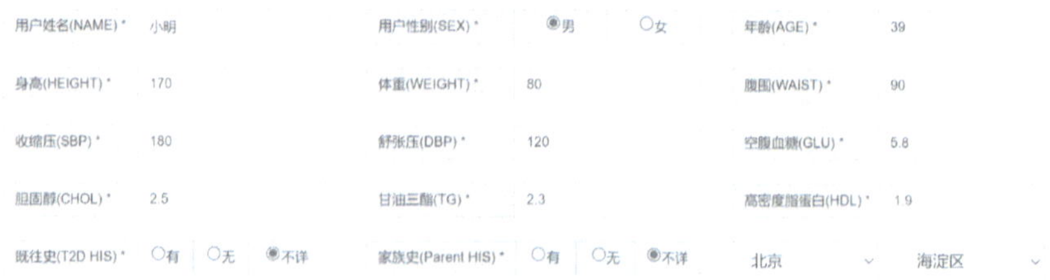

风险评估 重置

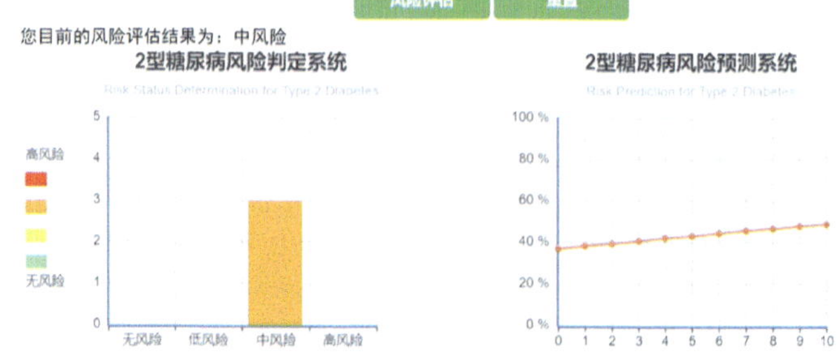

风险评估结果说明：

1. 糖尿病患病风险程度从小到大分为四种。既无风险、低度风险、中度风险和高度风险，分别对应的颜色和数值为蓝(0)、黄(1)、橙(2)和红色(3)。
2. 根据您目前的状态，未来十年内您患糖尿病的几率显示在右侧图上。

Fig. 6.14 Risk assessment system for type 2 diabetes produced by Beijing Forest Studio

In contrast to the risk classification study, the risk prediction is based on a cohort study. The individual disease situation is affected by different environments, such as time, region, and heredity. Hence, the type and quantity of variables used in this study are more than those of risk classification. According to the different types of variables used in risk prediction, related research can be divided into the prediction model based on the non-invasive index, biochemical index, and genetic information index.

Prediction models based on non-invasive indicators mostly use demographic and physical fitness indicators, and some also include factors such as diet, exercise-related factors, or family history. The AUC of these models is generally in the range of 0.7–0.8, but related studies show that the AUC is less than 0.7. These studies basically use only 3–4 indicators for prediction. For the non-invasive indicators used, the number of factors selected is still related to the model's accuracy. Among all the models, only the FINDRISC (The Finnish Diabetes Risk Score) model and the German Diabetes Risk Score model have an AUC greater than 0.8. However, the FINDRISC model is aimed at people who are treated with drugs as the model input, excluding people who are not taking drugs, and the analyzed population has a certain limitation. The German model uses only categorical variables as model input, and the risk scores are all integers, which reduces the discrepancy of the model to a certain extent.

The prediction model based on biochemical indicators uses non-invasive indicators and includes metabolic syndrome indicators, such as cholesterol, triglycerides, high-density lipoprotein, etc. Studies have shown that adding metabolic syndrome indicators to the risk prediction model can improve the prediction accuracy. For example, after adding lipid indicators to the atherosclerosis risk score model, the model AUC can be increased from 0.71 to 0.80 ($P < 0.001$); the Framingham model can be improved from 0.72 to 0.85 by adding high-density lipoprotein and triglyceride based on age, gender and BMI (body mass index). Generally, based on non-invasive indicators, adding biochemical indicators can improve the model's prediction effect.

Compared with the above two aspects, few related studies on the prediction model are based on the genetic information index, which aims at the single nucleotide polymorphism and multi-nucleotide polymorphism in risk prediction. The results of the Malmo project, Botina study, and Rotterdam study showed that single nucleotide polymorphism did not improve the accuracy of risk prediction. On the contrary, polynucleotide polymorphism had a certain effect on improving the model's accuracy based on age, gender, and other non-invasive indicators. The prediction model of genetic information can find the key factors from the perspective of diabetes, but it has little significance for population screening.

In conclusion, different factors of model selection will reduce the practicability and accuracy of risk prediction in the process of use. The selection of non-invasive indicators can facilitate the use of a risk prediction model. It can be realized only through a questionnaire survey without a routine physical examination or computer-related calculation and discrimination, which is easy to implement. However, the accuracy and practicability of the model are the opposite. If the practicability is guaranteed, the accuracy of the model will be reduced, and vice versa.

Next, a long-term prediction method of type 2 diabetes incidence probability based on the Markov model is introduced. In this method, a naive Bayes algorithm is used to calculate the probability vector of the risk level, and the prediction model is constructed by selecting the attribute subset after attribute selection, such as fasting blood glucose, body mass index, cholesterol, triglyceride, waist circumference, gender, family history of diabetes, and age. The comparison between the experimental results and the prediction results of the Archimedes model shows that the model can predict type 2 diabetes incidence in a long period, and the prediction method is accurate and straightforward.

The Markov-based prediction method for the incidence probability of type 2 diabetes is shown in Fig. 6.15.

The Naive Bayes algorithm is used to calculate the probability vector of risk level, which requires that each attribute is independent of each other. Experimental results

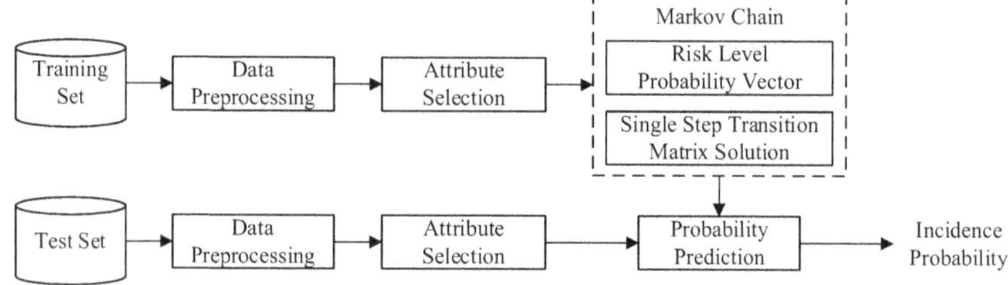

Fig. 6.15 Principles of Markov-based method for predicting the incidence of type 2 diabetes

show that the attribute subset with the highest accuracy of Naive Bayes classification includes 8-dimensional attributes of fasting blood glucose, body mass index, cholesterol, triglyceride, waist circumference, gender, family history of diabetes, and age. These 8-dimensional attribute sets are relatively independent of each other and are the optimal subsets for calculating the probability vector of the risk level. Markov chain is widely used in prediction models in various fields because it has no aftereffect and does not require much historical data. Markov model does not need to find the rules between the various factors from the prediction factors. It only needs to consider the evolution characteristics of the event's historical situation, so the Markov model is practical in predicting diabetes incidence probability. The formula calculates the probabilities $P_n^T = P_0^T \times Q^n$ of each state after several transitions, where P_n represents the state probability vector after n transitions; P_0 represents the initial state probability vector, and Q represents the one-step state transition matrix. The onset process of type 2 diabetes is divided into four states: no risk, low risk, high risk, and pathogenetic. The initial state probability vector P_0 is a 4×1 matrix vector, and the state transition matrix Q is a 4×4 matrix.

The prediction method based on the Markov chain can effectively predict the incidence rate of type 2 diabetes, so as to achieve the purpose of early detection, early attention, and early intervention, which can fully reduce (or delay the occurrence of type 2 diabetes, greatly save the cost of treatment; at the same time, for different intervention programs (such as changing diet habits, strengthening physical exercise, etc.), after some time once again, the model can be used to quantitatively evaluate the intervention effect. The intervention effect can then be improved by adjusting the intervention scheme. In addition, this method is based on the assumption that the initial probability is constant. With the development of society, people's eating habits will change, and their initial probability and transfer probability will also change. Therefore, the development of online learning model will help track the development trend of diseases and improve the effectiveness of risk prediction.

(2) **Establishment of Guidance Scheme**

1) **Intervension Simulation**

Accurate selection of intervention indicators can make the intervention measures more targeted and improve the intervention effect. The selection and verification

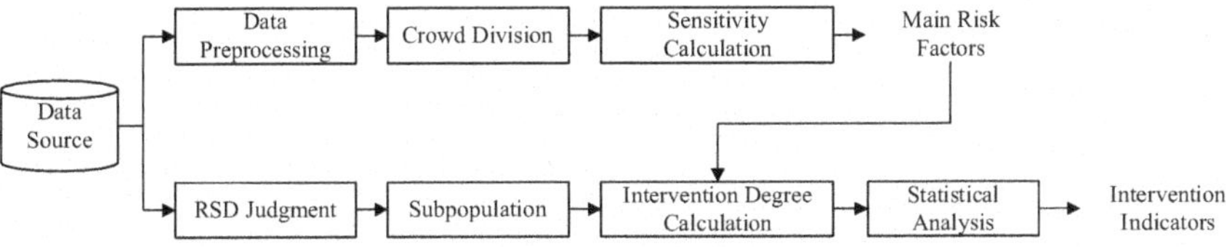

Fig. 6.16 Principles of intervention index selection method

method of intervention indicators for high-risk cross-section data is through the combination of qualitative analysis and quantitative calculation, respectively, training the selection and validation model of intervention indicators for high-risk groups. In constructing the model, machine learning and mathematical statistics are used to construct the subpopulation attribution model by using the H-SVM (Hierarchical Support Vector Machine) algorithm. The intervention indexes are adjusted quantitatively in different subpopulations. The transition probability of each subpopulation from high-risk state to low-risk state (medium-risk, low-risk, and no risk) is calculated. The intervention indicators are verified. The principle is shown in Fig. 6.16.

The RSD system is used to determine the risk status of input data, and four risk levels of high-risk, medium-risk, low-risk, and no-risk groups are obtained from the output of the system high-risk groups were screened out and further analyzed. To further refine the high-risk groups, the H-SVM algorithm is used to construct the subpopulation attribution model. According to the general characteristics of high-risk groups and combined with the sensitivity of different indicators, the high-risk groups were sub-divided into four categories: severe high-risk (high-risk A), moderate high-risk (high-risk B1 and high-risk B2), and mild high-risk (high-risk group C). The intervention aims to reduce the high-risk populations' risk level and reasonably adjust the intervention indicators in each subpopulation. In practical intervention, changing the lifestyle can improve the intervention index and reduce type 2 diabetes incidence. Therefore, to reduce the risk level of high-risk groups in the simulation, it is necessary to reasonably adjust the intervention index. By adjusting the size of intervention indicators by simulation, the proportion of maintaining high-risk and transferring to other risk levels (medium-risk, low-risk, and no-risk) is calculated, respectively. The number and percentage of the subpopulation with high risk or transferred to low-risk state (medium-risk, low-risk, or no-risk) after the adjustment of intervention indicators were statistically analyzed.

2) **Prediction of Intervention Effect**

The incidence rate of type 2 diabetes is difficult to cure. Intervention before high-risk groups can effectively reduce the incidence and improve the quality of life. However, under normal circumstances, most of the relevant studies are follow-up studies, and it is necessary to wait until the end of the intervention cycle to know how the intervention effect is. The intervention cycle is generally half a year or even longer. If the intervention effect is not ideal after the intervention, the disease may be delayed.

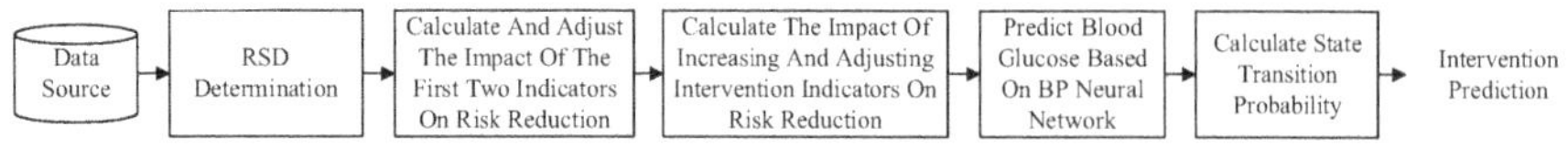

Fig. 6.17 Principles of simulation prediction method for intervention effect of high-risk population

Therefore, predicting the intervention's impact at the early stage of intervention can save time and improve intervention efficiency.

Intervention effect prediction simulates and predicts the intervention effect of the high-risk population of type 2 diabetes. Combined with the physical condition of a high-risk population, the whole intervention cycle's effect can be predicted at the beginning of the intervention, which is conducive to the effective use of resources and improves the intervention efficiency. The simulation prediction model of the high-risk population was trained by quantitative and statistical analysis, respectively. In the model construction, machine learning and mathematical statistics are used to build a model to calculate the state transition probability. The indexes with adjustment sensitivity in the first two places and the increasing adjustment intervention index on risk reduction are calculated. Then, the blood glucose value is predicted based on BP neural network algorithm. Finally, the state transition probability is calculated, and the intervention prediction results are given. The principle of the intervention effect simulation prediction method for high-risk population is shown in Fig. 6.17.

The risk level determination and intervention consequence prediction analysis method can provide technical support for high-risk population with diabetes. Help to understand both the population risk level and intervention effect from the whole. It is convenient for effective use of resources and improvement of intervention efficiency.

3) **Construction of Intervention Guidance Program**

The intervention analysis for a high-risk population of type 2 diabetes mainly includes three parts: intervention guidance, intervention effect evaluation, and formulation of intervention guidance scheme. The schematic diagram is shown in Fig. 6.18.

First, collect the physical examination data, and then conduct intervention guidance based on the collected data, which mainly includes: RSD determination, subclass attribution of high-risk groups, determination of intervention indicators, and calculation of intervention degree of intervention indicators. Then, the intervention effect is evaluated. Finally, two kinds of intervention guidance programs are formulated; one is for the intervention instructor, the other is for the intervention object.

Store the input physical examination data in the database and conduct high-risk screenings through the RSD system. An individual requires intervention if identified

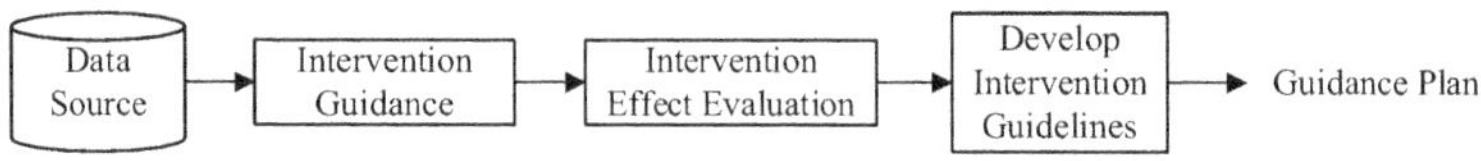

Fig. 6.18 The schematic diagram of constructing intervention guidance programs for high-risk groups of diabetes

as high-risk, and no intervention is needed if they are not. For high-risk individuals, perform subcategory attribution, determine their subpopulation, select intervention indicators, and calculate the degree of intervention for these indicators. Then calculate the transfer probability of the high-risk person to reduce the risk. Finally, provide intervention guidance suggestions. For non-high-risk individuals, give guidance suggestions without direct intervention.

Based on the study of intervention indicators and intervention effects of high-risk populations, combined with the practical application needs, the intervention guidance scheme for the high-risk population of type 2 diabetes has been constructed. It can be used for high-risk screening, subclass attribution, determination of intervention indicators, calculation of the intervention degree, and transfer probability of risk reduction. By integrating this scheme with personal exercise and dietary habits, an individualized intervention guidance program can be formulated to achieve effective intervention.

(3) **Implementation of Intervention Program**

When intervening with individuals at high risk for type 2 diabetes, a dietary assessment system can be constructed to provide dietary structure, energy regulation, and personalized nutritional guidance.

(4) **Measurement of the Rationality of Dietary Structure**

Many researchers at home and abroad, have done a lot of effective investigation and research work in nutrition research, and have summarized many dietary assessment and guidance methods, such as HEI (Healthy Eating Index), DQI (Dietary Quality Index), etc. However, due to the complexity of Chinese people's dietary structure, which is different from that of European and American countries, many foreign dietary assessment and guidance methods are not suitable for China's national conditions. Most of the research work on dietary nutrition is carried out for the population in our country, which fails to provide personalized guidance according to individuals' actual situation. In addition, the survey found that there are few systematic tools for individual dietary assessment and advice. Most of the system tools still use evaluating single nutrients in an individual diet, which is not easy to operate in the actual diet. Single nutrient or food group index can only reflect one aspect of the diet's status and problems, not enough to explain the complex interaction between nutrients. Along with people's attention to nutritional issues gradually from malnutrition to the nutritional imbalance caused by related chronic diseases and the quality of the overall diet, research should consider the single-factor analysis and pay more attention to the analysis of multiple and interactive risk factors.

In consideration of the aforementioned circumstances, He [14] et al. proposed the Chinese Dietary Balance Index (DBI), an index based on dietary guidelines for Chinese people and the optimal proportions of various foods within a balanced diet, as delineated by the balanced diet pagoda. The proposed index was then evaluated through a combination of the dietary characteristics of the Chinese population and an assessment of the consumption of grains, potatoes, vegetables and fruits, livestock, and poultry. This method is appropriate for ordinary residents to assess their dietary

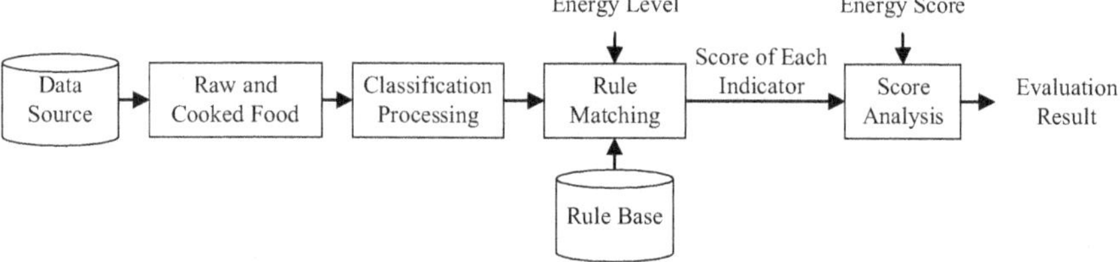

Fig. 6.19 Principles of DBI evaluation

intake. Additionally, combined with the Food Exchange List portion method (FEL) in nutritional diet, it can provide suggestions on the quantity of various kinds of food for individuals, thereby facilitating the evaluation and guidance of individual diets, and guiding the development of individual diets in a healthy direction.

The Dietary Quality Distance (DQD) was calculated for individuals with different energy levels in accordance with the corresponding evaluation rules, thereby obtaining their corresponding DBI score. The DBI represents the dietary balance index, with higher scores indicating a more optimal dietary structure. The principle of DBI evaluation is shown in Fig. 6.19. As the DBI indicators are based on the quantity of raw food, it is necessary to convert cooked foods into their raw equivalents prior to classifying diets according to the DBI index system.

(2) **Dietary Energy Assessment Combined with Individual BMI**

In dietary assessment, by combing the dietary energy assessment method with individual BMI, the corresponding dietary energy assessment standards were proposed according to the different energy intake of different individuals. The energy assessment results were scored and quantified using the energy nutrition curve, thereby establishing a single dietary energy evaluation standard. The average error of the standard was 1.69% when it was applied to the actual population of 1.69%. It can accurately evaluate individual dietary energy intake and provide certain guidance for dietary intervention. The principle of the individual dietary energy assessment method is illustrated in Fig. 6.20.

The application of individual dietary energy assessment methods in individual dietary energy assessment has a more accurate and individualized evaluation effect. It can not only take into account the individual energy intake but also guide the individual energy consumption to achieve different energy consumption goals.

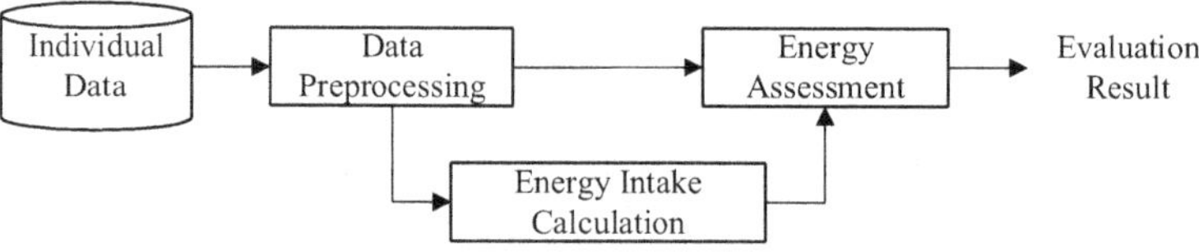

Fig. 6.20 Principles of individual dietary energy assessment methods

The overall rationality of an individual diet is based on energy. The individual dietary energy assessment method can effectively evaluate individual dietary energy and play an essential role in guiding individual dietary energy intake.

(3) **Personalized Dietary Guidance Based on FEL Ideal Recipe**

A nutritional diet is one of the common dietary guidance methods in nutrition. The nutritional diet refers to the design of a diet for one day, one week, or one month in combination with human needs and the content of various nutrients in food, to make the proportion of protein, fat, carbohydrate, vitamin, and mineral intake reasonable, to achieve a balanced diet. The purpose and significance of nutritional diet are to implement the dietary nutrient reference intake of various groups of people into the daily diet of diners so that they can eat enough energy and various nutrients according to their needs, and at the same time, prevent excessive intake of nutrients or energy. According to the needs of various nutrients, combined with the varieties of local food, production season, economic conditions, and kitchen cooking level, all kinds of food can be reasonably selected to achieve a balanced diet.

Considering the advantages and disadvantages of the food exchange portion method, an improved food exchange method combined with the DBI index system can generate a corresponding FEL template. Then, the previously obtained individual dietary evaluation results obtained before can be used to select the corresponding guidance mode by constructing a decision tree. The final dietary guidance scheme is generated by combining various kinds of FEL food exchange tables.

The schematic diagram of the dietary guidance algorithm based on FEL is shown in Fig. 6.21, which consists of three parts: energy requirement calculation, food exchange method improvement, and gguidance result generation.

The DBI index system is applied to the generation of the FEL template. The number of copies of various foods can be searched in the template by individual energy requirements. The corresponding guidance mode is then selected according to the decision tree constructed by previous individual dietary evaluation results to generate the final dietary guidance results. This is of great significance for a rational diet and promotion of healthy diet.

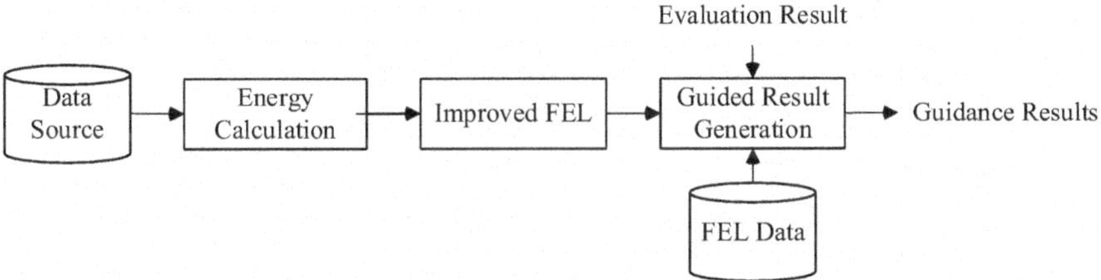

Fig. 6.21 Principle of dietary guidance algorithm based on FEL

6.5.3 *Comprehensive Health Assessment System for the Elderly*

A. **System Introduction**

Comprehensive health assessment for the elderly refers to a multi-dimensional assessment of the overall health level of the elderly. The comprehensive assessment of elderly health originated in a large chronic disease hospital in England in the 1970s. At that time, it was a process of collaborative diagnosis by a multidisciplinary team. It could assess the medical, psychological, and functional aspects of the elderly and put forward a comprehensive and coordinated treatment and long-term follow-up plan. This assessment method was quickly popularized in hospitals, and gradually extended to communities, nursing homes, and became an essential means in geriatrics and geriatric health care management. Different from the traditional medical assessment, the comprehensive evaluation of elderly health has two characteristics: one is "multi-dimensional" assessment, especially the assessment of physical function and mental state, which helps to achieve comprehensive assessment. The other is "multidisciplinary," in which traditional medicine and informatics teams work together to improve the assessment and implementation effect.

B. **Overall Structure**

In general, the comprehensive assessment of older people's health mainly includes four dimensions: physical function, disease state, psychological state, and living environment, as shown in Fig. 6.22.

C. **Function Features**

(1) **Assessment of Physical Function**

With increasing age, the physical function of the elderly will show a downward trend. Everyone can generally understand their physical state, but cannot quantify the evaluation of physical function. The physical role of the elderly is mainly reflected in the

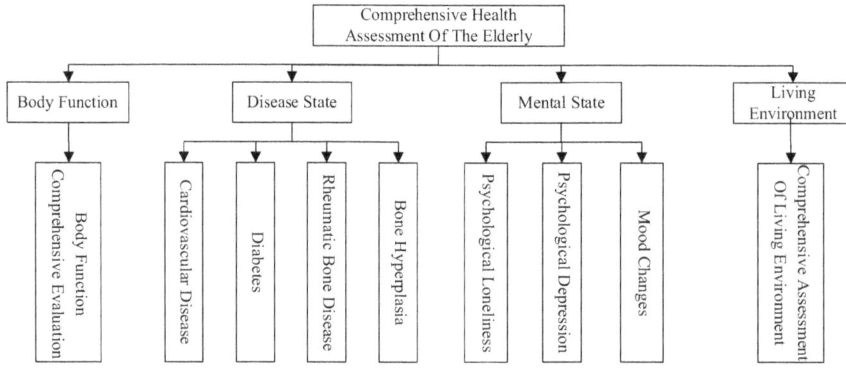

Fig. 6.22 Overall framework of comprehensive health assessment for the elderly

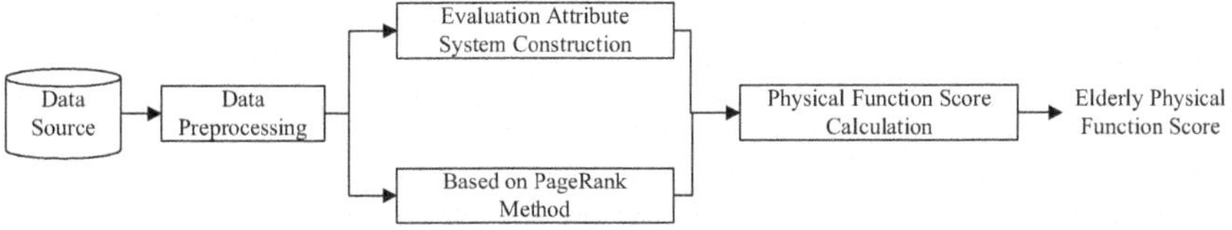

Fig. 6.23 Framework of the algorithm of the elderly body function evaluation model

physical function, which includes walking speed, cognition, sleep, excretion, chronic diseases, hearing, and vision. The 7-dimensional attributes form an almost complete whole from the limbs, five senses, thinking, and physical and mental state. Therefore, the evaluation of the physical function of the elderly through the 7-dimensional attributes can truly reflect the physical state of an older adult.

This section mainly introduces a physical function evaluation method for the elderly based on the PageRank algorithm. The algorithm framework is shown in Fig. 6.23.

The evaluation attribute system consists of seven dimensions: walking speed, cognition, sleep, defecation, chronic disease, hearing, and vision. The specific conditions and results of each dimension attribute are as follows: two defecation (urination and defecation difficulty are 0, only one of them is 1, neither is 2), walking speed ($0 \sim 0.65$ is 0, $0.66 \sim 1$ is 1, $1.01 \sim +\infty$ is 2), cognitive function (MMSE score $0 \sim 9$ is 0, $10 \sim 20$ is 1, $21 \sim 26$ is 2, $27 \sim 30$ is 3), sleep quality (no good value is 0, general value is 1, good value is 2), chronic disease (with value is 0, no value is 1), visual impairment (with value is 0, no value is 1), hearing impairment (with value is 0, no value is 1).

PageRank essentially outputs the probability distribution of web pages, which is used to represent the probability of randomly clicking on a page to jump to other pages. If there are only four pages A, B, C, and D in the whole network, each page has an initial probability value of 0.25. Web pages B, C, and D can be linked to web page A, as presented in Fig. 6.24.

The PageRank score of the web page can be obtained from the graph:

Fig. 6.24 Webpage link diagram

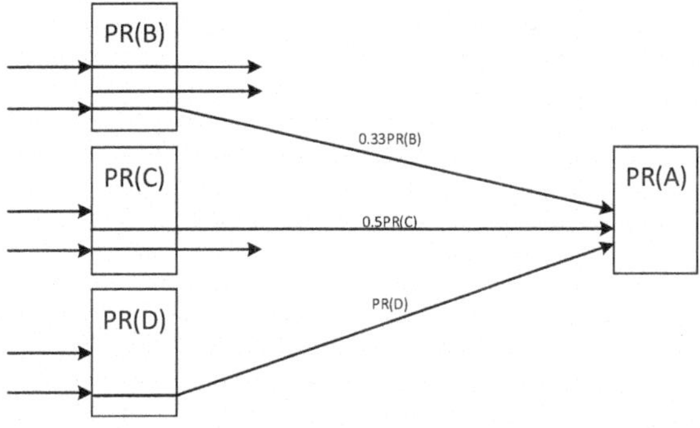

$$PR(A) = \frac{PR(B)}{3} + \frac{PR(C)}{2} + \frac{PR(D)}{1}$$

However, the number of web pages in the real network is huge, so the main problem of the transfer matrix is the sparsity of the matrix. The PageRank value of some pages in all pages becomes larger and larger through each iteration calculation, while the PageRank value of most pages gradually tends to 0. In order to solve the above problems, the method of smoothing is adopted, and the parameter β is introduced to adjust the PageRank score of the webpage. The formula is modified as follows:

$$PR(A) = \frac{\beta}{4} + (1-\beta)(\frac{PR(B)}{3} + \frac{PR(C)}{2} + \frac{PR(D)}{1})$$

If the PageRank values of all web pages are put together, a one-dimensional vector is formed, which is represented by letter A and matrix B is the link relationship between web pages. The elements in matrix B represent the number of links b_{ij} from the web page i to page j. The results of all the above pages can be generalized to a single page. The calculation formula of PageRank can be expressed as follows:

$$A_i = \left[\frac{\beta}{N} \cdot I + (1-\beta)B\right] \cdot A_{i-1}$$

i represents the number of iterations. If the difference between sum A_i and A_{i-1} is small, that is to say, the difference between the two iterations is close to 0, the iteration process of the whole algorithm is terminated. Based on the concept of the PageRank algorithm, the covariance matrix is used to represent the correlation between attributes as the input of the algorithm. Therefore, the physical meaning of the column vector is the weight of attribute importance. If the relationship between the trait and other attributes is strong, the attribute is given a higher weight; if the correlation between the attribute and other attributes is low, the attribute is given a lower weight so that the attribute weight will affect the score of personal physical function. After transposition, the covariance matrix is used as the input of the whole algorithm, which is represented by B.

The formula for calculating the comprehensive score of body function is as follows:

$$score = \sum_{i=1}^{n} \frac{x_i}{\max(x_i)} * W_{x_i},$$

where n represents the number of attributes, x_i represents the score value of each dimension, $\max(x_i)$ represents the maximum value of each dimension attribute, and W_{x_i} represents the weight of each dimension attribute.

(2) **Assessment of Disease Status**

With the aggravation of social aging, more and more attention has been paid to the health problems of the elderly, and cardiovascular and cerebrovascular diseases are the primary diseases threatening the health of the elderly. The number of cardiovascular and cerebrovascular diseases is the highest in the world. The deaths caused by cardiovascular and cerebrovascular diseases has accounted for 1/4 of the total deaths. The high incidence rate, high mortality rate, high disability rate, and high recurrence rate have put enormous economic pressure on families, society, and even the country. Disease status assessment assesses the risk of diseases that the elderly have suffered and may suffer soon. Research on the cardiovascular morphology and function of the elderly is a very representative and important content in the disease status assessment of the elderly. The following is a brief introduction to the aging analysis of cardiovascular morphology and function in the elderly.

The data used in this case were collected by echocardiography and carotid ultrasound. These data were used to analyze the characteristics of aging in depth. The algorithm framework is shown in Fig. 6.25. Initially, the data undergo preprocessing to ensure accuracy. Subsequently, a univariate analysis is conducted on the attributes derived from echocardiography and carotid ultrasound, utilizing age groups as the independent variables. This approach allows for a detailed examination of the relationship between age and cardiovascular attributes. Following this, the K-means clustering method is employed to segment the data into distinct groups. Ultimately, a comprehensive analysis of the outcomes from both methodologies is performed. This analysis serves to confirm the aging characteristics associated with each morphological attribute of cardiovascular function.

In univariate analysis, the mean, standard deviation, or percentage of continuous data and discrete data in cardiovascular morphological and functional attributes are described with aging, and the changing trend of age and each attribute is obtained. One-way Analysis of Variance (ANOVA) was used to test for significance.

The K-means clustering method is used to make up for the data's missing values and analyze them after normalization. The order of the change degree of each attribute with aging is obtained. Compared with the single factor age group's rigid classification, there is no rigid age condition in the clustering process. Simultaneously, the

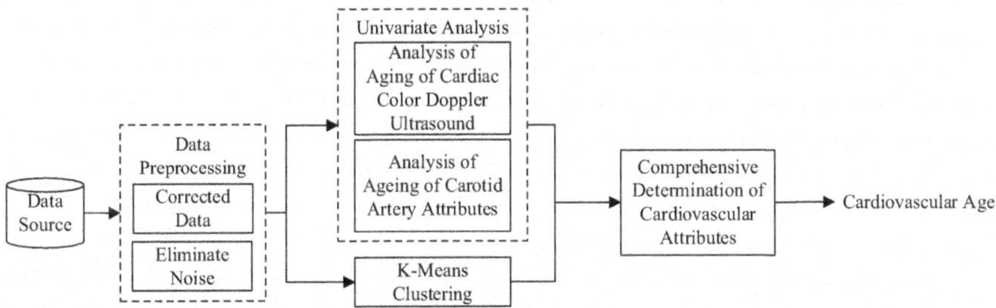

Fig. 6.25 Aging analysis algorithm framework of cardiovascular morphology and function of the elderly

order of each cardiovascular attribute's aging change degree is obtained by comparing the absolute value of each attribute's mean difference after normalization, thus eliminating the influence of dimension on the calculation results. In particular, the clustering method's use is for aging analysis, so it differs from other k-means algorithms in two aspects: the initial value selection is set according to the age, and the initial centroid value is determined by the sample value selected by the age. The K-means normalization attribute does not include age, so increasing age in the clustering process is important to avoid the small difference of different ages after clustering cannot be used for aging analysis.

(3) **Mental State Assessment**

The health of the elderly is the physical health of the elderly and the mental health of the elderly. Senile depression is a common mental disorder among the elderly. It has a high incidence rate and a high disability rate, which seriously affects the quality of life of the elderly. Taking the risk assessment of depression in the elderly as an example, this paper briefly introduces the psychological state assessment of the elderly.

The algorithm framework of elderly depression risk assessment is shown in Fig. 6.26, which includes the training module and test module. The training model needs to pre-process the data first, and then use the logistic regression method to build the risk assessment model of elderly depression. The model is tested under the test set to verify and analyze the model.

The logistic regression model was constructed and trained on the selected training data to obtain the risk assessment model of depression in the elderly. The calculation method of the risk assessment value is that the value of each attribute is multiplied by corresponding coefficients of the risk assessment model, and the assessment value when the area under the receiver operation characteristic curve (ROC) curve is the largest is selected as the cut-off point.

(4) **Assessment of Living Environment**

The living environment is an essential factor for human survival, and its importance for the health of the elderly is self-evident. The World Health Organization (WHO) suggests that the main indicators for assessing the health of the elderly are not only

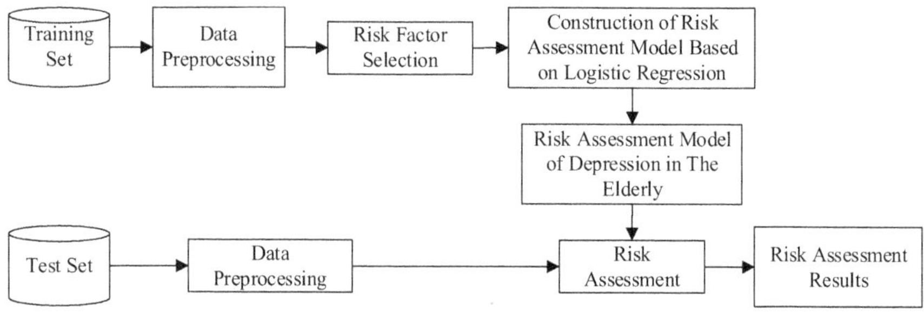

Fig. 6.26 Algorithm framework for risk assessment of depression in the elderly

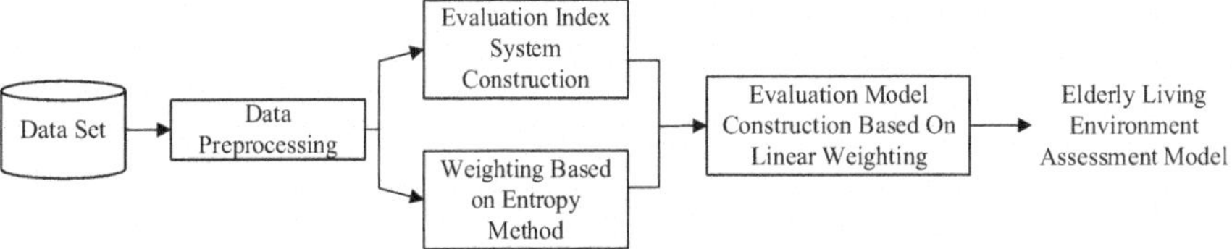

Fig. 6.27 Elderly living environment assessment model

death and disease. The study has shown that the living environment is closely related to the physical and mental health and quality of life of the elderly, and the assessment of the living environment of the elderly plays an important role in the comprehensive assessment of the health of the elderly.

This chapter introduces an assessment method based on linear weighting for the elderly living environment assessment. The structure of the assessment model is shown in Fig. 6.27. First, based on the elderly health management database, combined with expert opinions, to build the evaluation index system. Then, using the entropy weight method to weight the evaluation attributes. Finally, using a linear weighting method to build the living environment assessment model.

6.5.4 *Telemedicine*

A. **System Introduction**

Telemedicine, which integrates modern medicine, computer technology, and communication technology, is a new medical treatment form. Its primary purpose is to improve the level of diagnosis and medical treatment, reduce medical expenses, and better meet people's needs in health care. Telemedicine technology combined with the different needs of grass-roots hospitals to build a regional health information system platform to promote the smooth flow of a career. Information and data channels of tertiary and secondary hospitals and primary community medical service institutions, so as to realize the sharing of EMRs and examination information. Concurrently, when doctors in central hospitals need it, they can connect with experts from higher hospitals to share patients' conditions. Furthermore, remote surgery can guide the operation of patients without entering the operating room, and all kinds of problems encountered in operation can be handled at any time. Also, distance medical teaching can be implemented to train medical staff with high quality and efficiency to improve primary care workers' service level and professional ability.

Telemedicine is mainly composed of business supervision, telemedicine service, operation, and maintenance service. The public network collects all kinds of images and information in consultation workstations, expert consultation networks, and expert consultation workstations. Through the mobile medical cloud, remote medical data center, exchange system/video conference management system, combine all families and communities, primary hospitals and general hospitals in the local area into a set of data platform, thus forming the integration of resources, so that the province, even the whole country, can get the best medical services.

We should give full play to the advantages of medical technology and medical equipment of large hospitals or specialized medical centers, and carry out remote diagnosis, treatment, and consultation for the patients in remote areas, islands, or ships with poor medical conditions, so as to improve the level of diagnosis and medical treatment, reduce medical expenses, and meet the health care needs of the masses. Different application scope and telemedicine can be divided into global, international, domestic, regional, hospital, community, and family telemedicine.

Telemedicine technology has developed from the initial television monitoring, remote telephone diagnosis to high-speed network for digital, image, voice integrated transmission, and achieving real-time voice and high-definition image communication, providing a broader development space for the application of modern medicine.

B. **Overall Structure**

The overall architecture of big data-based telemedicine is shown in Fig. 6.28. Generalized telemedicine is divided into telemedicine approved by the health and family planning commission and other parts not included by the health and family planning commission. Among them, the telemedicine approved by the Health and Family Planning Commission includes remote consultation, remote case diagnosis, remote medical imaging diagnosis (including imaging, ultrasound, nuclear medicine, electrocardiogram, electromyography, electroencephalogram), remote monitoring, remote pathological discussion, and remote outpatient consultation; other parts include online consultations (no prescription).

C. **Function Features**

(1) **Basic Functional Specification**

With the wide application of telemedicine service, the national level has made clear regulations on the management standards, implementation procedures, responsibility identification, supervision, and telemedicine management. Relying on information technology to provide telemedicine services is an effective way to improve the level of primary medical services and solve the medical treatment problems of people in the grassroots and remote areas. The Central Committee of the Communist Party of China and the State Council have put forward clear requirements on deepening the reform of the medical and health system, the 12th Five-Year Plan for the Development of Health Services, and several opinions of the State Council on promoting information consumption and expanding domestic demand.

The "Basic Functional Specification of Telemedicine Information System" stipulates the functional composition, functional requirements, and overall requirements of the telemedicine information system.

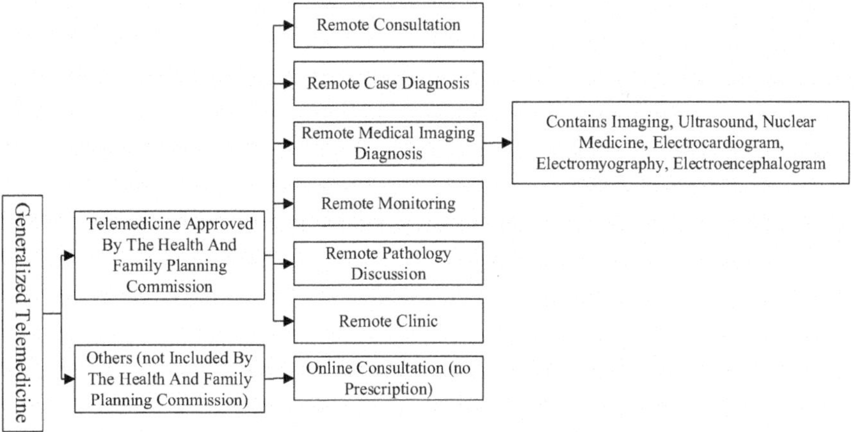

Fig. 6.28 Diagram of the overall telemedicine architecture

The functions of the telemedicine information system include basic business functions, extended business functions, and system management functions.

The telemedicine information system's basic business functions include six types of remote consultations, remote appointments, remote two-way referrals, remote image diagnoses, remote ECG (electrocardiograph) diagnoses, and remote medical education. All telemedicine information systems must have these functions, and none of them are indispensable.

The extended business functions of the telemedicine information system include three types: remote intensive care, remote surgical teaching, and remote pathological diagnosis. Hospitals can choose to implement some or all these functions if they are able to do so.

The system management functions include the management of basic data and business data. It is the unified management of medical institutions, medical personnel, and patient information resources at all levels. It is connected with other functional subsystems to realize the storage, exchange, and sharing of basic data and business data: update, share, and backup. The system management functions include nine categories: authority management, medical and health institution data management, department data management, expert data management, medical record data collection and storage, follow-up management, financial management, statistical analysis, functional collaboration, and data interaction.

(2) **Characteristics of Telemedicine System**

Telemedicine is one of the rapidly developing high-tech application fields in the world. It has been widely valued and applied in the global healthcare industry, and has gradually become a new medical service model generally accepted by the government, hospital managers, medical experts, patients, and their families. Telemedicine cannot only improve the popularity of hospitals and achieve excellent social benefits, but also bring significant economic benefits to hospitals. Some people call it one of the seven most promising industries in the twenty-first century.

The implementation of telemedicine services is closely related to the development of modern communication technology. This technology appeared in the United States in the late 1950s. From the early 1960s to the mid-1980s, this technology was regarded as the first generation of telemedicine. However, the development of telemedicine technology was relatively slow at this stage. With the continuous improvement of modern communication technology, telemedicine has entered the second generation since the late 1980s. The United States is an early country to conduct telemedicine research, and the first developed telemedicine system is used for non-invasive monitoring of astronauts and first aid for battlefield wounded and sick. Since then, medical institutions began to use telemedicine, and gradually carried out remote consultation, remote consultation, remote transmission of medical images, remote control surgery, and other projects. Other countries, such as Western Europe, Japan, and Australia, also attach great importance to the telemedicine development and have invested a lot of money in the research and development of telemedicine information technology. Telemedicine abroad is mainly used to carry out remote consultation and treatment, using various communication lines (such as ATM (Asynchronous Transfer Mode), ISDN (Integrated Service Digital Network), PSTN (Public Switched Telephone Network).) to provide medical services with the help of video conference or other communication systems; carrying out computer management and networking of medical data, and sharing medical data. Some Western European countries have developed and tried to use IC cards containing basic medical information, so that any networked hospital can use it to get the latest treatment information about the patient. Since the 1980s, telemedicine has been explored in China, and many telemedicine networks and institutions have been applied.

With the development of medical technology and computer network communication technology, medical informatization is limited to the provision of telemedicine and extended to the scope of telemedicine and digital health. In contrast, telemedicine mainly refers to implementing the diagnosis and treatment of patients from the physiological aspects. Telemedicine refers to the treatment of patients with physiological diseases and includes the maintenance of patients' mental health, with a broader scope. Recently, the concept of digital health has been proposed, which mainly refers to the comprehensive electronic and information-based medical services. Digital health faces patients and uses the most advanced information, and information communication technology provides the most convenient and fastest medical service for the majority of patients and is oriented to the public, aiming to provide convenient and cheap medical information service for all.

With the construction of the hospital information system, the information service is developing from the hospital internal information management networks to the external remote network. The increasingly perfect internal LAN of the Internet can access the Internet-wide area network, which can be connected anytime and anywhere. At the same time, with the continuous improvement of network technology and the continuous popularization of network, it is possible to establish a comprehensive and three-dimensional telemedicine health service system including VSAT

(Very Small Aperture Terminal), DDN (Digital Data Network), FR (Front Engine, Rear Drive), ISDN (Integrated Service Digital Network), PSTN (Public Switched Telephone Network), LAN (Local Area Network), ATM (Asynchronous Transfer Mode), Internet networks and video phones, which can fully make telemedicine build a bridge in the air and serve people across time and space. In the Omni-directional three-dimensional telemedicine system, people in cities and remote areas can use telemedicine very conveniently. Patients pour out their illness to doctors through the network, and doctors provide diagnosis and treatment plans through the network, and the whole process does not need to meet with patients.

In addition to building an omni-directional and three-dimensional telemedicine system to solve the problem of people seeking medical treatment in remote areas, the development of urban community telemedicine services can not only improve the work efficiency of doctors and expand the coverage area of experts' personal services, but also avoid the expansion and growth of investment sites and equipment costs.

6.6 Summary

Medical information processing is a key research avenue in the field of big data analytics, and is of great value in the study of disease risk factors and the development of treatment schemes. Characterized by its knowledge-intensive nature, diversity, practical applicability, timeliness, and public relevance, medical information differs from other forms of data processing in terms of complexity and specificity. As medical information processing technologies become more sophisticated and medical applications expand, the field faces new challenges that require innovative solutions and adaptive strategies.

From the perspective of basic resources, medical information processing in the era of big data is rich in all kinds of resources, such as PDB, SGD, gene data resources represented by the international thousand people genome project, such as MedPix, ABIDE, cancer imaging medical image resources represented by the archive, EHRs defined by MIMIC, STRIDE, and CDC, and medical voice records represented by TORGO, NKI-CCRT, and EMOTASS.

Medical information processing is a cornerstone in the field of big data, with most state-of-the-art technologies incorporating machine learning and deep learning methodologies. These approaches often integrate reasoning processes and expert knowledge to build sophisticated models for handling medical data. For instance, in gene sequence analysis, predictive models are developed using algorithms such as AdaBoost and SVM. In the domain of medical imaging, CNN U-Net architecture, is employed for the segmentation of biomedical images.

The adaptability of medical information processing to big data scenarios is evident in numerous applications. A prime example is the diabetes health promotion system, which encompasses disease risk assessment, establishment of guidance schemes, and evaluation and provision of intervention strategies. This system not only screens and predicts disease risks but also offers tailored advice on diet, exercise, and medication, thereby highlighting the significant value and potential of big data processing in medical informatics. The rapid evolution of big data analytics and its application in medical information processing is poised to significantly enhance scientific medical decision-making and elevate the overall quality of healthcare.

6.7 Exercises

1. Please list the specific application scenarios of medical information processing.
2. How to select the appropriate sampling rate for specific biomedical signals?
3. Please list and explain the types of medical images, imaging principles, and differences.
4. Please briefly describe the difference between the electronic medical record and paper medical record, as well as the advantages of the electronic medical record.
5. Please try to summarize the relationship between the digital hospital and big data analysis and briefly describe it.
6. Please try to design the digital hospital at the system level and explain the relationship between its modules.
7. What are the characteristics of medical information big data? What are the obvious differences from other types of big data?
8. What are the purposes, tools, and methods of gene sequence analysis?
9. What are the key technologies of medical image data mining?
10. Please elaborate on the application process of big data analysis in precision medicine.
11. What are the specific functions of the diabetes risk assessment system?
12. What are the dimensions of the comprehensive assessment of elderly health?

References

1. Otoom AF, Abdallah EE, Kilani Y, et al. Effective diagnosis and monitoring of heart disease. Int J Softw Eng Appl. 2015;9(1):143–56.
2. Kunwar V, Chandel K, Sabitha A S, et al. Chronic Kidney Disease analysis using data mining classification techniques. In: Cloud system & big data engineering. Piscataway: IEEE;2016. p. 300–5.
3. Bejnordi BE, Veta M, Van Diest PJ, et al. Diagnostic assessment of deep learning algorithms for detection of lymph node metastases in women with breast cancer. J Am Med Assoc. 2017;318(22):2199–210
4. Ogden PJ, Kelsic ED, Sinai S, et al. Comprehensive AAV capsid fitness landscape reveals a viral gene and enables machine-guided design. Science. 2019;366(6469):1139–43.
5. Bin YU, Shan LI, Cheng CH, et al. Human LncRNA big data gene prediction based on ensemble learning. J Qingdao Univ. Sci. Technol. (Natural Science Edition) 2018;39(1):106–13.

6. Sahiner B, Chan HP, Wei D, et al. Image feature selection by a genetic algorithm: application to classification of mass and normal breast tissue. Med Phys. 1996;23(10):1671–84.
7. Ginneken VB, Setio AAA, Jacobs C, et al. Off-the-shelf convolutional neural network features for pulmonary nodule detection in computed tomography scans. In: 2015 IEEE 12th International symposium on biomedical imaging. Piscataway: IEEE;2015. p. 286–9.
8. Yoo Y, Tang LYW, Brosch T, et al. Deep learning of joint myelin and T1w MRI features in normal-appearing brain tissue to distinguish between multiple sclerosis patients and healthy controls. Neuroimage Clin. 2017;17(3):169–78.
9. Pereira S, Pinto A, Alves V, et al. Brain tumor segmentation using convolutional neural networks in MRI images. IEEE Trans Med Imaging. 2016;35(5):1240–51.
10. Ronneberger O, Fischer P, Brox T. U-Net: convolutional networks for biomedical image segmentation. In: Medical image computing and computer-assisted intervention. Berlin: Springer;2015. p. 234–41.
11. Hripcsak G, Albers DJ. Next-generation phenotyping of electronic health records. J Am Med Inform Assoc Jamia. 2013;20(1):117–21.
12. Ho J, Ghosh J, Sun J, et al. Marble: high-throughput phenotyping from electronic health records via sparse nonnegative tensor factorization. In: International conference on knowledge discovery & data mining. New York: ACM;2014. p. 115–24.
13. Rosdi F, Salim SS, Mustafa MB. An FPN-based classification method for speech intelligibility detection of children with speech impairments. Soft Comput. 2019;23(7):2391–408.
14. He YN, Zhai FY, Ge KY. Establishing a Chinese dietary balance index. J Hygiene Res. 2005;34:208–11.

Chapter 7
Big Data and Internet of Things (IoT)

7.1 Introduction

The Internet of Things (IoT) is experiencing a period of rapid growth in the number of connected devices, including sensors, smartphones, vehicles, and data centers. The vast quantity of sensing devices is engaged in the continuous collection, generation, and transmission of data over time for various applications and sectors. This results in an extremely huge and fast/real-time data stream, commonly referred to as "big data." The management and control of such enormous datasets present significant challenges for the research community at large. The analysis of big data to discover, predict, and control decisions is of critical and inevitable importance for the improvement of the quality of services, experience, and life, as well as the generation of numerous benefits from data. It is essential to highlight that the integration of big data analytics over IoT networks can successfully develop IoT. Since big data and IoT are independent of each other, the integration still presents a challenge to the research community, which must propose new techniques in order to speed up the research advances and business models of IoT.

This chapter examines the role and the feasibility of big data analytics in the context of IoT networks. Furthermore, it examines the interrelationship between big data and IoT, as well as the existing related works and applications. In conclusion, this chapter identifies the open challenges and issues that have been encountered in the combination of big data analytics and IoT.

S. Luo and L. Pan, *Big Data Analysis*, Big Data Management,
https://doi.org/10.1007/978-981-95-5725-7_7

7.2 Fundamentals of IoT

7.2.1 Basic Concept

Over the past two decades, the Internet of Things (IoT) has emerged as a prominent area of interest within both industry and academia. Hence, the IEEE IoT [1] also furnishes two baseline definitions, the first of which pertains to the small scenario, and the second of which pertains to the large scenario:

In the Small Environment Scenario

"An IoT is defined as a network that connects uniquely identifiable 'things' to the Internet. These 'things' have sensing/actuation capabilities and, in some cases, potential programmability. Through the exploitation of unique identification and sensing, information about the 'things' can be collected and their state can be changed from any location and at any time by any authorized entity."

In the Large Environment Scenario

"An IoT is defined as a self-configuring, adaptive, complex network that connects 'things' to the Internet using standard communication protocols. The interconnected things have a physical or virtual representation in the digital world, a sensing/actuation capability, a programmability feature, and are uniquely identifiable. The representation contains information including the thing's identity, status, location, or any other business, social, or privately relevant information. The things offer services, with or without human intervention, by exploiting the unique identification, data collection, communication, and actuation capabilities. The service is delivered through the use of intelligent interfaces and is made available anywhere, anytime, and to anything, with security in mind."

7.2.2 IoT Phases and Related Technologies

Things in the IoT can be humans and intelligent machines that are connected in three different types, namely machine-to-human, human-to-human, and machine-to-machine communication. In fact, this connection is based on three main phases: (i) data acquisition and transmission, (ii) data processing, and (iii) action and utilization. These phases and their corresponding elements are shown in Fig. 7.1.

Data Collection and Transmission. There are devices widely used in IoT networks, including sensors, controllers, RFID tags, barcode/Quick Response Code (QR Code) tags, and other wireless/wired devices that collect, process, and convert data from physical and digital worlds into a signal. These devices connect to the gateways with different technologies such as WiFi (IEEE 802.11), Bluetooth and Bluetooth Low Energy (IEEE 802.15.1), 5G, LTE, and Low-power Wireless Personal Area Networks (6LoWPAN).

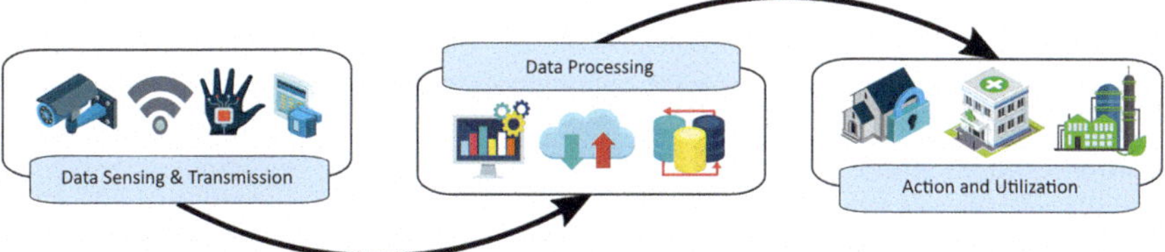

Fig. 7.1 IoT phases and related technologies

Data Processing. The purpose of this phase is to collect, send, and store data for data processing and to perform real-time actions to produce useful information from it. Various manipulation techniques, such as classification, sorting, and computation, are used in this phase. Furthermore, there are different technologies used to store and process the data, such as cloud computing and big data analytics. Cloud computing is one of the most favorable platforms in IoT data processing, where the data can be collected and sent to the cloud services for processing, storing, and analysis. Big data analytics [2] is another important technology in IoT data processing. It processes and stores a large amount of data in real time using various tools such as Hadoop MapReduce or Spark to improve decision-making.

Action and Utilization. This phase provides high-quality intelligent services to meet the customers' needs. It also includes different vertical markets such as smart home, smart building, and smart healthcare. This phase is also based on the semantics, which extracts the knowledge to recognize and analyze smartly, and make the right decision to choose the exact service.

7.2.3 Applications of IoT

Internet of Things is involved in various applications, including:

(1) **Smart city**

Smart cities encompass many sectors and services such as smart homes, traffic management, city lighting, weather systems, surveillance systems, smart energy, smart grids, and smart healthcare, which enable a smart economy and environment, smart people, and living. In fact, this brings all the requirements and challenges of smart cities that we have discussed in the previous section. For example, smart cities require large-scale addressing, high security, and privacy, as well as interoperability between devices, protocols, and applications. Cost reduction, efficiency, and resource optimization are crucial drivers that underscore the importance of smart cities in our lives. For instance, Smart Lighting System (SLS) is one of the most prevalent applications in smart cities, aiming to reduce energy consumption and achieve an intelligent and more efficient lighting management system.

(2) **Smart home**

A smart home integrates IoT sensors, gateways, protocols, cloud computing, and databases. Sensors are employed to collect data on home conditions such as lighting, climate, entertainment systems, alarm systems, and appliances, then send the collected data to home controllers for monitoring and control. Cloud computing enables remote access to the latest status of the home and is divided into a platform-as-a-service (PaaS) and an infrastructure-as-a-service (IaaS). Common applications used in smart homes are lighting control, fire detection, temperature monitoring, and voice control.

(3) **Healthcare**

Healthcare applications are the most important and prevalent IoT applications that are used to monitor and track patients. Medical sensors and wearables are used to connect to patients and collect data from physiological signals or body movements (e.g., electrocardiogram, oxygen consumption, heart rate, etc.), allowing doctors to monitor and control their status in real time and outside of the hospital. Other sensors are used to improve the environmental conditions for patients, such as temperature, light, humidity, and carbon dioxide levels.

(4) **Smart transportation**

IoT plays an important role in the management of vehicular traffic in large cities. It can improve a lot of services in vehicular networks, such as displaying the routing condition to avoid congestion, calculating the time of arrival, controlling the speed of the vehicle, transferring information in real time, and using sensors in smart parking systems to improve mobility in urban areas.

(5) **Smart grid**

Smart grid uses IoT to improve and optimize the energy consumption of homes and buildings. It consists of collecting, analyzing, controlling, and managing resources to connect service providers to detect faults, increase efficiency, and improve service quality.

(6) **Agriculture**

Agriculture or smart farms use IoT to automate farming techniques to improve quality and quantity, and minimize risk and waste. Sensors are used to collect data on soil content, rainfall, wind speed, and humidity, which helps farmers make better decisions and control.

7.2.4 Characteristics and Requirements of IoT

IoT constitutes a sophisticated network that interweaves various domains and sectors of the real world, each endowed with distinct characteristics and properties. The key features of this network are illustrated in Fig. 7.2 and are delineated as follows.

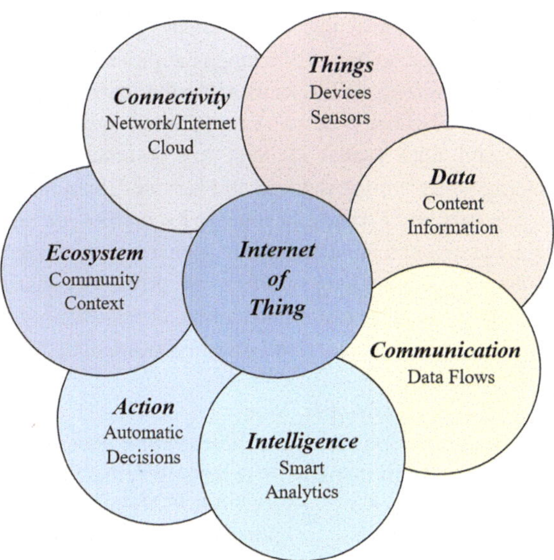

Fig. 7.2 Internet of things characteristics

Connectivity. Various technologies are used to connect to IoT devices, enable service accessibility and compatibility in the things, exchange information globally, and communicate between different infrastructures.

Sensing. The enormous amount of sensing devices in the IoT interact with different environments to measure changes, collect, and generate status data. Sensing technologies enable devices to communicate with the physical world and people around them. Sensing capabilities can be used in different IoT sectors such as healthcare, industrial control, traffic monitoring, and control.

Scalability. Billions of data points generated by a huge number of sensing devices, smartphones, and actuators make the management of IoT networks more challenging in terms of scalability. IPv6 is one of the solutions that can serve devices by enabling a larger address space. However, addressing is not the only scalability issue; scalable IoT architectures and solutions are required to support billions of devices and manage big data generated in an efficient way.

Intelligence. IoT devices collect, generate, and transmit data over time in different domains. They can also use various algorithms and technologies, such as big data analytics, artificial intelligence, software, and hardware, to make decisions, enable smart data analytics, and enable devices to interact intelligently.

Dynamic changes. IoT network is characterized by a dynamic topology in which devices are connected and disconnected according to their status, such as sleeping, waking, and moving. The growing number of devices also makes the IoT topology more dynamic. Hence, serious design solutions that support seamless mobility are required to ensure anytime, anywhere connectivity and data availability.

Addressing. IoT comprises billions of devices, each needing a unique and consistent address. IPv6 provides a vast address space, but IoT networks also require addressing schemes that are efficient, with minimal processing demands and the ability to adapt to network changes without losing continuity.

Heterogeneity. IoT is marked by the heterogeneity of its components, including hardware platforms, networks, operating systems, devices, communication protocols, and data types. Hence, this complex ecosystem is required to support heterogeneity among devices and services in an interoperable manner.

Security and Privacy. In some IoT scenarios, devices can generate highly sensitive data. For example, in a healthcare scenario, any attack that changes the data will result in danger to human lives or affect user privacy by sharing the information. Hence, IoT security should provide authorization, confidentiality, and integrity. Privacy must also be safeguarded through strict access controls that determine who can access sensitive information.

Resource constraints. Many IoT devices face limitations in battery life, storage, and computing power. These constraints necessitate the development of energy-efficient solutions to maintain connectivity and functionality despite these challenges.

7.3 Enabling Big Data Analytics Via IoT

7.3.1 IoT Data Characteristics and Analytics Requirements

Many IoT applications generate a huge volume of continuous data. The data can be streaming data captured in a short period of time that needs to be analyzed quickly to make fast decisions at the right time. For example, vehicular networks, particularly those with emergency response systems, necessitate real-time data processing capabilities. In the event of an accident, the system must generate responses within a critical window ranging from milliseconds to a few seconds.

Another poignant real-time application of streaming data in IoT is seen in remote patient monitoring—a widespread practice within healthcare. This application streams data for vital signs such as heart rate and blood pressure, which must be processed with immediacy to ensure that patients receive timely and appropriate care based on their current health status.

In contrast, other IoT applications, such as consumer product analytics, generate massive data that is stored in the big data sources. This data is used in various analytics, such as business intelligence and transactional analysis, where the results do not require real-time response and can be delivered after a few days. Consequently, these datasets have distinct analytical demands that necessitate tailored treatment and analysis methods. Moreover, the sheer volume of data can result in massive datasets that surpass the storage, management, and analytical capabilities of conventional hardware and software platforms.

Leveraging analytics on IoT-generated big data offers numerous benefits, particularly for businesses. It enables the extraction of valuable insights for more informed decision-making and trend forecasting. This includes understanding customer behavior, enhancing products and services, and anticipating various events and actions. Therefore, it is essential to comprehend the unique characteristics and structure of this big data to harness its full potential effectively.

7.3.2 *Big Data Analytics Architecture for IoT*

Drawing from pertinent research and endeavors [3], as depicted in Fig. 7.3, the overarching framework for IoT-driven big data analytics can be systematically segmented into five distinct stages: Data Acquisition, Data Preprocessing, Data Storage, Data Analytics, and Data Visualization. Further details about these stages have been discussed in the following sections. It is important to emphasize that we have also discussed big data analytics frameworks and platforms for each IoT application in Sect. 7.4.

Data Source. IoT weaves a vast network of interconnected devices, each a conduit for a plethora of information and data. The data source is varied due to the diversity of things and applications in the IoT network, so it can be sensors (such as temperature sensors and accelerometers), Radio-Frequency Identification (RFID), smartphones, GPS, and cameras. The variety of data sources brings great challenges to IoT because most of the generated data is under different types, structured, semi-structured, and unstructured data, and in a real-time manner.

Data Acquisition. This step refers to the data collection by obtaining the raw data from various sensors and devices, and then transmitting these huge volumes of data to the data storage infrastructure. However, the transmission of a massive volume of data brings many challenges, such as high bandwidth consumption and low energy efficiency. Various efforts have been proposed to address these problems by collecting and processing the data on Mobile Edge Computing (MEC) servers.

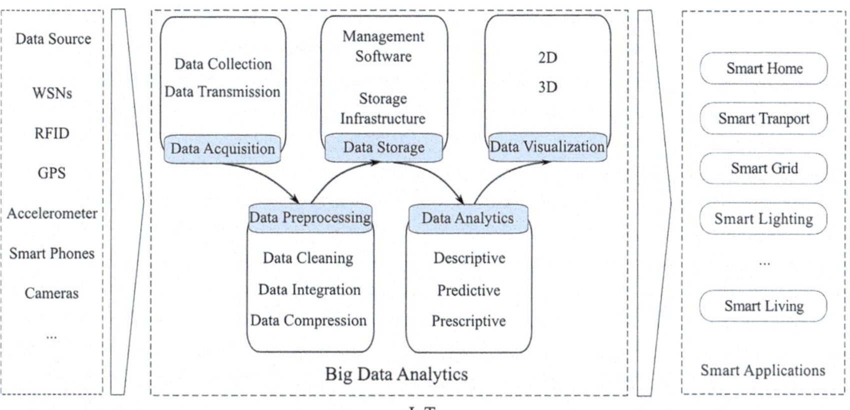

Fig. 7.3 Big data analytics in IoT

Data preprocessing. Data obtained from the previous step is characterized by variety, noise, and duplication. Various data are collected in the format of text, image, audio, video, etc., with structured, semi-structured, and non-structured types. The generated data can also be noisy or duplicated due to communication losses, transmission, and collection failures. Consequently, these characteristics make the cleaning and compression of difficult tasks. Various data cleansing and compression solutions have been proposed, such as regression models, probabilistic models, and outlier detection models.

Data Storage. Data storage is a crucial step in data analysis. This step can be divided into storage infrastructure and data management software. Storage infrastructure mainly includes persistent storage devices, such as micro-SD cards, and temporary storage devices, such as Random Access Memory (RAM). At the same time, data management software refers to file systems such as Google File System (GFS) and database management systems such as NoSQL databases.

Data Analytics. Data analytics methods can be divided into three types:

- Descriptive methods: use existing datasets to reveal the features of data. Various techniques in this category, such as clustering, descriptive statistics, and association rule mining, are used in network performance and content analysis.
- Predictive analytics: use historical data to predict forthcoming events and actions. This type of analytics is subdivided into classification, regression, supervised learning, unsupervised learning, and deep learning algorithms, and is used in traffic prediction and consumer behavior prediction.
- Prescriptive analytics: extends the results of both descriptive and predictive analytics for decision-making. This category includes simulation, optimization, and decision-making, and is used in network optimization and planning.

Data Visualization. This step plays an important role in presenting the results to help users take the right actions in real time. Various techniques have been used in data visualization, such as 2D and 3D visualization.

7.4 IoT Applications Powered by Big Data

7.4.1 Big Data Analytics in a Smart City

The smart city is the most potent IoT application, composed of various sectors and domains, such as smart transportation, smart healthcare, smart education, smart grid, smart surveillance, smart lighting, water management, waste management, urban security, and environmental monitoring. Smart cities make cities more livable, vibrant, and improve the quality of life by solving most of the existing problems faced by people in cities, such as pollution, traffic congestion, and energy shortage. By installing various IoT sensors and devices throughout the city, anyone can easily find a free parking spot in real time, find nearby charging stations, detect accidents in real time, predict traffic jams, and detect issues in the power system. However, a large amount of data is generated from these devices and applications, which has increased over time with the development of smart cities and technologies, leading to several issues and problems in the storage and analysis.

Big data analytics is the most suitable solution for smart cities to address these issues. Taking examples for the impact of big data analytics on real-world smart city applications, such as ensuring citizens' safety by using predictive analytics to predict the time and location of crimes in a real-time manner, making the city a much safer place. Another example in the transportation sector is the analysis of traffic data, which can help to reduce traffic congestion and accidents. Big data analytics can also help to manage and monitor cities more intelligently by analyzing cities' growth and future changes in various fields.

Figure 7.4 illustrates the implementation of big data analytics in the context of smart cities. A substantial quantity of real-time data is generated and collected from a multitude of sources, including automobiles, mobile phones, healthcare sensors, air quality sensors, and camera surveillance systems, for analytical purposes. This big data can be stored in three principal locations: cloud storage, which can drive data by removing redundant data and encrypting data transmission; edge computing, which enables the computation and data storage close to the users to improve the response time; and hybrid data storage systems, which combine the benefits of cloud and edge storage. The data processing phase encompassed a number of distinct phases, including preprocessing, correlation, regression, forecasting, classification, clustering, and visualization. Each phase employed a different analytics method, such as the Support Vector Machine (SVM), regression trees, the Bayesian classifier, k-means, and others, in order to extract useful information from the input data. The results of the analytics are used to inform intelligent decision-making in accordance with the specific field in question.

Many platforms and frameworks have been proposed to integrate big data analytics in smart cities. Osman et al. [4] proposed a big data analytics framework for smart cities called Smart City Data Analytics Panel—SCDAP. This framework consists of three layers, including the platform layer, security layer, and data processing layer. The platform layer includes hardware, operating systems, and communication protocols. At the same time, the security layer is responsible for authentication and data access. The data processing layer contains various processing functions, including data acquisition, data preprocessing, model manager, and aggregation. The functionalities of the data processing layer are summarized as follows:

- **Data acquisition**: this layer collects and captures data from various sensors, devices, social media networks, and sources.
- **Data preprocessing**: this layer contains various functionalities, such as data cleansing and transformation, to transform input data into an analysis-ready format.
- **Online and real-time analytics**: performs stream data processing for real-time applications.

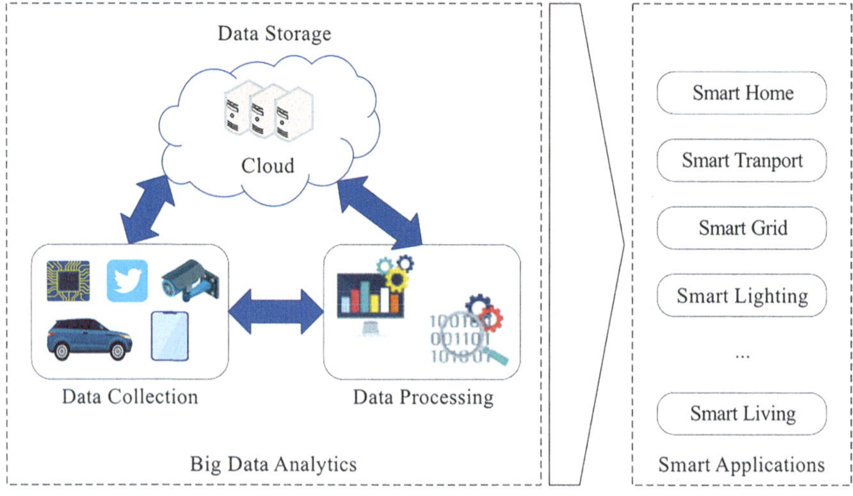

Fig. 7.4 Big data analytics in smart cities

- **Batch data repository**: represents the data storage management system (e.g., Hadoop HDFS, NoSQL database management systems).
- **Batch data analytics**: applied to applications that support latency.
- **Model management**: the result of data analysis can be persisted, retrieved, or deleted.
- **Model aggregation**: applied for higher-level and more complicated analytics.
- **Smart application**: built based on data analysis results.
- **User interface**: provides various tools allowing access, reporting, and ad hoc queries on persisted and/or aggregated models.

A business model and architecture for managing big data in the smart city has been proposed in [5]. In this model, the data collected and stored in the database is used in the business intelligence and big data analytics model to predict future actions. The results of the analysis can be presented in the form of a report or an alert. This model also includes a data science layer that comprises data sources, business intelligence, and analytics to extract insights and useful information from structured and unstructured big data. Similarly, Cheng et al. [6] developed a big data platform for smart cities called CiDAP. CiDAP handles both historical and real-time data. It contains four main modules: (i) IoT-broker and IoT-agents, which collect different data formats from different data sources; (ii) big data repository, which is used to store the data and transform into the new structured data; (iii) big data processing, which is used for intensive data analytics; (vi) and CityModel server, which is used to communicate with external applications. These platforms' main drawbacks are a lack of security and the inability to be handled by other big data analytics platforms.

Many research efforts and solutions have also been proposed in each area of smart cities. Further details of these efforts for each smart city domain are presented in the following sections.

7.4.2 Big Data Analytics in Smart Transportation

The transportation system has received much attention from industry and academia with the goal of improving driving safety, efficiently managing traffic conditions, and enhancing the user experience while driving. To achieve these goals, many solutions, protocols, and applications have been proposed, such as Dedicated Short-Range Communications (DSRC) and Wireless in Vehicular Environments (WAVE). However, several challenges still exist, such as the big data generated from multiple and diverse sources. With the development of the transportation system, the amount of these complex data generated is rapidly increasing from the Terabyte level to the Petabyte level, in which the traditional data processing methods are inefficient, and cannot meet the data analysis requirements.

Big data analytics can provide new techniques and methods to process and analyze such a huge amount of data efficiently and in real time. It can also solve three main problems, which are data storage, data analysis, and data management. Big data analytics can also provide fast data collection and analysis in real time for current and previous traffic data to predict events and provide useful information to prevent accidents, identify asset problems, prevent failure states, etc. Furthermore, it enhances the experience of driving by helping users to reach their destination in the most suitable route and in the shortest possible time.

The data generated in the transportation system is increasingly complex and has big data features, which are described by three Vs: volume, variety, and velocity. Volume indicates the amount of data generated from different sources, such as smart cards, GPS, sensors, video detectors, social media, and so on. This data is increasing and growing significantly over time and is generated by tracking transponders, infrastructure, environmental and meteorological monitoring, etc. Variety refers to the types of data, which increase notably as the number of users and applications. Velocity refers to the speed of processing the data in a short period. The speed of monitoring and processing has increased due to new communication technologies.

A. **Big Data Analytics Architecture for Transportation Systems**

The architecture of big data analytics for transportation systems can be divided into three layers, including the data collection layer, the data analysis layer, and the application layer, as shown in Fig. 7.5.

After collecting various structured, semi-structured, and mixed data from different sources, these data are transferred to the data analysis layer to classify and clean the data, remove duplicates, and apply analysis techniques, mainly including descriptive analysis and predictive analysis, to extract useful information. The analysis results are used in the application layer to predict future traffic and make decisions.

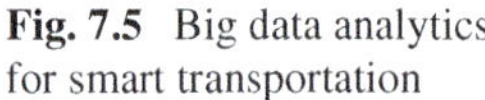
Fig. 7.5 Big data analytics for smart transportation

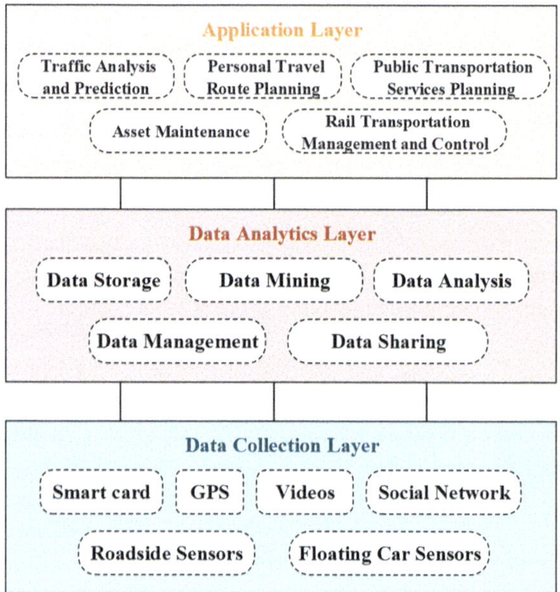

Data Collection Layer. It is the most important layer in this architecture. Data is collected from people, vehicles, and various sources, resulting in an increase in the complexity, diversity, and volume of data collected. The data is sourced from various channels within transportation systems, as detailed below:

- **Smart card** is one of the main data sources that collects the information of passengers traveling by bus or train, such as travel time and origin-destination information. These cards introduce a huge amount of data records every day, especially in big cities. For instance, at London metro stations, Transport for London (TfL) processes smart card data from approximately 8 million trips every day. This makes the smart card data an important data source for planning and managing public transportation services.
- **GPS** is another major source of data collection with location tracking for use in travel mode detection, delay, and traffic monitoring.
- **Video cameras** are also widely used in transportation systems to collect information on vehicle identification, traffic flow detection, etc.
- **Sensors** are installed in various components of transportation systems to collect data related to vehicle speeds, vehicle density, traffic flows, travel times, etc. The data can be collected from three types of sensors: roadside sensors collect data from sensors located along the roadside, including magnetometers, vehicle detectors, infrared systems, ultrasonic and acoustic sensor systems, and so forth; floating car sensors collect data from different detectors installed in vehicles, such as automatic vehicle identification and license plate recognition; and wide-area sensors collect traffic flow data using various tracking sensors, such as photogrammetric processing, sound recording, and video processing.

- **CAV and VANET**. Connected and Autonomous Vehicle (CAV) represents a cutting-edge technology that integrates the radical vehicle capabilities with the road infrastructure, generating a huge amount of real-time data, such as speed, coordination, acceleration, and safety data. Vehicle Ad Hoc Network (VANET) is a dynamic mobile network that collects a large amount of data from different vehicles and infrastructures, creating a rich dataset. Real-time results are a challenging task for analyzing such large amounts of data. Thus, big data analytics can overcome most of the VANET challenges, such as congestion and traffic prediction.
- **Passive collection** refers to the data that is passively collected for use in the research, such as mobile phone data, Internet access data, and social media data.

Data Analytics Layer. This layer applies various big data analytics techniques to the collected data to perform data storage, management, mining, analysis, and sharing. The most common big data analytics technique used in transportation systems is machine learning. Four pivotal models of machine learning are notably deployed in this sector:

- **Supervised Learning**: This model employs a mapping function, utilizing new input data to predict output variables for specific variables. Most models adapted in different transportation system scenarios are linear regression in traffic flow prediction, traffic speed estimation, etc., decision trees in traffic accident detection, accident severity analysis, etc., Artificial Neural Networks (ANN) in traffic flow prediction, travel time prediction, etc., and Support Vector Machines (SVM) in travel time prediction, bus arrival time prediction, etc.
- **Unsupervised Learning**: This model deals with unlabeled data and works on its own to discover structure in its input. K-means is the most common unsupervised learning model used in highway traffic planning, travel time prediction, etc.
- **Reinforcement Learning**: It is the training of models to make decisions by intelligent agents to maximize reward in an appropriate situation. This model is applied in the transportation system to formulate control and optimization problems in the transportation system, and is basically used in traffic signal control scenarios.
- **Deep Learning**: It is based on ANN with representation learning, which can achieve better performance than traditional machine learning models. Recurrent Neural Network (RNN) and Convolutional Neural Network (CNN) are the most widely used architectures, which are applied in transportation system scenarios, such as predicting traffic congestion, mobility prediction, and traffic flow prediction.

Application Layer. It used the analytics results in different transportation system scenarios, including traffic analysis and prediction, personal travel route and public transportation services planning, asset maintenance, etc. More details regarding these applications and research efforts can be found in the subsequent section.

B. **Research Efforts**

A variety of applications and solutions have been put forth for use in transportation systems that employ big data analytics. In the context of the traffic accident analysis application, Naseer et al. [7] proposed a framework based on using deep learning techniques to predict and classify road accident data with the objective of identifying potential hotspots and subsequently reducing the number of accidents in the future. Similarly, to solve the limitation of the installed sensors among roads, Dogru et al. [8] developed an intelligent traffic accident detection system using machine learning methods, including Artificial Neural Networks (ANN), Support Vector Machines (SVM), and Random Forests (RF), to analyze the collected data from VANET in a manner that takes into account the position and velocity values of each vehicle. Hence, other vehicles that have received this information can readily detect accidents by applying machine learning methods to analyze vehicles' behaviors over the previous few seconds; and Peng et al. [9] introduced an identification model for risky driving behaviors, including overclose car-following and lane departure, based on a neural network and Bayesian filter.

The field of traffic flow prediction has been significantly enriched by the application of Big Data analytics, with numerous studies contributing to its advancement. Cheng et al. [6] put forth a deep learning-based time series model for forecasting the traffic flow, comprising two principal components. The initial component employed deep learning to extract the traffic flow feature, while the subsequent component used the time series function to predict the traffic flow over a specified interval and to identify the fluctuations therein. Similarly, Yi et al. [10] introduced a deep learning neural network model with supervised learning. The proposed model is based on TensorFlow for the analysis and estimation of traffic conditions using real-time transportation big data. In the context of public services and personal route planning, authors in [10] present a methodology for predicting the average daily origin-destination trips using mobile phone data from millions of users. The data are then clustered based on their respective locations. Regarding transportation management and control, the work presented in [11] an architectural framework based on big data analytics for real-time traffic control. This framework employs a systematic analysis of the traffic control systems' requirements.

Conversely, numerous prominent big data analytics platforms have been used in the transportation sector. For instance, Apache Hadoop has been utilized to analyze a multitude of data types, including smart card data, sensor data, social media data, GPS data, etc. Apache Spark is another proper platform that adapts to machine learning methodologies and is utilized in the transportation system.

7.4.3 Big Data Analytics in Smart Healthcare

In the present era, numerous challenges persist within the healthcare sector, contributing to elevated health risks and mortality rates for a multitude of reasons. One of the primary factors contributing to this situation is the lack of adequate treatment. Conversely, numerous public and private healthcare organizations generate a massive volume of structured and unstructured data for each patient on a daily basis, derived from a range of sources including sensors, radiography, and social media. Consequently, valuable insights can be derived from big data through the application of analytical techniques. The application of big data analytics on IoT will facilitate improvements and developments in numerous fields, particularly in the healthcare system. In essence, the strategic application of big data analytics in healthcare is poised to transform the way care is delivered, offering a more comprehensive, patient-centered approach that prioritizes quality, efficiency, and patient satisfaction. This shift toward data-driven healthcare promises to not only save lives but also enrich the overall health and well-being of communities.

The seamless integration of big data analytics into healthcare opens new avenues for studying intricate aspects of human behavior, such as physical actions, personality, temperament, emotions, and social interactions, which have heretofore been regarded as challenging to quantify. Hence, this will facilitate an enhanced comprehension of the disease, its dissemination, its impact on the community level, and the nature of human interaction. Furthermore, a more reliable and appropriate decision can be made by conducting a thorough analysis of the available data at the right time. Moreover, the application of big data in healthcare has the potential to save millions of lives and significantly reduce treatment costs, revolutionizing the way we approach health and wellness.

The healthcare system generates and processes complex, heterogeneous, and voluminous datasets pertaining to clinical data, images, and records in a rapid and meaningful manner. To effectively interpret these diverse and voluminous datasets, sophisticated tools and solutions are essential. The objective of big data analytics is to process and analyze such a massive amount of data that is heterogeneous in format and generated from diverse sources, with the aim of facilitating the acquisition of knowledge that can inform decision-making processes in real time. Employing advanced techniques such as data mining, statistical analysis, and leveraging platforms like Hadoop, MapReduce, and Spark, big data analytics not only processes and classifies the data but also visualizes it to facilitate informed decision-making, as depicted in Fig. 7.6. This comprehensive approach enables healthcare professionals to make more accurate diagnosis decisions, ultimately enhancing patient care and outcomes.

A. Level of Healthcare Systems

Big data analytics in healthcare has captured a lot of attention from both industry and academic researchers. Hence, various research areas within the healthcare system are reaping the benefit of big data analytics in the healthcare system, including:

(1) **Bioinformatics**. This field is heavily dependent on the molecular data generated from human bodies. Many types of data are used in bioinformatics research, including those of a voluminous and considerable magnitude, such as gene-expression data, DNA, RNA, protein sequence data, protein–protein interaction, and human disease networks.

    Bioinformatics is witnessing an exponential growth in data volume every year. In 2014, the European Bioinformatics Institute (EBI) reported an estimated 40 petabytes of genetic, proteomic, and molecular data, a significant leap from

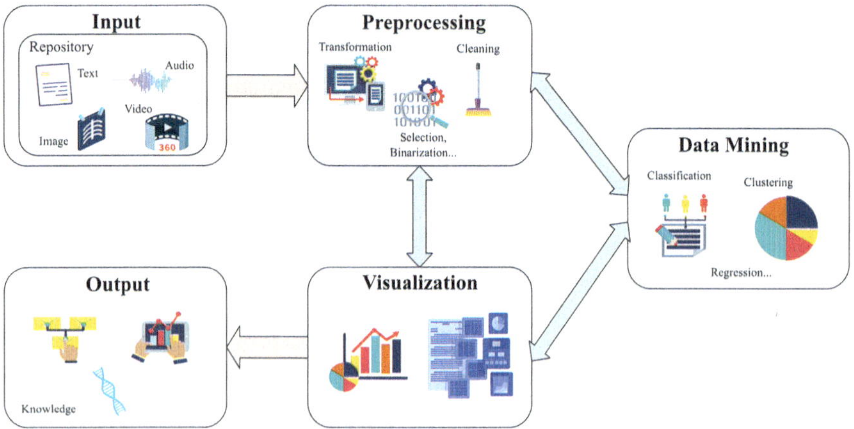

Fig. 7.6 A framework for big data analytics in a diagnostic system

the 18 petabytes recorded in 2013. Consequently, the vast quantity of data has given rise to a multitude of issues. These issues are related to the vast and heterogeneous bioinformatics data, which is generated from a multitude of sources and exhibits considerable diversity both in type and in size. To address these challenges, organizations have turned to cloud computing technology. Beijing Genomics Institute (BGI), a leading genome sequencing center, has installed a cloud-based analysis platform named Gaea that uses Hadoop tools to perform large-scale genome analysis. Similarly, Stanford University and UC Berkeley have implemented Bina box, a cloud-based genome analysis tool designed to reduce data size in the cloud. However, significant challenges remain in the field of bioinformatics research, particularly with regard to the management of large-scale microarray datasets, which encompass a vast array of expression values across diverse temporal and pathological contexts. It is therefore evident that the development of advanced big data analytics technologies is a necessity in order to address these issues.

Various approaches have been used in the field of bioinformatics big data analytics, including machine learning methods, mathematical optimization, and statistical techniques. SVMs are the popularly used supervised learning method in the domain of big data analytics for bioinformatics. Other mathematical techniques have also been applied for the purpose of dimensionality reduction, with the objective of overcoming the problem of big data volume in the context of big data. These include techniques such as principal component analysis, singular value decomposition, and kernel principal component analysis. At the same time, other algorithms for distributed data analytics have been proposed to handle the large-scale computing problem.

(2) **Medical Image Informatics**. Often referred to as medical image processing, this domain is a cornerstone in the healthcare industry. It harnesses imaging data generated from different information to facilitate the computer-aided medical

diagnostics and decision-making processes, with the aim of detecting diseases, including brain and lung tumors, artery stenosis, and aneurysms, among others. The data is also notable for its substantial volume and the heterogeneity of its size, dimension, and type, which contribute to the formation of "Big Volume." Imaging data can be Magnetic Resonance Imaging (MRI), Computed Tomography (CT), Positron-emission tomography (PET), and so forth. The analysis of these massive medical images can assist in the reduction of errors and the provision of information to monitor patients in real time.

Various research works within the field of big data analytics have been proposed to enhance medical image processing. Istephan et al. [12] put forth a framework for querying unstructured medical images in the field of epilepsy, with the aim of enhancing the reliability of patient care. Their framework is a two-step process: the first step focuses on structured data, while the second leverages the Hadoop platform to handle unstructured data in a distributed way, thereby facilitating more efficient medical image queries. Whereas Kouanou et al. [13] proposed an optimal workflow for managing and analyzing biomedical images using big data analytics. They designed two distinct architectures for the image classification phase. The first is based on the Hadoop framework, while the second employs Spark, which is more suitable than the proposed Hadoop architecture because it allows for the classification of a large number of medical images. In addition, the authors employed in each step of their workflow a suitable method and algorithm from the literature review in each step of their workflow to obtain the final architectures.

Research in neuroinformatics is conducted at the tissue level, whereby brain image data is analyzed to gain insights into brain function and elucidate the relationships between the information contained within brain images. Thus, the research in this subfield is most closely aligned with the data level, rather than with the subject matter itself. Hence, further research is needed to examine the whole brain MRI datasets using a variety of prediction methods, such as machine learning, in order to help physicians in making accurate diagnoses.

(3) **Clinical informatics**. This field is fundamentally centered on leveraging predictive analytics through the examination of patient data, thereby equipping physicians with the tools to make precise decisions regarding patient care. Clinical data, predominantly unstructured, encompasses a range of clinical documents, laboratory tests, and reports detailing patients' diseases and health issues, making the research in this subfield a big value. The application of big data analytics enables the storage and organization of clinical data into a structured format, thereby facilitating enhanced retrieval, extraction, and improvement in clinical decision-making, patient assessment, and the formulation of treatment plans.

Electronic Care Records (EHR) and Electronic Medical Records (EMR) are used to store and record the medical and clinical data obtained from patients' physical and mental health situations. The principal benefits of these records are an enhancement of service quality, a reduction in healthcare costs, and a decrease in medical errors. Additionally, they facilitate more effective healthcare for patients by providing a more reliable classification of their health.

Public health informatics. This domain harnesses the power of population data sourced from hospitals, epidemics, and even social media platforms. Characterized by their sheer volume, rapid accumulation (velocity), and diverse nature (variety), these datasets present a rich source for health-related insights. By applying big data analytics, valuable information can be extracted from social media, revealing hidden patterns that inform medical understanding. This can lead to more effective strategies for preventing illness, predicting, and diagnosing diseases. Public health informatics leverages various datasets, such as the Human Mortality Database, which provides comprehensive data on population and a country's mortality. Online Patient Network Database is another valuable resource, offering information related to the disease and symptoms. Additionally, research has been conducted using social media platforms like Twitter and Facebook to detect and predict various diseases such as depression, flu, and influenza. These studies employ different probabilistic and statistical models to analyze the data, demonstrating the potential of digital media in enhancing healthcare.

Medical Signal Analytics. In the realm of healthcare, monitoring devices like Physiological signal monitoring and Telemetry devices are indispensable. They generate data in the form of signals, but the sheer volume and velocity of this data pose significant challenges, particularly during streaming data collection. Issues such as bandwidth limitations, scalability, and cost-effectiveness are common hurdles that need to be addressed. Therefore, the important steps of big data analytics in streaming data are the acquisition of signals and the storage of the data generated from monitoring devices using big data analytics tools such as Hadoop, MapReduce, MongoDB, and so forth. Various datasets have been used in numerous research works, including the Multi-parameter Intelligent Monitoring in Intensive Care II (MIMIC-II) clinical database, which contains physiological waveforms, minute-by-minute records subsets, and the ECG-ID database, which contains EEG signal recordings, among others.

B. **Research Efforts**

The integration of big data analytics in healthcare has been a subject of extensive research, with studies exploring its potential and the challenges it faces. Andreu-Perez et al. [14] addressed the challenges associated with the processing of massive heterogeneity of structured and unstructured data, emphasizing the importance of ensuring data privacy and security, and highlighting that data must be obtained from authorized sources and through secure sensors to ensure accurate diagnoses. This underscores the importance of data integrity and the ethical considerations inherent in healthcare data management. Khan et al. [15] examined the potential of big data analytics in the context of m-Health systems, exploring the capabilities of various AI-based algorithms and big data methods. Also, they proposed a model based on AI and big data analytics for m-Health. Notwithstanding the aforementioned considerations, a number of limitations persist with respect to a large section of the population, privacy, and security.

Similarly, Rehman et al. [16] delineated five sub-disciplines included in the healthcare system within the domain of big data analytics: medical image analysis, bioinformatics, clinical informatics, public health informatics, and medical signal analytics. Furthermore, the authors emphasized the significant challenges and issues associated with the complexity and accessibility of data.

One of the most compelling aspects of big data analytics in healthcare is its real-time capabilities, particularly when integrated with the Internet of Things (IoT). Kario et al. [17] have demonstrated this through the development of a multisensory blood pressure monitoring system that collected and analyzed different formats of data on blood pressure and environmental parameters, then used this system for cardiovascular disease. The real-time analytics facilitated by IoT sensors allow for the seamless transmission of patients' physiological data to the cloud, enabling health professionals to access and respond to this information promptly.

From the perspective of life care, the quality of life for patients with severe chronic illnesses who are confined to their homes can be improved and managed with the integration of existing wearables technology and analytical techniques. A lifecare recommendation model for lifecare based on peer-to-peer networking has been developed in [18]. The system collects data from various wearable devices, including heart rate and body temperature, and analyzes it using a decision tree. Similarly, Yacchirema et al. [19] proposed an IoT-based system aimed at detecting falls among the elderly and alerting the relevant care groups. This system used 6LowPAN wearable devices to collect data in real time from the movements of elderly individuals, subsequently analyzing said data by using a decision tree-based big data model and cloud computing.

In other words, Verma et al. [20] proposed an IoT cloud-based m-Healthcare monitoring and disease diagnosing framework to predict illness and its severity level. This framework is based on cloud technology and is designed to facilitate the prediction of diseases. Similarly, Kumar et al. [21] put forth a cloud- and IoT-based framework for disease prediction. Their framework is underpinned by a novel classification algorithm known as the Fuzzy Rule-based Neural Classifier.

7.4.4 Big Data Analytics in Smart Grid

The European Union defines a smart grid as an electricity network that integrates the actions of all generators and consumers in a manner that enhances the reliability, security, and efficiency of the electric system through information exchange, distributed generation, and storage resources.

A multitude of data sources contribute to the smart grid, including market data, equipment data, geographic data, and power system data. These data are utilized in analytics to predict future occurrences, provide situational awareness, and issue advanced alerts. Additionally, they are employed in the analysis of grid stability, the detection of faults, and other critical functions. However, a massive amount of heterogeneous data is generated in a relatively short period from these smart grid

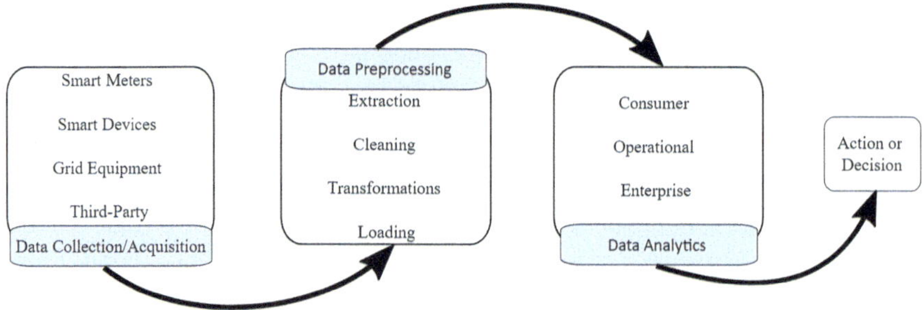

Fig. 7.7 Big data analytics for smart grid

sources. For instance, one million smart meters in a smart grid can produce 35.04 billion records, equivalent to 2920 TB of data with a sampling rate of 4 times per hour. Therefore, data analytics plays a pivotal role in enhancing the intelligence, efficiency, and productivity of the grid.

Figure 7.7 shows the application of big data analytics in the context of the smart grid. It is divided into the following sections for detailed explanation.

Data Collection/Acquisition. The data are generated from a variety of smart grid entities, including power generation unit(s), transmission lines, distribution stations, electric vehicles, and end-users (including industrial, commercial, and residential consumers). This data is collected and transmitted using intelligent devices, such as Advanced metering infrastructure (AMI), Phasor measurement unit (PMU), Wide-area monitoring system (WAMS), Remote terminal unit (RTU), Supervisory control and data acquisition (SCADA), and Intelligent electronic device (IED).

Data Preprocessing. Once the data are collected, the data undergo preprocessing to ensure quality and reliability. In this phase, the data are subjected to cleansing in order to remove any missing values and erroneous data that could potentially impact the quality of the resulting analytics. Subsequently, the data are then transformed to provide a unified view of the dataset, which is then loaded into a repository.

Data Analytics. This step is applied to process the preprocessing data to inform actions and decisions. The primary objectives of this step are to facilitate learning and responsiveness by enabling the sharing of smart grid status between utilities and consumers, thereby ensuring and improving the stability of the smart grid. The analytics can be divided into three types: consumer analytics, operational analytics, and enterprise analytics. Consumer analytics encompass energy forecasting, consumption analysis, and theft detection. Operational analytics include asset maintenance, outage management, and distribution optimization. Enterprise analytics encompasses real-time grid awareness and the visualization of data. Various analytics algorithms have been used in this step, including supervised learning, unsupervised learning, correlation, and dimensionality reduction.

A. **Research Efforts**

Various frameworks and solutions have been integrated with big data analytics for use in smart grids. Munshi et al. introduced a big data framework for smart grids, aiming at storing and processing a large amount of data in near real time. The framework includes a series of analytical procedures pertaining to the acquisition, storage, processing, querying, and analysis of big data. Authors used Hadoop's HDFS to archive the data pertinent to the smart grid, employing Hive and Impala to facilitate the retrieval of these datasets for further analysis and the generation of selected data. Furthermore, the framework was implemented on a secure cloud-based platform. In other work, Zhang et al. [22] discussed the applicability of big data analytics to the smart grid, including the collection of data, existing population techniques, and current applications of data analytics in the smart grid. The authors also addressed the challenges associated with this integration, including the heterogeneity of data and data sources, security and privacy concerns, and the limitations of the proposed frameworks and strategies from researchers.

Big data analytics have been pivotal in different smart grid applications, such as fault detection, predictive maintenance/condition-based maintenance, transient stability analysis, electric device state estimation/health monitoring, power quality monitoring, topology identification, renewable energy forecasting, load forecasting, load profiling, load disaggregation, and non-technical loss detection.

The primary constraints associated with the proposed platforms and solutions in the smart grid based on big data analytics pertain to security, privacy, and data quality. It is notable that the majority of the proposed solutions do not incorporate considerations pertaining to security and privacy. While alternative solutions employ diverse algorithms to ensure privacy, such as encryption algorithms, this approach often results in heightened complexity and latency in the analytics process.

7.4.5 *Big Data Analytics in a Smart Home*

The potential benefits of a smart home are numerous and include the ability to control and monitor various home attributes, such as lighting, climate, entertainment systems, alarm systems, and more. Such systems ensure family safety and security, improve efficiency and intelligent energy consumption, and enhance the quality of life for individuals. Hence, a massive amount of data can be produced by a multitude of IoT devices and appliances. By conducting real time or offline analysis of this big data, we can derive actionable insights that have a significant impact on societal safety, health, and economic well-being. For example, the monitoring and analysis of appliance usage can help identify the health status of residents within a smart home, potentially detecting health issues early on. Furthermore, companies may analyze energy consumption data to provide consumers with recommendations for reducing their electricity bills.

A few efforts have introduced solutions and frameworks for IoT-based smart homes that leverage big data analytics, incorporating both fog and cloud computing. Data collected from different IoT devices and appliances is transmitted through IoT gateways to either fog or cloud platforms for preprocessing, analytics, classification, prediction, and visualization. Most of these works are focused on analyzing smart home energy consumption, while others concentrate on monitoring the activities of individuals within the smart home for healthcare purposes. We will delve into these initiatives in the subsequent section.

A. **Research Efforts**

Yassine et al. [23] presented a platform for the analysis of large datasets from IoT in smart homes. This platform employs a combination of fog and cloud computing to facilitate the management, analysis, and extraction of valuable insights from energy consumption data. Fog computing is used to facilitate the acceleration of analytics services. The system comprises a number of functions, including data preprocessing, pattern mining, classification, prediction, and visualization. Cloud computing serves as the backbone for historical data analytics, providing extended storage capabilities and supporting the core infrastructure for smart home management. The authors further proposed a model that harnesses the power of big data analytics in a smart home for healthcare purposes. This model employs pattern mining, cluster analysis, and prediction techniques to analyze temporal energy consumption, thereby learning and discovering human activities.

Likewise, Sultan et al. [24] put forth a concept for a self-learning and adaptive smart home. In this system, data is captured through IoT devices and subsequently stored and analyzed using machine learning and big data analytics. Chen et al. [25] introduced a hybrid model that leverages big data analytics to enhance the living environment. This model utilizes the optimal regressive features of the Kalman Filter to predict the actions of users in the future, thereby improving the overall predictive capabilities and responsiveness of the smart home system.

7.4.6 Big Data Analytics in Smart Agriculture

The integration of big data analytics and IoT is revolutionizing the agriculture sector by enhancing productivity, optimizing crop management, refining risk assessments, and streamlining the agricultural supply chain. A plethora of IoT devices and sensors generate an enormous amount of dynamic, complex, and spatial data. This data is collected by monitoring and measuring various parameters such as soil nutrients, weather, air temperature, rainfall, wind speed, and relative humidity. The data encompasses a wide range, including historical data, sensor data, live-streamed data, business data, and market-related data. It varies in structure, from structured to unstructured, and in format, including text, images, audio, and video. The sheer volume and variety of data necessitate robust storage and processing solutions.

Cloud computing is one of the existing solutions used to store and manage such massive data. Big data analytics technologies complement these efforts, offering storage and analysis capabilities. For instance, HDFS is used to store a large amount of data, while MapReduce facilitates the distributed processing of data. Various methods have been used in the analytics stage, such as classification, clustering, prediction, and association rule mining. Applying these methods to agricultural data can yield significant benefits, such as predicting environmental changes, managing farm operations, and providing early warnings to farmers about potential disease or extreme weather conditions. They can also assist in predicting fire risks and managing overall farm activities.

There is a lack of solutions and proposed platforms for the integration of IoT in agriculture and big data analytics. Most of these works[26] have focused on the applicability of big data analytics in agriculture and smart farms, emphasizing the benefits and challenges that remain to be addressed. In contrast, a few other works[27] used big data analytics tools, such as MapReduce, for the processing of various agricultural data.

7.5 Summary

Big data analytics has emerged as a crucial component in IoT, enabling the extraction of valuable insights that facilitate more reliable decision-making and accurate trend predictions, including understanding customer behavior, improving products and services, and predicting various events and actions.

In this chapter, we focused on IoT from the perspective of big data analytics. More precisely, we discussed and presented the role of big data analytics in realizing IoT, drawing on existing literature, solutions, and considerations for future research and implementation. We commenced with an overview of IoT applications, outlining their characteristics and examining the various issues and challenges affecting IoT communications. We then provided a comprehensive account of big data analytics and its architectural elements, elucidating the interconnections. Additionally, we discussed the suitability and applications of big data analytics in IoT.

Throughout this chapter, we have reviewed the state-of-the-art literature regarding the existing solutions, architectures, and platforms in the context of IoT and big data analytics. We concluded by identifying and discussing the research gaps and challenges that the academic community might address in order to develop a robust architecture for the big data analytics-based IoT.

Many challenges persist in the integration of big data analytics with IoT, including those related to data quality, security and privacy, analytics, storage, and visualization. We believe that further research is essential to establish a common framework for the application of big data analytics in IoT.

7.6 Exercises

1. Briefly describe the basic characteristics of IoT.
2. What are the IoT data characteristics and requirements for analytics?
3. Briefly describe the relationship between IoT and big data analytics.

4. What are the benefits and advantages of the applicability of big data analytics through IoT?
5. How many stages in big data analytics-based IoT? Briefly describe each stage.
6. Briefly describe the big data analytics-based IoT applications
7. What are the main challenges and issues that faced the integration of big data analytics through IoT?

References

1. Minerva R, Biru A, Rotondi D. Towards a definition of the internet of things (IoT). IEEE Internet Initiat. 2015;1(1):1–86.
2. Hariri RH, Fredericks EM, Bowers KM. Uncertainty in big data analytics: survey, opportunities, and challenges. J Big Data. 2019;6(1):1–16.
3. Dai H-N, Wong R C-W, Wang H, et al. Big data analytics for large-scale wireless networks: challenges and opportunities. ACM Comput Surv. (CSUR) 2019;52(5):1–36.
4. Osman AMS. A novel big data analytics framework for smart cities. Futur Gener Comput Syst. 2019;91:620–33.
5. Hashem IAT, Chang V, Anuar NB, et al. The role of big data in smart city. Int J Inf Manag. 2016;36(5):748–58.
6. Cheng B, Longo S, Cirillo F, et al. Building a big data platform for smart cities: Experience and lessons from Santander. In: 2015 IEEE international congress on big data.
7. Naseer A, Nour MK, Alkazemi BY. Towards deep learning based traffic accident analysis. In: 2020 10th Annual computing and communication workshop and conference (CCWC).
8. Dogru N, Subasi A. Traffic accident detection using random forest classifier. In: 2018 15th Learning and technology conference (L&T).
9. Peng J, Shao Y. Intelligent method for identifying driving risk based on V2V multisource big data. Complexity 2018;2018.
10. Yi H, Jung H, Bae S. Deep neural networks for traffic flow prediction. In: 2017 IEEE international conference on big data and smart computing (BigComp).
11. Amini S, Gerostathopoulos I, Prehofer C. Big data analytics architecture for real-time traffic control. In: 2017 5th IEEE international conference on models and technologies for intelligent transportation systems (MT-ITS).
12. Istephan S, Siadat M-R. Unstructured medical image query using big data--an epilepsy case study. J Biomed Inform. 2016;59:218–226.
13. Kouanou AT, Tchiotsop D, Kengne R, et al. An optimal big data workflow for biomedical image analysis. Inform Med Unlock. 2018;11:68–74.
14. Andreu-Perez J, Poon CCY, Merrifield RD, et al. Big data for health. IEEE J Biomed Health Inform. 2015;19(4):1193–208.
15. Khan ZF, Alotaibi SR. Applications of artificial intelligence and big data analytics in m-health: a healthcare system perspective. J Healthc Eng. 2020;2020(1):8894694.
16. Rehman A, Naz S, Razzak I. Leveraging big data analytics in healthcare enhancement: trends, challenges and opportunities. Multimedia Syst. 2021;1–33.
17. Kario K, Tomitani N, Kanegae H, et al. Development of a new ICT-based multisensor blood pressure monitoring system for use in hemodynamic biomarker-initiated anticipation medicine for cardiovascular disease: the national IMPACT program project. Prog Cardiovasc Dis. 2017;60(3):435–49.
18. Yoo H, Chung K. Mining-based lifecare recommendation using peer-to-peer dataset and adaptive decision feedback. Peer-to-Peer Netw Appl. 2018;11(6):1309–20.

19. Yacchirema D, de Puga JS, Palau C, et al. Fall detection system for elderly people using IoT and big data. Procedia computer science. 2018;130:603–10.
20. Verma P, Sood SK, Kalra S. Cloud-centric IoT based student healthcare monitoring framework. J Ambient Intell Humaniz Comput. 2018;9(5):1293–309.
21. Kumar PM, Lokesh S, Varatharajan R, et al. Cloud and IoT based disease prediction and diagnosis system for healthcare using Fuzzy neural classifier. Future Gener Comput Syst. 2018;86:527–34.
22. Zhang Y, Huang T, Bompard EF. Big data analytics in smart grids: a review. Energy Inform. 2018;1(1):1–24.
23. Yassine A, Singh S, Hossain MS, et al. IoT big data analytics for smart homes with fog and cloud computing. Future Gener Comput Syst. 2019;91:563–73.
24. Sultan M, Ahmed KN. SLASH: self-learning and adaptive smart home framework by integrating IoT with big data analytics. In: 2017 computing conference.
25. Chen H, Xie X, Shu W, et al. An efficient recommendation filter model on smart home big data analytics for enhanced living environments. Sensors. 2016;16(10):1706.
26. Bendre M R, Thool RC, Thool VR. Big data in precision agriculture: weather forecasting for future farming. In: 2015 1st International conference on next generation computing technologies (NGCT).
27. Ghofrani M, Steeble A, Barrett C, et al. Survey of big data role in smart grids: definitions, applications, challenges, and solutions. Open Electr Electron Eng J. 2018;12(1).

The manufacturer's authorised representative in the EU is Springer Nature Customer Service Centre GmbH, Europaplatz 3, 69115 Heidelberg, Germany. If you have any concerns regarding our products, please contact ProductSafety@springernature.com

Printed and bound by CPI Group (UK) Ltd, Croydon, CR0 4YY
07/07/2026
02160909-0004